NEC® Q&A

Questions & Answers
on the *National Electrical Code*®

Noel Williams

Jeffrey S. Sargent

JONES AND BARTLETT PUBLISHERS
Sudbury, Massachusetts
BOSTON　　TORONTO　　LONDON　　SINGAPORE

World Headquarters
Jones and Bartlett Publishers
40 Tall Pine Drive
Sudbury, MA 01776
978-443-5000
www.jbpub.com

Jones and Bartlett Publishers Canada
6339 Ormindale Way
Mississauga, Ontario L5V 1J2
Canada

Jones and Bartlett Publishers International
Barb House, Barb Mews
London W6 7PA
United Kingdom

National Fire Protection Association
1 Batterymarch Park
Quincy, MA 02169
www.NFPA.org

Jones and Bartlett's books and products are available through most bookstores and online booksellers. To contact Jones and Bartlett Publishers directly, call 800-832-0034, fax 978-443-8000, or visit our website www.jbpub.com.

Substantial discounts on bulk quantities of Jones and Bartlett's publications are available to corporations, professional associations, and other qualified organizations. For details and specific discount information, contact the special sales department at Jones and Bartlett Publishers via the above contact information or send an email to specialsales@jbpub.com.

ISBN-10: 0-7637-4473-5

Production Credits

Chief Executive Officer: Clayton E. Jones
Chief Operating Officer: Donald W. Jones, Jr.
President, Higher Education and Professional Publishing:
    Robert W. Holland, Jr.
V.P., Sales and Marketing: William J. Kane
V.P., Production and Design: Anne Spencer
V.P., Manufacturing and Inventory Control: Therese Connell

Publisher, Public Safety Group: Kimberly Brophy
Acquisitions Editor: Chuck Durang
Production Editor: Susan Schultz
Director of Marketing: Alisha Weisman
Composition: Modern Graphics
Text Design: NFPA
Cover Design: Kristin E. Ohlin
Cover and Text Printing: Courier Corporation

Copyright © 2007 Jones and Bartlett Publishers, Inc.

All rights reserved. No part of the material protected by this copyright notice may be reproduced or utilized in any form, electronic or mechanical, including photocopying, recording, or by any information storage and retrieval system, without written permission from the copyright owner and publisher.

Publication of this work is for the purpose of circulating information and opinion among those concerned for fire and electrical safety and related subjects. While every effort has been made to achieve a work of high quality, neither the publisher, the NFPA, the authors, or the contributors to this work guarantee the accuracy or completeness of or assume any liability in connection with the information and opinions contained in this work. The publisher, NFPA, authors, and contributors shall in no event be liable for any personal injury, property, or other damages of any nature whatsoever, whether special, indirect, or consequential, or compensatory, directly or indirectly resulting from the publication, use of or reliance upon this work.

This work is published with the understanding that the publisher, NFPA, authors, and contributors to this work are supplying information and opinion but are not attempting to render engineering or other professional services. If such services are required, the assistance of an appropriate professional should be sought.

Library of Congress Cataloging-in-Publication Data
Williams, Noel.
  NEC Q and A : questions and answers on the national electrical code / Noel Williams and Jeffrey S. Sargent.
     p. cm.
  ISBN 0-7637-4473-5 (pbk.)
  1. Electric wiring, Outdoor. 2. Buildings—Power supply—United States. 3. National Fire Protection Association. National Electrical Code (1996) I. Sargent, Jeffrey S. II. Title.
  TK3221.W55 2006
  621.319′24—dc22

2006022674

6048
Printed in the United States of America
10 09 08 07 06   10 9 8 7 6 5 4 3 2 1

Contents

Chapter 1 NEC Scope and General Requirements 1

Question 1.1	Does the *NEC* say who can do "low-voltage" work covered by the *NEC*? 2
Question 1.2	Are all broadband installations covered by Article 830? 3
Question 1.3	Does the *NEC* cover wiring on transit buses or trains? 4
Question 1.4	Are electrical substations covered by the *NEC*? 6
Question 1.5	What is the difference between wet and damp locations? 7
Question 1.6	Does the *NEC* require listed equipment? 9
Question 1.7	Are all switches required to have a short-circuit or interrupting rating? 12
Question 1.8	How do terminal ratings affect conductor and ampacity selection? 13
Question 1.9	Where can I find information on protective electrical apparel? 17
Question 1.10	How are the requirements for working spaces determined? 19
Question 1.11	What working spaces apply to HVAC equipment? 26
Question 1.12	Are enclosed panels facing each other considered to be "exposed live parts on both sides"? 28
Question 1.13	Do transformers require working space for servicing while energized? 30
Question 1.14	What clearance is required for equipment that is serviced only while de-energized? 32

Chapter 2 Wiring and Protection 35

Question 2.1	Is gray permitted for any grounded conductor or just certain voltages? 36
Question 2.2	Where is GFCI protection required? 36
Question 2.3	What types of devices are permitted for providing AFCI protection? 41
Question 2.4	Why must single receptacles match the rating of the circuit but not duplex receptacles? 44
Question 2.5	Where are receptacle outlets required in dwelling units? 45
Question 2.6	Do unit load values for general lighting include factors for continuous loads? 54
Question 2.7	May a building be supplied by sources in addition to a service? 56
Question 2.8	How many services are permitted for a building? 58
Question 2.9	Are goosenecks permitted to be located below service drop conductors? 62
Question 2.10	What size service is required for temporary power for construction? 63
Question 2.11	Are "gang-operated" isolating switches required ahead of primary metering equipment? 65
Question 2.12	Where is an overcurrent device required to be located in a circuit? 66
Question 2.13	May I terminate a grounding electrode conductor in a meter base? 73
Question 2.14	What is meant by a "parallel path"? 76
Question 2.15	How are grounding electrode conductors sized? 80
Question 2.16	Must water piping and steel be bonded where there is no building? 84

Question 2.17	Is a lighting pole considered grounded by a direct connection to an electrode? 85	
Question 2.18	Is metal roofing, siding, and metal veneer required to be grounded or bonded? 87	
Question 2.19	Where can I find flexible metal conduit that is listed for grounding? 89	
Question 2.20	Should a service lateral include a grounding or bonding conductor? 91	
Question 2.21	Is a separate electrode system permitted to establish a "clean ground"? 93	
Question 2.22	Where must isolated grounding conductors terminate? 97	
Question 2.23	Is a 250-volt DC system required to be grounded? 98	

Chapter 3 Wiring Methods and Materials 101

Question 3.1	Is wiring restricted from sharing a penetration with plumbing? 102
Question 3.2	Is air tubing permitted in a raceway for conductors? 103
Question 3.3	Are sleepers laid on a roof adequate for support of exposed raceways? 104
Question 3.4	When can raceways or cables be attached to ceiling grids or supports? 105
Question 3.5	Must neutrals be pigtailed at devices supplied by multiwire branch circuits? 107
Question 3.6	Are receptacles permitted in plenum ceilings? 109
Question 3.7	What wiring methods and equipment are permitted in ducts, plenums, and other air-handling spaces? 110
Question 3.8	When are solid conductors allowed in raceways? 113
Question 3.9	Is a listing number adequate for identifying a manufacturer? 115
Question 3.10	What color coding does the *NEC* require for conductors? 116
Question 3.11	Where can I find application information for conductors 119
Question 3.12	Do the special conductor sizes for dwellings apply to 208/120-volt, single-phase systems 121
Question 3.13	How are ampacity calculations coordinated with termination and continuous load requirements? 123
Question 3.14	Do the ampacity tables of Article 310 apply to both AC and DC conductors? 129
Question 3.15	What are the differences in requirements for boxes in combustible and noncombustible materials? 130
Question 3.16	Are box extensions the only way to shield combustible material at boxes? 132
Question 3.17	Where is Type NM cable permitted? 135
Question 3.18	When and where is protection required for cables in attics and roof spaces? 138
Question 3.19	Do the extended support lengths for RMC apply where three-piece couplings or unions are used? 141
Question 3.20	Are aluminum conduits and fittings permitted to connect to steel enclosures? 142
Question 3.21	In what occupancies are cable trays permitted? 143
Question 3.22	What type of single conductors or multiconductor cables may be used in cable tray? 145

Chapter 4 Equipment for General Use 149

Question 4.1	Are green or white conductors in flexible cord permitted to be re-identified? 150
Question 4.2	Where can I find information on the proper and improper uses of flexible cords and extension cords? 151
Question 4.3	Is marking required for GFCI-protected receptacles in new installations? 154
Question 4.4	When is a service panel required to have a main overcurrent device? 155
Question 4.5	Are luminaires permitted to be connected by a cord and plug? 158
Question 4.6	Must a low-voltage lighting system come entirely from one manufacturer? 160
Question 4.7	Must all disconnects be readily accessible? 162
Question 4.8	Is a pool pump motor treated as an appliance or as a motor? 163
Question 4.9	What is the purpose of code letters on motors? 165
Question 4.10	How is overcurrent protection provided for motors and motor circuits? 167
Question 4.11	Do small motors such as damper motors require local disconnects? 171
Question 4.12	When are motors permitted to be connected by a cord and plug? 173
Question 4.13	How are conductors sized for motors over 600 volts? 175
Question 4.14	May the circuit breaker for an air-conditioning unit also serve as the disconnect? 176
Question 4.15	Can a circuit for HVAC equipment be tapped to separate integral circuit breakers? 177
Question 4.16	What restrictions does the *NEC* impose on the locations of transformers? 181

Chapter 5 Special Occupancies 187

Question 5.1	Where do I find the classification of a hazardous location? 188
Question 5.2	Must areas where solvents, fuel oil, or diesel is used be classified? 190
Question 5.3	How is equipment identified as suitable for hazardous locations? 192
Question 5.4	Is there any length limitation on flexible conduit runs or flexible fittings in Class I areas? 196
Question 5.5	Is "explosionproof flex" permitted between an enclosure and a required seal? 198
Question 5.6	What types of facilities are and are not covered by Article 511? 200
Question 5.7	What type of portable lighting is permitted in garages and aircraft hangars? 204
Question 5.8	Is there a way to use ordinary equipment in a hazardous area? 205
Question 5.9	Are optometrists' or psychologists' offices patient care areas? 207
Question 5.10	Is special wiring or other special equipment required for patient care areas? 209
Question 5.11	What loads are permitted on the equipment branch of an essential electrical system in a hospital? 213
Question 5.12	What requirements apply to temporary wiring at trade shows and conventions? 214

Chapter 6 Special Equipment, Conditions, and Systems and Related Standards 217

Question 6.1	When does Article 645 apply? 218
Question 6.2	Why are special boxes required for connections to pool luminaires? 220
Question 6.3	Where are receptacles required for pools, spas, and hot tubs? 222
Question 6.4	How is bonding to be accomplished for hydromassage bathtubs? 226
Question 6.5	May the power supply to a fire pump originate in a service panel switchboard? 228
Question 6.6	What physical separation is required for a fire pump service disconnect? 231
Question 6.7	Which conductors for fire pumps require special routing and protection? 233
Question 6.8	How are emergency loads identified by the *NEC*? 237
Question 6.9	Are transfer switch control wires considered to be emergency wiring? 238
Question 6.10	Is more than one emergency system permitted in a building? 240
Question 6.11	Is emergency power required to be supplied when a branch circuit fails? 244
Question 6.12	Is an outside generator for an emergency or standby system a "separate structure"? 245
Question 6.13	How do I know if a circuit is Class 1, Class 2, or Class 3? 247
Question 6.14	What are the bonding and grounding requirements for control circuits? 249
Question 6.15	How are rules for data and communications circuits applied to a Class 2 "interbuilding circuit"? 253
Question 6.16	When or where are "low-voltage" cables required to be in conduit? 257
Question 6.17	Are separations required between power and instrument or control wiring? 259
Question 6.18	What is the purpose of "CI" cable? 261
Question 6.19	What wiring method should be used to interconnect residential smoke detectors? 263
Question 6.20	Are communications cables permitted on a roof without a raceway or overhead clearance? 264
Question 6.21	What raceways or equipment are required to be grounded in communications systems? 266
Question 6.22	Does the *NEC* specify minimum spacing between conductors for communications and power? 267
Question 6.23	Are communications and power system grounding electrodes required to be interconnected? 269
Question 6.24	How is raceway fill determined for multiconductor cables? 271

Preface

Where these *NEC* Questions and Answers Came From

Many, though not all, of the questions and answers that comprise this book come from the advisory service archives of NFPA's Electrical Engineering Department. The questions have been selected from among the thousands of advisory service inquiries that were submitted to the electrical engineering staff from people who use the *NEC* in their respective roles in the electrical industry. Engineers, designers, installers, inspectors, and others use this membership benefit as a job aid to better understand and apply the *NEC*. The advisory service is offered as a benefit of NFPA membership and is also offered at no cost to those who enforce the *NEC* at the federal, state, county, city, or other level.

NFPA employs a full-time technical staff in its engineering division and one of the duties of the technical staff is to provide advisory service on the NFPA codes and standards to which they have been assigned. This advisory service is provided via telephone, letter, fax, and e-mail. The answers provided by staff members, regardless of the means of communication, are considered to be informal staff opinions based on their expertise in the subject matter covered by the particular code or standard about which the question has been asked. No direct response from technical staff can be considered as a formal interpretation, and all written and e-mail responses contain a disclaimer to that effect.

Formal interpretations on NFPA codes and standards are processed under the requirements established in Section 6 of the NFPA Regulations Governing Committee Projects. There are very strict conditions under which a formal interpretation request will be considered and sent on to the responsible *NEC* Code-Making Panel and *NEC* Technical Correlating Committee. For more information on the formal interpretation process, the *NFPA Regulations Governing Committee Projects* can be downloaded at no charge at: *www.nfpa.org/assets/files/ PDF/NFPADirectory2006.pdf*.

NFPA's Electrical Engineering Department, under the direction of the chief electrical engineer, has a technical staff that performs the myriad of responsibilities associated with developing and supporting the *NEC* and its associated products, as well as the other codes and standards assigned to the electrical department. The answers in this book are based on staff-developed responses to advisory service questions. Many of the actual answers are far more concise because it is not the function of the advisory service to provide tutorial responses. This type of enhanced information is the role of the *National Electrical Code Handbook*, and other products offered by NFPA including this book.

The role of an advisory service response is to provide a direct, accurate response to questions on the requirements of the *NEC* or other NFPA codes and standards. Engineering solutions on how to achieve compliance with code requirements is the role of engineers or expert consultants. A listing of experts and consultants who provide this type of service is compiled in the *NFPA Buyers Guide*. Archival research is a service provided by the staff of NFPA's Charles Morgan Library. This research is performed for a nominal cost.

The expertise of NFPA's technical staff is enhanced by their ability to draw on the collective knowledge base of over 6,000 technical committee members who are recognized experts and leaders in their respective professions. The *NEC* committee is comprised of over 400 code-making panel members representing all facets of the electrical industry. The strength of the *NEC* is the volunteers who serve on the code-making panels. These dedicated men and women are the true subject matter experts and it is through interaction with these individuals that NFPA staff members are able to provide the expertise that users of NFPA's advisory service rely on.

Introduction

How to Use This Book

This book consists of a series of questions and answers. Each item in the series is headed by a summary question that is related to some area of the *National Electrical Code (NEC)*. The questions and answers in this book are organized in the order of the primary *NEC* references that are used to answer the summary questions. The summary questions and answers are also grouped by chapter. Chapter 1 addresses questions that are based on articles in Chapter 1 of the *NEC*. Similarly, the summary questions in Chapters 2 through 5 of this book are based on articles in same-numbered chapters of the *NEC*. Chapter 6 of this book includes questions based on articles in Chapters 6 through 8 of the *NEC*. The summary questions are listed by chapter in the table of contents.

In most cases, more than one reference or section of the *NEC* must be used to arrive at an answer or adequately discuss related issues, or references to documents other than the *NEC* may be needed to support an answer. The other significant references that are used in an answer are listed with each summary question along with the main code reference for the question. The other references may be discussed in the answer, or they may simply be mentioned as related or additional requirements. One index in this book is based on the code references.

Each summary question may have related questions that are listed with the summary question and answered in the discussion. The subjects mentioned in the discussion from both the summary questions and the related questions are summarized as keywords that are also listed with each summary question. The second index in this book is based on the keywords.

As the reader of this book, you may choose to simply browse through the summary questions or the related questions for subjects in which you are interested. Or, you may use the list of summary questions in the table of contents to focus your reading on a chapter or subject area. If you are a reader who is looking for answers or discussions related to specific subjects or code references, the indices based on keywords and references will help to shorten your search.

If you know the subject you are interested in, you may find that subject listed in the keyword index. Or, you may look through the keyword index to pick a keyword that is related to your area of interest. The keywords may be the topic of a discussion or a term used in a discussion. Some keywords will appear in many questions and answers, and others may be discussed only once. If your subject is addressed in one of the answers you look at, or if it is not, other keywords related to that answer may provide a path to a more complete discussion or to discussions of similar or related issues.

Code references can also be used to locate the answers that deal with a specific subject. If you have questions about application or interpretation of a specific code section, you can locate questions and answers that refer to a specific section by finding that section in the reference index. Like the keywords, lists of additional references with each question and answer will also lead the reader to other answers that may cover or include the same or related references to specific code rules. These other answers and explanations may cover a different aspect of a rule or may provide examples of the application of the same rule in a different context.

The questions and answers in this book are based on the answers to questions asked of NFPA advisory service. However, most such questions and answers are very specific, and the questions and answers in this book have been generalized to be more broadly applicable. This book is based on the 2005 *National Electrical Code*. Some rules may change in subsequent codes or may be different in earlier codes. This book should be used in conjunction with a copy of the *NEC*. The actual text of the code must be the basis for any code interpretation, and this book is only meant to provide commentary, explanation, and examples of the application of the *NEC*.

Chapter 1

NEC Scope and General Requirements

Question 1.1 Does the *NEC* say who can do "low-voltage" work covered by the *NEC*?

Keywords

License
Communications
Data
Controls
HVAC

NEC References

90.1, 90.2, 90.3, 90.4, Article 422, Article 424, Article 430, Article 440, Article 725, Annex G

Related Questions

- Does the *NEC* require that communications wiring or data network wiring be installed by a licensed electrician?
- Have changes been made recently that would require licensed electricians where they were not required in the past?
- If conduit is run under Chapter 3, but the wiring is installed under Chapter 8, is licensing required?
- Does the *NEC* cover HVAC wiring and HVAC technicians?
- What wiring may be done by HVAC technicians and what wiring is required to be done by electricians?
- Is there a difference between the requirements for people working on control wiring and for people working on power wiring?

Answer

The *National Electrical Code* does not deal with the qualifications of installation personnel. As stated in Sections 90.1 and 90.2, the *NEC* deals with the requirements for electrical installations, including signaling and communications. However, licensing rules are usually promulgated at the state, county, or local level. A jurisdiction may or may not have licensing requirements, and those licensing rules may or may not include communications, remote control, signaling, or other so-called low-voltage work.

Most, but not all, areas in the United States have some form of licensing for electrical installers or electricians. Often, the statute that adopts the *NEC* and makes it a law for electrical installations also covers qualifications and licensing of electricians. Some jurisdictions (states, counties, cities, or the like) require all work covered by the *NEC* to be done by licensed personnel, but some jurisdictions have limited licenses for communications and other limited-energy circuit work, or licenses that are limited to certain types of occupancies. Other jurisdictions exempt wiring from licensing requirements if it does not present a shock or fire hazard, or they may exempt employees of an owner doing work on the owner's premises. Another of the many possible ways in which licensing is applied is that some jurisdictions require licensing for all raceway installations, but may not require licensing for those installing communications conductors in the raceways. With regard to heating, ventilating, and air-conditioning (HVAC) power wiring, many jurisdictions allow the wiring from an outlet or disconnect to be done by persons not licensed as electricians, but prohibit such persons from altering the fixed premises wiring. Other jurisdictions require all work covered by the *NEC* to be done by licensed electricians. An installer or other code user must check state or local laws to see if licensing applies to a specific job or facility.

The "changes" referred to in the questions are not really changes in the work covered by the *NEC*. In recent editions, some language has been added, primarily in Sections 90.3 and 90.4, to emphasize that communications installations *are* covered by the *NEC*, but this was not a change in intent. Previously, some people tried to read the *NEC* to say that communications installations were not covered, and the added language is simply meant to eliminate that misinterpretation.

Wiring for HVAC equipment is also covered in the *National Electrical Code*. The general rules of the first four chapters of the *NEC* apply, as well as specific articles. Article 422 applies to appliances generally, including gas and oil furnaces. Article 424 applies to electric space heating equipment, including electric furnaces, unit heaters, and duct heaters. Article 430 applies to motors such as those used in fans and pumps, and Article 440 applies to the hermetic refrigerant motor compressors that are widely used in air-conditioning and refrigeration equipment. Most electric temperature control systems are covered by Article 725 if they are not line voltage controls or communications. The *NEC* does not say who can do these types of work, only how it is to be done so it will be safe.

There is some relatively new language in Annex G that covers qualifications of electrical inspectors, but that language is not part of the *NEC* unless it is specifically adopted by the authority having jurisdiction. Annex G is meant to be a model for a statute covering enforcement of the *NEC* and is designed to be incorporated and completed in a statute so that it becomes law. However, as it appears in the *NEC*, it does not cover installers.

Are all broadband installations covered by Article 830?

Question 1.2

Related Questions

- What portions of a broadband or communications installation are covered or not covered by the *NEC*?
- What is meant by "exclusive control"?
- What cables are required to be listed?

Answer

Article 830 covers a specific method of providing broadband communications. Although all broadband signals provide some power, the systems contemplated by Article 830 provide enough power to supply the needs of the network interface unit (NIU) that splits the communications signal into video, audio, voice, and data signals. Some other broadband technologies, such as cable, satellite, or DSL, usually use power from the customer's facility to power the set-top box, modem, or equivalent. As shown in Figure 1.2, the building wiring for these systems may fall under Article 725, Article 770, Article 800, or Article 820, depending on the power source and the type of cable used. Where satellite dishes are used, the antenna and incoming cable are covered by Article 810 and the building wiring is covered by Article 820.

The query as to what portions of the network-powered broadband communications system are covered by the requirements of Article 830 of the *National Electrical Code* is answered in Section 90.2(B)(4). Section 830.1 has an explanatory note that references Section 90.2(B)(4). The purpose of this note is to direct users of the *NEC* to the overall scope of the document to better understand the delineation between what is covered and what is not covered. Section 90.2(B)(4) clearly indicates that communications equipment that is under

Keywords

Broadband
Communications
CATV
Network interface unit
NIU
Exclusive control

NEC References

90.2, 830.1, 830.3, 830.179

Broadband installations are covered by various *NEC* articles, depending on the technology.

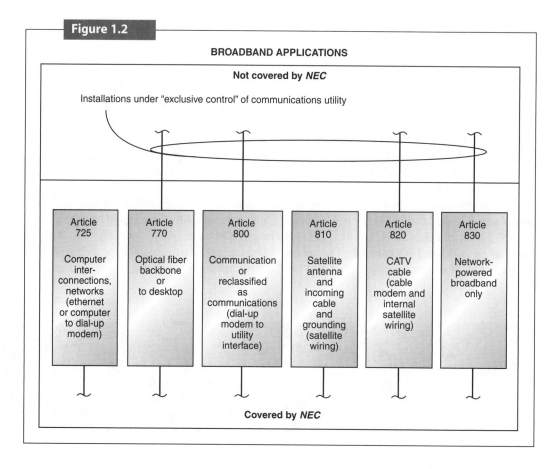

Figure 1.2

the exclusive control of a communications utility and is located either outdoors or in building spaces used exclusively for such installations is outside the scope of the *NEC*.

Another question concerns what is meant by "exclusive control." The intent is that the communications system equipment will only be serviced by the provider of the communications system; such equipment is not intended to be serviced, examined, or otherwise modified by the subscriber or end user. Typically, the wiring on utility poles, regardless of whether they are shared poles, is looked upon as being under the exclusive control of the communications utility and is therefore outside the *NEC*'s scope. With regard to indoor installations, the key is whether the cables are installed in spaces used exclusively for such installations. Generally, the cables run within a building are not looked upon as being installed in spaces used exclusively for such installations and therefore would be covered by the requirements of the *NEC*.

Section 830.179 requires that the cables used with a network-powered broadband communications system be listed for that purpose. The permission to use cables identified in Article 820 is for installations made prior to January 1, 2000. These cables can only be used for low-power circuits. However, the CATV cables covered by Article 820 may be used for output circuits from the NIU as provided in 830.3(D). All output circuit cables used inside buildings are required to be listed in accordance with their specific articles, which means that they may be listed as Class 2 cables, optical fiber cables, communications cables, or CATV cables.

As noted, cables installed within a building or structure are usually not viewed as being under the exclusive control of a communications utility. However, installations and cables that remain within a dedicated room or other portion of a building that is used only by a communications utility and are thus under the exclusive control of a communications utility are not covered by the *NEC*. Most communications rooms are used for other purposes as well, and are accessible to other than utility workers. Such rooms are covered by the *NEC*.

Question 1.3

Does the *NEC* cover wiring on transit buses or trains?

Related Questions

- Does the *NEC* cover wiring on passenger trains and ferries?
- Where do I find the things covered and not covered by the *NEC*?
- Does the *NEC* apply to aircraft?

Answer

The scope of the *NEC* is given in Section 90.2. Section 90.2(A) says what is covered, and Section 90.2(B) lists those things that are not covered. According to 90.2, the *NEC* "covers the installation of electric conductors, electric equipment, signaling and communications conductors and equipment, and fiber optic cables and raceways," followed by a listing of many types of installations and occupancies covered by the *NEC*. The *NEC* provides installation requirements for electrical circuits and systems in all the listed situations.

Section 90.2(B) lists a number of items that limit the scope of the *NEC*. Installations in "ships, watercraft other than floating buildings, railway rolling stock, aircraft, or automotive vehicles other than mobile homes and recreational vehicles" are specifically excluded. This section goes on to list a number of other types of installations related to mining, railways, communications utilities, and electrical utilities that are excluded from coverage by the *NEC*. The excluded items are generally covered by other standards that may be imposed by governmental jurisdictions or by the entities themselves.

The list of excluded installations is really just a statement of the intended coverage of the *NEC*. Some of the provisions of the *NEC* may be incorporated into the rules adopted by any given entity, such as rules adopted by the federal government or rules adopted by utilities. In fact, the 90.2(B)(1) Fine Print Note (FPN) says that some of the provisions of the *NEC* are included in federal rules governing ships. Certain parts of the *NEC* are also adopted by whoever has jurisdiction over safety in ferries and trains. Similarly, utilities often incorporate certain parts of the *NEC* regarding services into their own requirements, and may impose these requirements on any customers who want service from that utility.

Although the *NEC* may not be intended for some specific installation, that does not mean that it would not work to help provide a safe installation in a specific case. According to Section 90.4, the *NEC* is intended to be adopted by governmental jurisdictions and insurance inspectors. It must be adopted by those jurisdictions to have effect. If jurisdictions choose to adopt only parts of the *NEC* or apply it outside its intended scope, that is their prerogative.

Keywords

Train
Aircraft
Jurisdiction
Scope
Covered
Not covered

NEC References

90.2, 90.4

Question 1.4
Are electrical substations covered by the *NEC*?

Keywords

Substation
Overhead power line
Utility
Covered
Not covered
National Electrical Safety Code (NESC)

NEC References

90.2, *NESC*

Related Questions

- Are overhead power lines covered by the *NEC*?
- Are control buildings in substations covered by the *NEC*?
- Do other codes apply to substations and overhead lines?

Answer

The scope of the *NEC* is found in Section 90.2. This section describes what is covered and what is not covered by the *NEC*. According to 90.2(A), the *NEC* "covers the installation of electric conductors, electric equipment, signaling and communications conductors and equipment, and fiber optic cables and raceways." This is followed by a list of more specific descriptions, including "public and private premises, including buildings, structures, mobile homes, recreational vehicles, and floating buildings" as well as "yards, lots, parking lots, carnivals, and industrial substations." Based on this selected language alone, substations and overhead power lines are covered. However, 90.2(B) lists installations that are not covered, including "installations under the exclusive control of an electrical utility . . . consisting of service drops or service laterals, and associated metering, or . . . located in legally established easements, rights-of-way, or by other agreements . . . or . . . on property owned or leased by the electric utility for the purpose of communications, metering, generation, control, transformation, transmission, or distribution of electric energy." Thus, a substation or overhead power line like that shown in Figure 1.4 is covered by the *NEC* unless it is under the exclusive control of a utility or owned by a utility. The *NEC* does not define a utility, because utilities are regulated by government agencies such as state utility commissions and they determine who will be considered to be a utility.

It should be understood that although a substation or an overhead power line that is customer owned rather than utility owned may be within the scope of the *NEC*, that does not mean that the *NEC* provides sufficient information for installing or inspecting such facilities. For example, many of the accepted practices for substations, such as the installation of grounding grids for equalizing step and touch potentials within and around a substation, are not even mentioned in the *NEC*. The *NEC* also says nothing about some of the necessary rules for spacing, sag, line tension, cross arms, climbing space, pole construction, and so on, that apply to overhead medium- and high-voltage installations. Other standards must be used for installations of this type. The usual code that is used for substation and overhead line work is mentioned in 90.2(A)(2), FPN: ANSI C2—*National Electrical Safety Code*. Some states and other jurisdictions use variations or their own versions of this standard or a similar standard that covers the issues not covered by the *NEC*.

Another common argument is that since the *NEC* does not cover a utility-owned substation, the buildings used for switching, generators, metering, and control that may be part of a substation are not covered either. According to 90.2(A)(4), these buildings are covered if they *are not* an integral part of a generating plant, substation, or control center. But this statement says nothing about the application of other codes, such as building, mechanical, or plumbing codes. The applicability of other standards is not determined by the *NEC*. The application of the building code or other codes is determined by the jurisdiction.

Figure 1.4

Overhead line not covered by *NEC* if owned or controlled by a utility.

What is the difference between wet and damp locations?

Question 1.5

Related Questions

- How are wet and damp locations defined?
- Is the interior of a raceway in a wet location also a wet location?
- Is an area considered wet due to the presence of a wet-pipe sprinkler system?
- Is the interior of a greenhouse a wet or damp location?

Keywords

Wet location
Damp location
Dry location
Raceway
Outdoors
Underground

Answer

The definitions of wet and damp locations are found in Article 100, along with the definition of a dry location. These definitions are found under "Location, Damp," "Location, Dry," and "Location, Wet." Clearly, a damp location is somewhere between a dry and a wet location, so to completely understand the definition of a damp location, that definition must be used along with the definitions of wet and dry locations. The definitions are as follows:

Location, Damp. Locations protected from weather and not subject to saturation with water or other liquids but subject to moderate degrees of moisture.

NEC References

Article 100, 225.22, 230.53, 300.5, 334.12, 344.10, 410.4, 501.15, 517.2

Examples of such locations include partially protected locations under canopies, marquees, roofed open porches, and like locations, and interior locations subject to moderate degrees of moisture, such as some basements, some barns, and some cold-storage warehouses.

Location, Dry. A location not normally subject to dampness or wetness. A location classified as dry may be temporarily subject to dampness or wetness, as in the case of a building under construction.

Location, Wet. Installations under ground or in concrete slabs or masonry in direct contact with the earth; in locations subject to saturation with water or other liquids, such as vehicle washing areas; and in unprotected locations exposed to weather.

Figure 1.5 illustrates common examples of wet locations as well as damp and dry locations. Note that the definition of a wet location refers to the *installation* underground or in certain other locations. In effect, the inside of a raceway is in the same location and part of the same installation as the raceway itself, so the wiring that is installed inside raceways in wet locations must be suitable for wet locations. With regard to underground raceways, the requirement for conductors listed for wet locations is explicitly stated in 300.5(B).

The complete exclusion of moisture from a conduit cannot be guaranteed. Conduit systems will breathe, and conduits outside or underground will be especially subject to varying temperatures that will likely cause condensation inside raceways and other enclosures. For

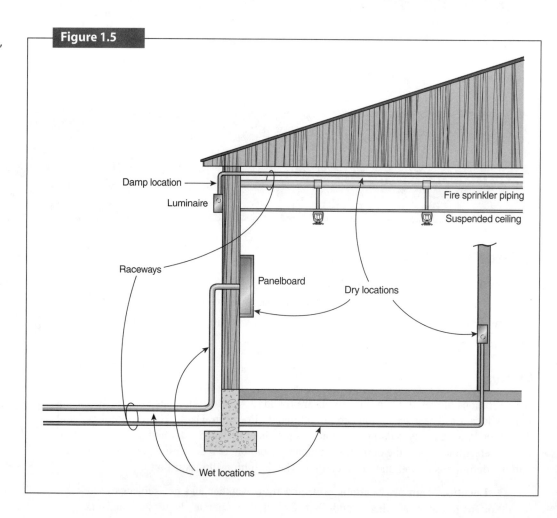

Examples of wet, damp, and dry locations as defined in Article 100.

this reason, Sections 225.22 and 230.53 require that raceways exposed outside or embedded in masonry be arranged to drain. Sections 300.5(G) and 501.15(F) are other examples of code rules that address this issue.

As noted, the definition of a wet location should be compared with the definitions of dry and damp locations to get a more complete picture of the intent and application of these definitions. On one hand, dry locations are areas "not normally subject to dampness or wetness," so the presence of a wet-pipe sprinkler system as shown in Figure 1.5 does not change a dry or damp location to a wet location. On the other hand, a greenhouse may well be a damp or wet location, depending on the way the irrigation system is set up. Many greenhouses used for commercial purposes are dry locations except near the floor, but others, such as those in public gardens and conservatories, may be designed to simulate a highly humid or wet outdoor location and may use overhead sprinklers or misters to create an interior damp or wet location.

The definition of a damp location provides many examples that help to distinguish between dry, damp, and wet locations. For example, not all outdoor locations are wet. Protected areas under eaves or porches are considered to be damp, as shown in Figure 1.5. Only some basements are considered to be damp, depending on the presence of or exposure to moisture.

The locations as defined in Article 100 are for use in applying wiring methods and equipment, so the definitions of dry, damp, and wet locations are significant for the proper selection of wiring methods. Many wiring methods are prohibited in wet or damp locations, whereas others are permitted in any type of location. For example, according to 344.10, Type RMC (rigid metal conduit) is suitable for all atmospheric conditions and occupancies, and the primary limiting consideration is corrosion. Consider Type NM (nonmetallic-sheathed) cable as another example. According to 334.12(B)(4), Type NM cable may not be used "where exposed or subject to excessive moisture or dampness," which would limit NM cable to areas that are not normally damp or wet, or, in other words, to dry locations. Thus, NM cable should not be installed in raceways underground or on the outside of buildings.

In some cases, a location may be more specifically defined or defined for other purposes. For example, in patient care areas of health care facilities, 517.2 provides a definition of a wet location that is concerned more with patient risk than with protection of wiring. Parts of the areas that are considered wet locations with regard to patient care may be damp or wet under the definitions of Article 100. In other cases, as in Section 410.4(A), specific requirements are applied to luminaires that are to be installed in wet or damp locations. In 410.4(D), a bathtub or shower area is defined, not so we know where the wet or damp areas are, but rather so we know where luminaires may not be installed at all.

1.6 Does the *NEC* require listed equipment?

Related Questions

- What alternatives to listing are recognized by the *NEC*?
- Is a large appliance "equipment" or a collection of equipment?
- Must an assembly of listed components be listed as an assembly?

Keywords

Approved
Listed
Labeled

Keywords

Identified
Equipment
Product standard
Machine
Appliance

NEC References

90.7, Article 100, 110.2, 110.3

Answer

Section 110.2 requires that all conductors and equipment be approved. The Fine Print Note that follows 110.2 refers to 90.7 and 110.3 and to the definitions of "approved," "identified," "labeled," and "listed." "Approved" is defined in Article 100 as "Acceptable to the authority having jurisdiction." Section 90.7 discusses the responsibility of the authority having jurisdiction (AHJ) to examine equipment for safety, and explains that listing by a qualified testing organization is intended to provide assurances that listed equipment meets a specific uniform standard. Section 90.7 also explains that the code intends for listing to relieve the AHJ from the task of inspecting internal wiring and internal parts of equipment in the field. For these reasons, listed products are often preferred, but the *NEC* does not require that *all* products be listed, only that they be approved.

Some items covered by the *NEC* are specifically required to be listed or are required to be listed by other standards; others are required to be identified for a use. In some cases, listing is used to identify suitable equipment, and in other cases certain items required by the code may not be available as listed products and need to be approved without a listing. Following are some examples to illustrate these points:

- According to 250.8, grounding connections must generally be made with listed devices, and rigid metal conduit must be listed in accordance with 344.6.
- The *NEC* does not say that fire alarm equipment must be listed, but NFPA 72, *National Fire Alarm Code,* does require listing.
- The fittings used to attach a multiconductor service cable to a building must be *identified* for use with service conductors according to 230.27, but these straps and many others intended for use with raceways, such as one-hole straps for electrical metallic tubing (EMT), are not listed.
- Section 500.8(A) requires equipment for use in classified areas to be identified, and says listing or labeling is one means of identifying suitable products.
- Nonmetallic sheathed cables are required to be "supported and secured by staples, cable ties, straps, hangers or similar fittings designed and installed so as not to damage the cable." These support methods are required by 334.30, but many are not available as listed products, so they must simply be approved.

Annex A includes a list of many product standards that apply directly to equipment that is required to be listed under the *NEC*. This is not a complete list of the standards that are used or available. Where listed products are not available, Section 110.3 provides a partial list of factors that should be considered in judging equipment for approval. These are many of the same factors that are considered in the product standards that are used in the "examinations for safety" mentioned in 90.7. Obviously, listing by a qualified third party to a recognized product standard is a much easier, and usually more thorough, way for a user or inspector to determine suitability of equipment than conducting separate and individual evaluations based on a list of generalized criteria. It should not be surprising that so many approvals are based on listing and labeling of equipment. Where the *NEC* does not specifically require listing, the AHJ must decide on what basis equipment should be approved, or whether listing should be required in a certain case.

The term "equipment" is defined in Article 100 as "A general term including material, fittings, devices, appliances, luminaires (fixtures), apparatus, and the like used as part of, or in connection with, an electrical installation." (Based on this definition, conductors are also equipment, even though the *NEC* mentions both in some contexts.)

Appliances are specifically mentioned as items of equipment. Where an appliance is listed as such, the listing indicates compliance with a product standard that covers the entire appliance. Many large "appliances" are not listed as appliances, but may be listed under some other standard or may just be constructed of listed components. A large machine or control panel or an entire process system may be judged in much the same manner as some portion of a building—as a group of items that can be inspected under the provisions of the *NEC*. However, if a product standard exists for an item, the AHJ may well insist that it be independently evaluated by a listing organization to determine if it meets the applicable standard. This is a decision that is left to the AHJ, because the AHJ carries the ultimate responsibility for judging and approving equipment under the *NEC*. An example of a small, listed, customized, field-assembled, control panel composed of listed components with other associated listed wiring and equipment is shown in Figure 1.6. Article 409 was added to the 2005 *NEC* to provide a basis for inspection and approval of industrial control panels similar to the one illustrated in Figure 1.6.

Section 110.3(A) should not be taken as criteria that should be used only by an AHJ. For the designer or installer of custom equipment and equipment for which there is no product standard, the list of criteria in 110.3(A) may also be used to document the reasons why the equipment should be approved, to aid the AHJ in making an evaluation.

Figure 1.6

Example of an approved assembly of listed components.

Question 1.7: Are all switches required to have a short-circuit or interrupting rating?

Keywords

Short-circuit current rating
Interrupting rating
Fault current
Fault level
Switch
Disconnect

NEC References

110.9, 110.10, 430.7, 430.109

Related Questions

- What short-circuit or interrupting rating applies to nonfused disconnects?
- Is the typical 10 kA rating of heavy-duty disconnects adequate for loads that are within the horsepower or other load current ratings of the switch?
- What devices are expected to "interrupt current at other than fault levels"?

Answer

The 10 kA short-circuit current rating of a nonfused disconnect or other disconnect is not necessarily adequate in all situations. Where the *NEC* refers to "fault levels" in 110.9, it is referring to the levels of current that could occur due to a short circuit or ground fault. Where the code refers to "levels other than fault levels," it is referring to normal running currents and overload levels of current, such as the locked-rotor current of a stalled motor. In the case of a nonfused disconnect switch used for a motor circuit, the nonfused disconnect is not required to interrupt short-circuit or ground-fault current. However, a nonfused disconnect switch used as a disconnecting means is required to interrupt ordinary full-load current and likely overload currents, such as locked-rotor current.

Section 110.9 breaks down into two separate and distinct requirements. The first paragraph of 110.9 applies to equipment incorporating automatic overcurrent devices such as circuit breakers and fuses (excluding motor overload devices or relays) that must be able to interrupt current at the levels occurring in short circuits or ground faults. The second paragraph applies to equipment that does not incorporate automatic overcurrent devices and is not intended to interrupt short-circuit or ground-fault currents but is intended to make or break normal current levels. Most nonfused switches fall under the requirements of the second paragraph of 110.9. Figure 1.7 shows examples of devices that are and are not intended to interrupt fault currents.

For most motors, the current that might be interrupted when the motor is in a locked-rotor condition is somewhere around 500–600 percent of the normal full-load current. Section 430.7 provides additional information for estimating the locked-rotor current based on locked-rotor code letters. Since locked-rotor current is proportional to horsepower (for a given code letter), the horsepower rating on a safety switch or other device with a horsepower rating is an indication of the levels of current that can be safely interrupted. Certain disconnecting means for large or higher-voltage motors are only intended for isolation, are not intended to operate under load, are required to be so marked, and are not required to have horsepower ratings according to 430.109(A) and (E).

The 10 kA short-circuit current rating of a nonfused disconnect switch is most directly related to Section 110.10. This section requires all system components (including the nonfused disconnect switch) to "ride through" the occurrence of a short circuit or a ground fault until the upstream overcurrent device clears or opens the fault current. Where the upstream available short-circuit current exceeds the short-circuit rating of the switch or other component, the overcurrent protective device upstream from this component must generally limit the let-through energy to the rating of the protected electrical component.

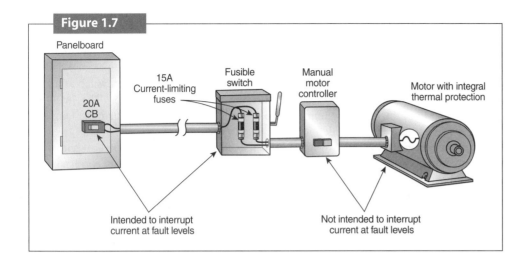

Elements of a circuit that are and are not intended to interrupt current at fault levels.

In other words, nonfused switches and other similar components (such as relays, contactors, and general-duty snap switches) must, in general, be rated at not less than the fault current to which they are exposed. This fault current may be as large as the let-though current of the upstream overcurrent protective device, but is usually limited further by the additional circuit impedance between the upstream overcurrent device and the device under consideration.

These requirements are generally applicable, but are really somewhat idealized because not all aspects of fault current are predictable. Therefore, the *NEC* does not require complete and comprehensive protection for all downstream components. The requirement of 110.10 is that the overcurrent device must clear a fault "without extensive damage to the electrical components of the circuit." Certainly, if the fault occurs in a relay, for example, that component is likely to be damaged and may have to be replaced. For that matter, not all components that interrupt current have short-circuit, interrupting, or withstand ratings. To clarify this point, the last sentence of 110.10 says, "Listed products applied in accordance with their listing shall be considered to meet the requirements of this section."

Question 1.8

How do terminal ratings affect conductor and ampacity selection?

Related Questions

- Do 75°C terminals require the use of 75°C conductors?
- How can conductors be specified if terminals are not available for inspection?
- When may I use a conductor at its 90°C ampacity?
- What is the use of 90°C conductors if most terminals are rated at 75°C or less?

Keywords

Terminal temperature rating
Termination
Conductor
Ampacity

Keywords

Ampacity selection
Adjustment factor
Correction factor

NEC References

110.3, 110.14, 240.4, 240.6, 310.15, Table 310.16

Answer

Section 110.14 covers electrical connections generally, including methods of terminating and splicing conductors. Section 110.14(C) is specifically concerned with the temperature of conductors where they terminate. The terminations may be part of another device, such as an appliance or circuit breaker, or may be a connection to another conductor. The intent of 110.14(C) is to limit the operating temperature of a conductor to no more than the temperature rating of the equipment to which it connects.

When an ampacity is selected from Table 310.16, that ampacity corresponds to one of the three temperature ratings included in the table. Different temperature ratings correspond to different ampacities for any given conductor. Ampacity is defined as the current that a conductor can carry continuously under specified conditions without overheating the conductor. In determining the ampacity of a conductor in any installation, other factors besides the load current must be considered. However, the load current is the most significant factor in determining the operating temperature of a conductor where it terminates on overcurrent devices or other equipment.

If a conductor is loaded to the ampacity taken from the 90°C column of Table 310.16 and is in an ambient temperature of 30°C in accordance with the heading of Table 310.16, we can reasonably expect the copper or aluminum in the conductor to operate at about 90°C. If the termination of that conductor is only rated for 60°C or 75°C, the termination and other connected equipment are likely to be overheated. Such overheating may result in damage to the equipment, equipment malfunction, deterioration of the connection, further increased heat, or other problems that could produce fire hazards and equipment failure. The ampacity of a conductor is not changed by the terminals to which the conductor is connected, but the amount of load carried by a conductor must be limited to no more than the current that would cause the conductor to operate at the rated temperature of the terminal.

Although ampacity is a defined term, it may be helpful to think of more than one kind of ampacity that may apply to a conductor. The first, of course, is ampacity as it is defined. For the purposes of this discussion, this will be called "defined ampacity." This is the ampacity determined by using Section 310.15 and the appropriate table or calculation, most commonly Table 310.16, and considering all the factors that are included in 310.15. Section 310.15 considers primarily conductor temperature rating, load current flow, ambient temperature, and the number of other current-carrying conductors in the same raceway or cable. Higher ampacities may apply to conductors run in free air or in certain configurations in cable trays and may be based on other tables referenced by 310.15, such as Table 310.17.

Another type of ampacity might be called the "usable ampacity." This ampacity is based on the maximum current that can be carried by a conductor where it is connected to an overcurrent device or other equipment or termination. The usable ampacity is based on the ratings and types of connected equipment. Section 110.14(C) addresses terminations and limits the ampacity that is selected for a conductor to the ampacity associated with the temperature rating of the terminal. Unless equipment is specifically marked otherwise, this usable ampacity for terminations and connected equipment is determined from Table 310.16, even where the defined ampacity is calculated in other ways or from other tables.

Overcurrent devices create another limiting factor for the usable ampacity of a conductor, especially where the conductor is used to carry continuous loads. In this case, the limit is based on the ability of a given size of conductor to carry heat away from the overcurrent device. These rules are about proper function of overcurrent devices and about using overcurrent devices as they are tested, not necessarily about overheating of terminals or conductors. Where loads are continuous, overcurrent device ratings may have to be increased to 125 percent of the continuous load, and conductor sizes will have to be increased to correspond to the rating of the overcurrent device. These rules are found in Articles 210, 215, and 230 for branch circuits, feeders, and service conductors, respectively, and are independent and separate from the "defined ampacity" and "usable ampacity" discussed here.

As an example, the defined ampacity of a 6 AWG copper conductor with 90°C insulation is 75 amperes, where the ambient temperature and number of conductors are as stated in the heading of Table 310.16 (not over 30°C ambient and not more than three conductors in a raceway or cable). However, the usable ampacity of a 6 AWG conductor depends on the terminals. If the terminals and equipment on each end of the conductor are rated at 90°C, the usable ampacity is still 75 amperes. This is a very unlikely situation unless the conductor is connected only to other conductors with terminals or connectors rated at least 90°C. Overcurrent devices rated 600 volts or less with 90°C terminal temperature ratings are unusual, if not rare.

Where the 6 AWG conductor is connected to an overcurrent device or other equipment, the assumed rating of the terminations is 60°C, according to 110.14(C)(1)(a). The usable ampacity of a 6 AWG copper conductor connected to terminals rated 60°C is 55 amperes. In this case, we would have to treat the conductor as if its ampacity were 55 amperes; that is, we would limit the load to 55 amperes and provide overcurrent protection for a 55-ampere conductor. If the terminals and equipment *on both ends* are rated at 75°C, we could choose a usable ampacity of 65 amperes from Table 310.16. Note, however, that according to 110.14(C)(1)(a), unless we have terminals with other actual markings, we must assume 60°C ratings for 6 AWG conductors or any other conductors of sizes 14 through 1 AWG. Where the conductor terminals are for conductors larger than 1 AWG and where overcurrent devices are rated at over 100 amperes, terminal ratings are assumed to be 75°C, according to 110.14(C)(1)(b). Section 110.14(C) provides the temperature ratings to be assumed for terminals when the terminals are not available for inspection, such as at the design stage. Although equipment may have terminals rated for 90°C or higher, usually because the load equipment itself produces heat, terminals rated over 75°C are very uncommon on overcurrent devices and panelboards.

Many of the most common conductor insulations now in use are rated 90°C. The ampacity selected for use with these conductors, or what we have called the usable ampacity, is limited by the terminal temperature ratings. However, Section 110.14(C) specifically states that higher-rated conductors may be used, and the ampacities of these higher-rated conductors may be used "for ampacity correction, adjustment, or both." Ampacity correction is the application of a factor to adjust for ambient temperatures that are other than 30°C. Ampacity adjustments are the percentages applied to conductors when the conductor shares a raceway or cable with more than two other current-carrying conductors.

To use the previous illustration, a 6 AWG copper conductor with 90°C insulation may have correction and adjustment factors applied to the 90°C ampacity of 75 amperes, but unless the connected equipment is marked otherwise, the final selected ampacity may not exceed the 60°C ampacity of 55 amperes. For example, assume the 6 AWG conductor is run with five other current-carrying conductors through an ambient that is up to 95°F, as shown in Figure 1.8(a). The correction factor in this case is 0.96 for 95°F from Table 310.16, and the adjustment factor is 0.80, or 80 percent from Table 310.15(B)(2)(a). The corrected and adjusted ampacity is then 75 A × 0.96 × 0.80 = 57.6 amperes. The defined ampacity is 57.6 amperes in this example, while the usable ampacity is 55 amperes. By selecting the 55-ampere value as the ampacity of the wire, the ampacity is "selected and coordinated" as required by 110.14(C) to avoid overheating at terminals and to ensure that the conductor is big enough and has a high enough temperature rating to avoid overheating of the conductor itself as required by 310.15. If the defined ampacity calculated according to 310.15 was lower than the usable ampacity or "terminal temperature ampacity" determined in accordance with 110.14(C), we would use the defined ampacity in order to satisfy both 110.14(C) and 310.15. In any case where the two values are different, we would use the lower of the two values in the selection of overcurrent protection, because both values are limits on the current allowed in the conductors. (The 60-ampere circuit breaker in the illustration is permitted by 240.4(B) since 55 and 57.6 are not standard sizes as listed in 240.6.)

Example of ampacity selection for both "usable ampacity" and "defined ampacity" as explained in the text.

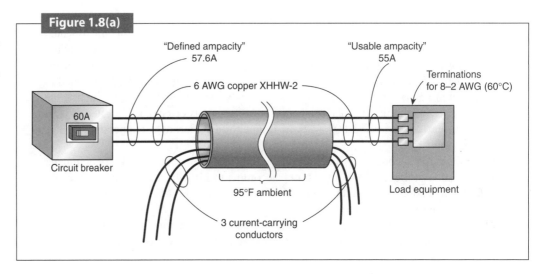

To this point we have been discussing the intent and use of 110.14(C) to ensure that terminations are not overheated. However, in some cases, equipment may be marked with a minimum rating for conductors that may be connected to the equipment. This is not the type of terminal rating that 110.14(C) is concerned with. For example, many types of equipment, such as luminaires, transformers, and electric heaters, produce heat. These types of

An example of terminal temperature markings on a circuit breaker.

Source: Based on *NEC Handbook*, 2005, Exhibit 110.6 (courtesy of Square D/Schneider Electric).

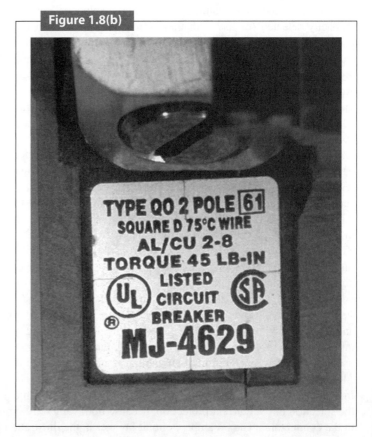

equipment may be marked with a minimum rating for conductors to keep the conductor insulation from being damaged by the operating temperature of the terminals or temperature in the area in which terminations are made. For example, many luminaires are marked to require 75°C or 90°C insulation on conductors. This is a minimum rating, so conductors rated 90°C could be used for either marking. Most circuit breakers in sizes of 20 to 100 amperes are marked 60/75°C, meaning the terminals may be used with conductors of either rating or a higher rating. However, if a circuit breaker is marked 75°C, like the one shown in Figure 1.8(b), wire rated 60°C is not suitable for termination on that device. The marking on the terminals of a circuit breaker or other devices that generate heat is both a minimum rating for insulation on connected conductors and a maximum operating temperature for those conductors. The minimum rating is part of the listing and labeling instructions that must be followed according to 110.3(B). The maximum operating temperature is the subject of 110.14(C).

Note that in Figure 1.8(b), the wire range of 2 to 8 AWG would normally be assumed to have 60°C terminals according to 110.14(C). However, the requirement for at least 75°C wire is not a significant restriction in new work because wire with 60°C ratings is very unusual in new work, and 60°C insulation types have not been widely used in the United States since the mid-1970s.

Question 1.9

Where can I find information on protective electrical apparel?

Related Questions

- What section of the *NEC* provides information on protective electrical apparel for different voltage classes and other characteristics?
- Is there a difference between clothing requirements for arc-flash and arc-blast protection?

Answer

The document that should be referred to for protective clothing requirements related to electrical safety is not the *National Electrical Code*, but rather NFPA 70E, *Standard for Electrical Safety in the Workplace*. Related to this standard are two additional National Fire Protection Association (NFPA) products that may be of interest. One is a textbook covering electrical safety issues in depth titled *Electrical Safety in the Workplace,* by Ray and Jane Jones. The other is a 20-minute video, *Working Safely with Electricity*. All of these products are available directly from the NFPA.

Under NFPA 70E, the selection of apparel and other personal protective equipment (PPE) is primarily for electrical shock and arc-flash protection. The selected apparel will also provide some protection from arc-blast hazards, but methods for assessing and providing protection from arc blast have not been fully developed yet. The selection of apparel for shock and flash protection is part of the required overall hazard-risk analysis of a particular task. In

Keywords

PPE
Protective apparel
Flash hazard
Incident energy

NEC References

110.16, NFPA 70E

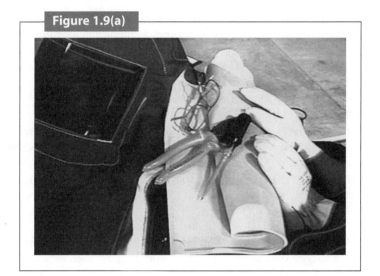

Examples of personal protective equipment, including switching suit and hood, voltage-rated tools and gloves, protective blankets, voltage tester, and safety glasses.

Source: NFPA 70E Seminar.

order to select apparel or other equipment, one must identify the task, determine the incident energy levels, select the apparel or equipment based on the task and incident energy, and then confirm the fire-resistive ratings of the apparel and the voltage ratings of some other tools and equipment. NFPA 70E provides a step-by-step method for performing this analysis. The only safe alternative, and really the preferred alternative, is to avoid working on energized equipment whenever possible. Figures 1.9(a) and 1.9(b) show some examples of personal protective equipment and applications.

Electrical worker clothed in personal protective equipment appropriate for the hazards involved.

Source: Based on *NEC Handbook*, 2005, Exhibit 110.7.

How are the requirements for working spaces determined?

Question 1.10

Related Questions

- What types of equipment require working space?
- How is the "nominal voltage to ground" determined?
- What is meant by "exposed live parts"?
- How is depth of working space determined for live parts that are not normally exposed?
- Is a wall constructed of gypsum wallboard considered to be grounded?
- Can doors to rooms or enclosures open into a working space?
- Can a required working space extend through a doorway?
- Are separate working spaces permitted to overlap?
- Is easily movable equipment allowed in required working space?
- When are two exits required?
- What is a "continuous and unobstructed way of exit travel"?

Keywords

Work space
Working space
Exposed live parts
Energized work
Large equipment
Entrances
Panic hardware
Exit travel

NEC References

110.26, 110.32, 110.33, 110.34

Answer

"Spaces About Electrical Equipment" is the title of Section 110.26. This section provides the general requirement that all equipment have "sufficient access and working space" as well as the explicit requirements for specific minimum dimensions where equipment is likely to be worked on while energized. (Requirements for dedicated space are discussed separately.) Very similar rules are found in 110.32, 110.33, and 110.34 for equipment over 600 volts. As may be expected, where equipment over 600 volts is likely to be worked on while energized, the rules are somewhat more restrictive and the required working spaces may be significantly larger.

Sections 110.26 and 110.34 mandate specific minimum dimensions of working space only when equipment is "likely to require examination, adjustment, servicing, or maintenance while energized." For the purposes of this discussion, we will shorten this to "energized work" or "worked on while energized," but all four conditions should be remembered. Simply using test equipment to check voltage or current levels on exposed parts is a type of "examination" while energized and therefore falls under what is termed here "energized work."

The *NEC* does not provide a list of types of equipment that are likely to be worked on while energized. That determination is left to the authority having jurisdiction. Nevertheless, there are certain types of equipment that are generally understood to be subject to energized work. These include panelboards, switchboards, motor control centers, control panels, and many types of disconnects and controllers. The mere possibility that somebody could work on something energized is not the criterion, however. The likelihood of energized work is the criterion for requiring working space. A judgment of this likelihood should be based on common and prudent work practices of trained persons, not the most risky or least likely work practices. For example, receptacles and snap switches in general should not and need not be subject to energized work, and checking voltage at a receptacle or inserting a plug into a receptacle is not intended to be considered exposure to live parts.

The requirements for working space are based on a person being exposed to live parts. The general installation rules of the *NEC* will make exposure to live parts unlikely under normal conditions because most live parts that present a shock or fire hazard are normally enclosed. However, when equipment covers or doors are removed or opened to do energized work, the live parts become exposed.

Article 100 defines the terms "live parts" and "exposed" (as applied to live parts). Live parts are simply "energized conductive components." "Exposed" as applied to live parts is defined as "Capable of being inadvertently touched or approached nearer than a safe distance by a person." It is applied to parts that are not suitably guarded, isolated, or insulated.

Working space is three-dimensional. The minimum depth, width, and height of the required working space are specified in Sections 110.26(A)(1), (2), and (3), respectively. The minimum depth is based on the conditions and the voltage to ground. The minimum width is the greater of 30 inches or the width of the equipment. The height may not be less than the required headroom, which is the greater of 6.5 feet or the height of the equipment. For equipment over 600 volts, the minimum width is 36 inches, and greater depths are required, but the depth is still based on the same three conditions and the voltage to ground.

According to 110.26(A)(1), the depth of the working space is measured from the exposed live parts, or from the enclosure if the live parts are enclosed. This language recognizes that exposed live parts may not be exposed normally, but the need for working space is based on conditions in which those live parts will be exposed by removing covers or opening doors. Also, if enclosed parts are on both sides of a working space and personnel are likely to be subject to energized work, it is assumed that both enclosures may be open at the same time, thus exposing the worker to Condition 3, "exposed live parts on both sides of the working space." This assumption is restated or reinforced by the language of 110.26(A)(1)(c), where, in existing buildings, restricted permission to use Condition 2 between facing equipment is provided. The three conditions are illustrated in Figure 1.10(a).

The depth of the working space is based on the "nominal voltage to ground." Article 100 provides a definition of the voltage to ground for both grounded and ungrounded systems. For grounded systems, it is "the voltage between the given conductor and that point or conductor of the circuit that is grounded." For ungrounded systems, the actual voltage to ground is not always known and may fluctuate. Nevertheless, in order to apply *NEC* rules to ungrounded systems, the voltage to ground is defined as "the greatest voltage between the given conductor and any other conductor of the circuit." Table 110.26(A)(1) also uses the term "nominal," and nominal voltage is also defined in Article 100. A nominal voltage is a convenient way of designating the voltage class of a system, and may be used even where the actual voltage varies "within a range that permits satisfactory operation of equipment." For example, the voltage may vary somewhat in a nominal 120/240-volt system, but the voltage to ground of this system is still defined as 120 volts. In a similar manner, the voltage of a 480-volt ungrounded delta system may vary as well, and the measured voltage to ground may be significantly more or less than 480 volts, but the voltage to ground of the system is *defined* as 480 volts. This definition provides a consistent basis for application of the working space rules and other provisions of the *NEC*. The definitions quoted previously refer to "the given conductor," so determining the nominal voltage to ground requires applying the definitions to each of the ungrounded conductors.

The three conditions illustrated in Figure 1.10(a) are based on the types of materials on the side of the working space that is opposite to the exposed live parts. One might think of the parts opposite to the live parts as being those behind the person performing energized work. In Condition 1, there will be no live or grounded parts "behind" the worker or opposite to the exposed live parts on which work is being done. Condition 2 covers situations in which the parts on the opposite side of the space are grounded. In Condition 3, there will be exposed live parts on both sides of the working space, that is, in front of and in back of the worker.

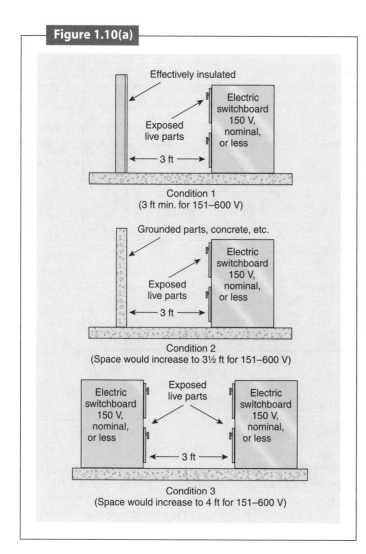

Figure 1.10(a)

Illustrations of three conditions described in Table 110.26(A) and Table 110.34(A).

Source: Based on *NEC Handbook*, 2005, Exhibit 110.9.

Where enclosures containing live parts face each other, Condition 3 applies because it is assumed that the enclosures could be open on both sides at the same time. However, a consideration of what is on the opposite side of the working space should include only those items that are at the boundary of the working space. An enclosure for live parts that is beyond the working space depth for Condition 3 should be treated separately. In other words, if live parts are on both sides of an otherwise open area, but separated by more than the Condition 3 working space depth, they should be considered separately as two independent spaces based on Condition 1, as illustrated in Figure 1.10(b). This rule of thumb is modified when using doubled working space depths in lieu of two entrances, as discussed later in this answer.

The description of Condition 2 states that "concrete, brick or tile walls shall be considered as grounded." ("Tile" in this context is intended to refer to ceramic tile or similar products, not linoleum tile, vinyl tile, carpet tile, or the like.) The *NEC* is very specific about these three types of construction products, because they might otherwise be considered to be nonconductive, and therefore might otherwise be interpreted as creating Condition 1. Other wall finishes that are nonconductive, such as drywall or wood, typically fall under Condition 1. Metallic enclosure surfaces or covers that are not removable fall under Condition 2.

Three working spaces that are allowed to overlap, with two items that are facing each other but separated by more than Condition 3 working space, treated independently as Condition 1 working spaces.

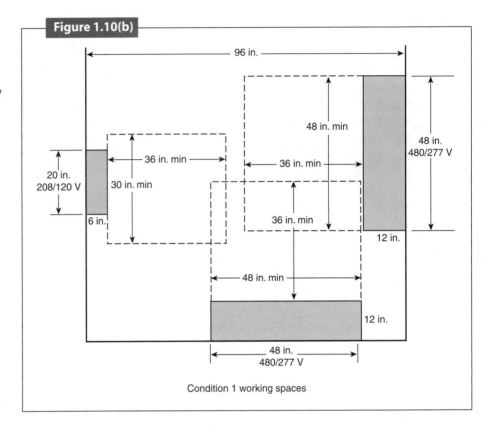

Figure 1.10(b)

Condition 1 working spaces

Where many items of equipment are installed side by side or on adjacent or opposite walls, the individual working spaces are permitted to overlap as shown in Figures 1.10(c) and 1.10(d). Stated another way, two or more pieces of equipment that are subject to energized work may share working space. This is true simply because the *NEC* does not say otherwise, and requiring individual, nonoverlapping working spaces for many items, such as time clocks, small safety switches, and controllers, would create an absurdity. The *NEC* does not anticipate that each item of equipment will have separate persons working on them simultaneously, so discrete working spaces are not required.

The 30-inch width of the working space is not required to be separate for each item of equipment or centered on each item of equipment, but space must be sufficient for safe operation and maintenance of each item of equipment.

Source: Based on *NEC Handbook*, 2005, Exhibit 110.12.

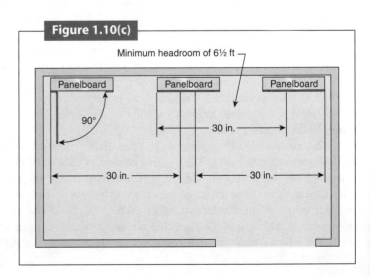

Figure 1.10(c)

According to Section 110.26(A)(2), the width of the working space (and the arrangement of equipment) must be such that equipment doors and hinged panels can be fully opened to at least a 90-degree angle. In effect, this says that equipment doors may open into the working space because the working space must be adequate for such opening. What about other doors, such as doors into a room? Can a door into a room open into the working space, or can the working space be arranged in such a way that a door must be either open or closed to provide the working space? The *NEC* does not address this issue directly. The direction of the swing of a door is addressed only for "large equipment," as discussed later in this answer. The *NEC* does not prohibit a panel from being located behind an open door, nor does the *NEC* require that doors be either open or closed to provide working space. In many cases, equipment such as panelboards may be safely and conveniently located in equipment closets or otherwise behind doors that can be fully opened to provide working space. Conditions such as these should be addressed on a case-by-case basis by the authority having jurisdiction. An installation that might be safe in one case might not be safe under other circumstances. The use of the door, the amount of traffic through the area, the likelihood that the door would be obstructed, and other issues should be considered.

Once a working space is defined, that area must be kept clear and may not be used for storage. This rule is found in 110.26(B), which also requires that a working space in a passageway or general open space be guarded while being used as a working space. This section implies another important point: that working space may be part of a passageway such as a hall or an aisle in an industrial or manufacturing facility, or may be part of a larger open area. However, the working space may not be used for storage, regardless of how easily the stored materials may be moved. Also, while the equipment is open and live parts are exposed for energized work, the working space must be "suitably guarded." This guarding might be provided by temporary barricades or by permanent barriers in the form of bollards or other guards.

Working space must also be free of obstructions and other equipment, including other electrical equipment. According to 110.26(A)(3), "the work space shall be clear and extend from the grade, floor, or platform to the height required by 110.26(E)." Section 110.26(E) provides the minimum headroom: the larger of 6.5 feet or the height of the equipment. Within this height, other electrical equipment ("equipment associated with the electrical installation") that is located above or below the equipment for which working space is required may extend up to 6 inches into the working space. For example, a wireway, a small transformer, or a time clock that is up to 10 inches deep may be located above or below a panelboard that is 4 inches deep without violating the working space of the panelboard, as illustrated in Figure 1.10(d).

Another need that must be addressed once a working space is defined is the requirement for access (and egress) from the working space. Entrances to (and exits from) working space are covered by 110.26(C). The minimum requirement of 110.26(C)(1) is that each working space be provided with at least one entrance of sufficient area. ("Sufficient area" is not defined in the *NEC*, but may be defined in building codes, or may be determined by the authority having jurisdiction.) Specific minimum dimensions are provided only for "large equipment" and equipment over 600 volts. "Large equipment" is defined as equipment that is rated 1200 amperes or more and contains overcurrent devices, switching devices, or control devices. Typically this means panelboards, switchboards, and motor control centers. For equipment over 600 volts, there is no ampere rating in defining large equipment. Large equipment over 600 volts is simply a switchboard or control panel that is over 600 volts and over 6 feet wide.

The general rule for large equipment is that two entrances are required: one at each end of the equipment or working space, as shown in Figure 1.10(e). (The diagram shows working space at the ends of the panelboard, but this working space is required only if there are serviceable parts behind the access panels.) Although the rule is about entrances, the objective of two entrances, especially for large equipment, is to provide for safe and reliable means of

Associated electrical equipment is permitted to extend up to 6 inches into working space.

Source: 1999 *National Electrical Code Changes*, p. 13.

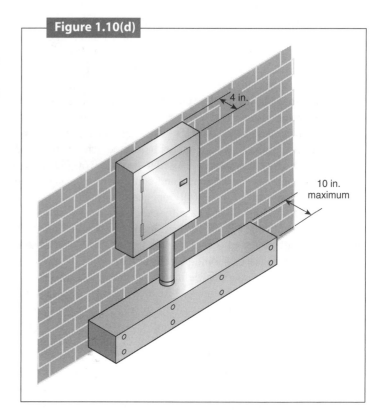

Figure 1.10(d)

exit. When accidents occur while working on large equipment, the resulting arcing or flame may engulf the equipment, obstruct the working space, or block an exit. Also, arcing faults on large equipment may not produce enough current to operate overcurrent devices, so the arcing may continue for relatively long periods of time. For these reasons, two exits are required from large equipment working spaces. Alternatively, a way may be provided to get out of the area without having to pass through the working space. Two ways to provide a way around the working space are to double the depth of the working space or to provide a "continuous and unobstructed way of exit travel" from the working space. The main rule and the two alternatives are found in 110.26(C)(2), or, for over 600 volts, in 110.33(A)(1).

The minimum size of the entrances for large equipment of any voltage or for any required working space for equipment over 600 volts is 2 feet wide and 6.5 feet high. The entrances

The basic rule for large equipment requires entrances at both ends of the working space.

Source: *NEC* Seminar.

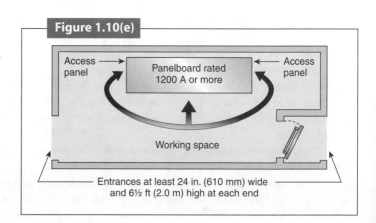

Figure 1.10(e)

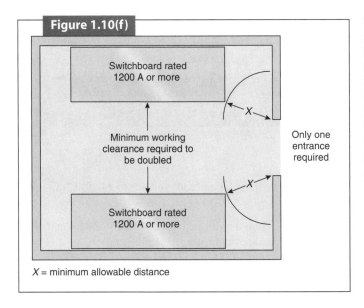

Where working space is doubled and one entrance is provided for large equipment, the distance from the edge of the large equipment to the entrance must be at least equal to the required working space before it is doubled.

Source: Based on *NEC Handbook*, 2005, Exhibit 110.18.

need not be doors, and could simply be openings to larger areas, or the equipment and the working space could be located in a larger open area. Where the entrances are doors, they must be equipped with "panic hardware" that allows the door to open under simple pressure (because workers who have been burned may be unable to operate knobs or levers), and the doors must open in the direction of exit travel. (The requirements covering door swing and hardware types were new in the 2002 *NEC* and are not retroactive, so they do not apply to existing installations, although they may require modifications when installations are upgraded.)

Where the option to provide doubled working space is used to permit the use of a single exit from large equipment working spaces, the edge of the door, opening, or other entrance area must be located at least as far from the end of the equipment as the normal (not doubled) working space depth required for the equipment and the condition, as shown in Figure 1.10(f). This means that the exit pathway will not be through the working space or closer to the equipment than the depth of the working space. A continuous and unobstructed exit pathway, as shown in Figure 1.10(g), will also provide an immediate exit from the working space, and will not require traveling through the working space to reach the entrance. This objective is easily met in large open areas, but must be carefully considered in enclosed rooms.

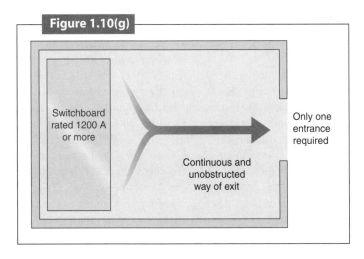

Large equipment with a continuous unobstructed exit pathway is not required to have two exits from the working space.

Source: Based on *NEC Handbook*, 2005, Exhibit 110.17.

Question 1.11 What working spaces apply to HVAC equipment?

Keywords

Clearance
Work space
Working space
Headroom
Grid ceiling
Energized
Sufficient access
Lock-out
Tag-out

NEC References

110.26, 430.14, 430.102, 430.107, 440.14

Related Questions

- Are working clearances required for HVAC equipment mounted above a grid ceiling?
- Are working spaces required for air-conditioning disconnects?
- Do the working clearances listed in Section 110.26 take precedence over the manufacturer's published or labeled clearance requirements?
- Do the clearance requirements of 110.26 apply to equipment located above suspended ceilings?
- Should equipment above dropped ceilings be treated differently due to the relatively small (low headroom) equipment spaces there?

Answer

This answer is based on the assumption that the specific HVAC electrical equipment mentioned is likely a blower motor, although other types of equipment could be involved. Section 430.14 of the NEC requires that motors be located so that there is adequate ventilation and enough space around the motor to perform the necessary maintenance. Similarly, 110.26 states that "sufficient access and working space shall be provided and maintained about all electric equipment to permit ready and safe operation and maintenance of such equipment." This rule applies to all types of electrical equipment that requires space for maintenance or operations, regardless of whether the equipment is energized while maintenance or service is performed. However, "sufficient access and working space" is variable, depending on the type and location of the equipment.

The working space described in 110.26(A) applies only to equipment that is "likely to require examination, adjustment, servicing, or maintenance while energized." (For the purposes of this discussion, this will be shortened to just "energized work.") The headroom requirements apply to equipment for which energized work will be done, unless, according to 110.26(E), the equipment is service equipment, switchboards, panelboards, or motor control centers, in which case the headroom must be provided regardless of whether energized work is expected.

The working clearances required by Section 110.26(A) are not required for the motors in question here, nor is the headroom from 110.26(E). However, a disconnecting means is required for this equipment in accordance with Sections 430.102 or 440.14, or both. Since the 2002 NEC, Section 430.102 requires that a motor must generally have a local disconnect "in sight" as defined in Article 100. The exemption for qualified persons using lock-out, tag-out programs is only applicable to industrial facilities. In some cases, a local disconnect may produce an increased hazard, and in such cases, the local disconnect is not required. Where only low-voltage HVAC or similar controls are likely to be worked on, reduced clearances are permitted by 110.26(A)(1)(b) because shock and flash hazards are greatly reduced or eliminated. Generally, however, a local disconnect is required. (Prior to the 2002 NEC, most motors did not require local disconnects if the disconnect at the controller could be locked in the off position.)

If the equipment and disconnecting means are not likely to require energized work, the specific working space described in 110.26(A) is not required, and in most cases, there is no specific headroom requirement. As noted, "sufficient access and working space" is always required. The manufacturer's instructions should be used in this case unless the applicable mechanical or plumbing code requires larger spaces. Typical rules from other codes require space adequate for removal and replacement of any serviceable parts, and specific clearances from top, bottom, sides, back, and front are often provided by the manufacturer. These clearances may sometimes be zero inches in some directions, such as on the sides or at the back of many furnaces. The furnace pictured in Figure 1.11 requires clearances from combustibles and clearances in the front that are specified by the manufacturer. The disconnect (toggle switch) shown on the wall does not require any specific clearance or working space, but the electrical equipment on the inside is likely to require service while energized, and working space should be provided in accordance with 110.26(A).

The final judgment of the likelihood of energized work rests with the authority having jurisdiction, and opinions vary about the disconnecting means themselves. Some may assume that, say, safety switches require working space, but toggle switches used as disconnects do not, or that fused switches require workspace, but nonfused switches do not. To provide working space, disconnects may be located below the grid ceiling in some cases, but opinions differ on whether the tiles in the grid interfere with the "in sight" rules, even though at least a few tiles will be removed to gain access. The question of ready access must also be addressed. For motors, at least one disconnect in the circuit must be readily accessible to comply with 430.107, and for refrigerant motors, a disconnect must be readily accessible from the equipment to comply with 440.14. Similar rules may apply to disconnects for appliances or other equipment. Some jurisdictions may want labels or signs

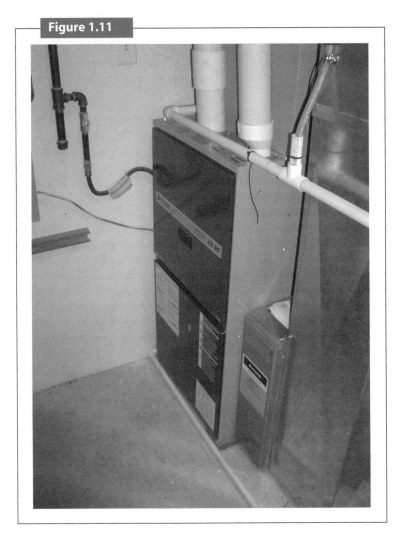

Figure 1.11 HVAC equipment that requires working space at the front for servicing of internal energized parts.

to indicate equipment that is not to be serviced while energized. The local authority should be consulted for this interpretation.

A decision about the likelihood that a disconnect will be serviced or examined while energized should consider the persons likely to use that disconnect. For example, a disconnect for HVAC equipment is very likely to be "examined" by a person doing HVAC service or maintenance simply because it is an easy place to expose parts and conductors to measure voltage or current or check for blown fuses. In some cases, service work by persons who are not electricians is confined to the portion of the circuit from the disconnect to the HVAC equipment by local laws or work rules. Where a disconnect is likely to be opened for such service, working space is required. Working space is not only for electricians.

Also, where renovations are concerned, jurisdictions will have different rules about which equipment or installations must be modified to comply with the present code, and which may remain as they are because they complied with previous codes. In general, the *NEC* is not intended to be retroactive.

Question 1.12 Are enclosed panels facing each other considered to be "exposed live parts on both sides"?

Keywords

Live parts
Exposed
Enclosure
Deadfront
Working space
Energized work
Voltage to ground

NEC References

Article 100, 110.26

Related Questions

- What is meant by "exposed live parts"?
- Would an installation consisting of two enclosed panelboards mounted on walls facing each other fall under Condition 3 of Table 110.26(A)(1)?
- How does Condition 3 in Table 110.26(A) apply where facing panelboards are of different voltages?
- How are depth measurements made?

Answer

Section 110.26 of the *National Electrical Code* requires sufficient access and working space to be provided and maintained about all electric equipment to permit ready and safe operation and maintenance of such equipment. Where electrical equipment is likely to be worked on (examination, adjustment, maintenance, or service) while energized, the specific clearances of Table 110.26(A)(1) are required. Although the panels are enclosed, live parts will be exposed while energized work is taking place. The assumption in the rule is that both panels could be open at the same time; thus, there will be "exposed live parts on both sides," and Condition 3 does apply.

Note, however, that Section 110.26(A)(1)(c) permits Condition 2 working clearances "between dead-front switchboards, panelboards, or motor control centers located across the aisle from each other" in existing buildings where electrical equipment is being replaced and "where conditions of maintenance and supervision ensure that written procedures have been adopted to prohibit equipment on both sides of the aisle from being open at the same time and qualified persons who are authorized will service the installation." Although these conditions do not apply in new installations, they do help to clarify the intent of the rule. That is, normally, panels facing each other are considered to be Condition 3 for the purpose of defining the depth of required working space unless the distance between the panels is more than the required depth of the working space. Furthermore, the language in 110.26(A)(1) says that measurements shall be made "from the enclosure or opening where the live parts are enclosed." This clearly recognizes that the live parts in question are likely to be exposed only while enclosures are open and energized work is taking place. At some distance, equipment facing each other is not close enough to be covered by Condition 3. The NEC is not specific on this point, but a reasonable interpretation of this is that where the distance between facing equipment is more than the depth required for any of the items of equipment, each equipment may be treated separately under Condition 1.

Consider the situation illustrated by Figure 1.12. Both of the panels (the 208Y/120 and the 480Y/277 panels) require working space. Larger working space is required for the 480Y/277-volt panel because the voltage to ground is 277 volts, or between 151 and 600 volts. Condition 3 for this panel requires 4 feet (1.2 meters). Condition 3 for the other panel is only 3 feet (900 mm) based on its voltage to ground of 120 volts. The two panels may share the same working space. From a code language standpoint, this is true simply because the rule never says otherwise. Separate, dedicated working space is not required for each item of equipment. Nevertheless, the working space that is required is based on the largest dimension that applies to that space, so in this case, the minimum dimension between the two panel enclosures is 4 feet. This space must also meet the requirements of 110.26(A)(2) for width, which is a minimum of 30 inches or the width of the largest equipment, along with the requirements of 110.26(A)(3) and 110.26(E) for headroom or the height of the working space. The height is not less than 6.5 feet (2.0 meters) or the height of the equipment if the top of the equipment is over 6.5 feet above the floor. If the panels were more than 4 ft apart, each might be considered as Condition 1, and in some cases, the working spaces might still overlap.

One other note: If a panel or other equipment is supplied by an ungrounded system, such as an ungrounded 480Y/277-volt system or an ungrounded 480-volt delta system, the voltage to ground may not be known or predictable in practice. However, for the purposes of *NEC* rules, the voltage to ground in an ungrounded system is defined in Article 100 as the maximum voltage between conductors.

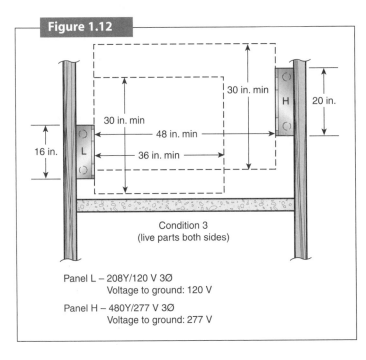

Figure 1.12 Panelboards with Condition 3 working space between them (exposed live parts on both sides of the working space).

Question 1.13

Do transformers require working space for servicing while energized?

Keywords

Transformer
Working space
Energized work

NEC References

110.26, 110.31, 110.34

Related Question

- What clearances are required for transformers?

Answer

According to Section 110.26(A), working space sufficient for ready and safe operation and maintenance is required for all electrical equipment. Such spaces may be very small, perhaps only big enough to remove the equipment or get access to the equipment. Spaces of the specific dimensions outlined in 110.26(A)(1), (2), and (3) are required only for equipment that is likely to require examination, adjustment, servicing, or maintenance while energized. Thus, working space of a specific minimum depth, width, and height is only required for transformers that will be examined, adjusted, serviced, or maintained while energized.

Since the dimensions of 110.26(A) are based on exposed live parts, the need for specified working spaces is based on the likelihood that live parts of transformers will be exposed while energized, and that likelihood depends on the type and use of the transformer. (From here on in this discussion, "working space" will be used to refer to the specific spaces required by 110.26(A), and "energized work" will mean examined, adjusted, serviced, or maintained while energized.)

Transformers may be one of a number of types. Probably the most common types installed under the *National Electrical Code* are dry-type transformers. These may be the small control type, larger enclosed encapsulated (nonventilated) types, or still larger enclosed and ventilated types. Oil-filled transformers may also be used, and these may be totally enclosed of the pad-mounted type or may be the types with some normally exposed live parts (bushings) that are mounted on poles, in vaults, or in substations. Each of these is different and should be treated differently with regard to working space. Certain spaces or clearances may be required for other reasons, as will be discussed later in this answer.

Many common control transformers are used to create Class 2 circuits. These transformers often have terminals that are normally exposed because they do not present a shock or fire hazard. There is little if any energized work that needs to be done on Class 2 transformers, and there is no shock or fire hazard anyway, so this type of transformer does not need working space. Other control power transformers are usually installed in control panels or controllers that themselves require working space, so working space is also provided for the control transformers.

Encapsulated dry-type transformers have separate wiring compartments with insulated leads. According to the note describing Condition 1 in Table 110.26(A)(1), insulated wire operating at not over 300 volts to ground is not considered a live part, so most such transformers don't have exposed live parts. Since there is nothing that can be accessed in the encapsulated windings, working space is unnecessary. Encapsulated transformers are usually not larger than about 15 to 25 kVA.

Enclosed ventilated dry-type transformers are commonly used to create separately derived systems that supply voltage systems that are different from the supply voltage or to isolate systems without changing voltage. For example, 208/120-volt systems are often needed in buildings that are otherwise supplied by 480/277-volt systems. These are very common in

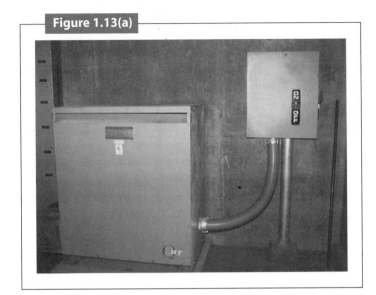

Figure 1.13(a) Dry-type transformer used for creating a separately derived 208Y/120-volt system from a three-phase, 480-volt system.
Source: *NEC* Seminar.

sizes from about 30 kVA to 112.5 kVA and are available in much larger sizes. Most are 600 volts or less, but some may be rated over 600 volts. A transformer of this type is shown in Figure 1.13(a).

Dry-type transformers have live parts that may be exposed by the removal of the cover, as shown in Figure 1.13(b). However, in most cases, the cover is not intended to be removed while energized and often cannot be removed safely while the transformer is energized. Usually, voltage and current measurements are more easily and safely made in the disconnecting means or at the overcurrent devices on the primary and secondary side rather than in the transformer itself, and tap changing is not done while the transformer is energized. In other words, such transformers are generally not subject to energized work, and therefore working space is not required.

One possible and increasingly common maintenance task may be done while a transformer is energized, however: thermographic (infrared camera) surveys. For this type of examination to be meaningful, the terminals in the transformer must be exposed while energized and while under normal load. Contact with terminals is not required, but exposure of termi-

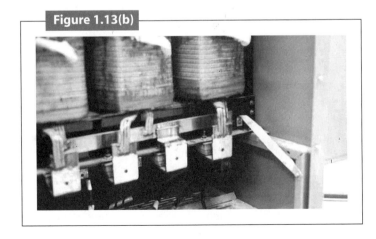

Figure 1.13(b) Dry-type transformer with cover removed, primary conductors terminated, and secondary yet to be connected.
Source: *NEC* Seminar.

nals to view is necessary. This task may require de-energizing the transformer, removing the cover, reenergizing the transformer, and waiting for normal loading conditions to be reestablished before the thermographic examination can take place. In other cases, the terminals may be exposed to view by partially removing the cover while energized. Either way, if this type of service is likely to be performed, working space is required.

Most pad-mounted, enclosed, oil-filled transformers have doors that can be opened while the transformer is energized. In many cases, the primary (over 600 volts) terminations are made using "load-break ells." These are designed to be connected and disconnected with an insulated tool or "hot stick." Other tasks, such as voltage or current measurements or service of metering, may be done in the secondary or primary compartments. Where such energized work is likely, the working space of the appropriate dimensions from 110.34 must be provided. The same rules apply to transformers in vaults, on poles, and in substations, where live parts are normally left exposed, but only to qualified persons. Table 110.31 also includes minimum dimensions to be provided between fences and live parts. These rules do not apply to utility transformers on the supply side of the service point, but similar requirements are found in the *National Electrical Safety Code*, which is the basis for the codes most commonly used by utilities.

This discussion has focused on working space for transformers. Other clearance requirements apply to most transformers. Most transformers must be readily accessible and have minimum required clearances for ventilation. Many require minimum clearances from combustibles. Oil-filled transformers may have to be installed in vaults or be provided with safeguards to protect fire escapes, windows, and door openings. These issues are all covered by Article 450 and are discussed in Chapter 4 of this book.

Question 1.14

What clearance is required for equipment that is serviced only while de-energized?

Keywords

Clearance
Work space
Working space
Energized
De-energized
Sufficient access
Non-electrical
Machine tool
Cranes
Hoists
Low voltage

NEC References

110.26, 670.1

Related Questions

- Is working space required for equipment in general even if no work is done while the equipment is energized?
- Are there any other specific rules covering clearances for machine tools or other special cases?

Answer

All electrical equipment is subject to the general requirement of Section 110.26, which states that "sufficient access and working space shall be provided and maintained about all electric equipment to permit ready and safe operation and maintenance of such equipment." For equipment that is not likely to be worked on or examined while energized, the *NEC* does not provide specific working clearance measurements. For example, conduit is "equipment," but it may usually be embedded or otherwise rendered inaccessible because access is not needed. Similarly, although boxes must be accessible, the amount of space required at a box is not specified. Certainly, adequate space must be provided to service, remove, or replace any parts that are serviceable or replaceable and to provide adequate ventilation where ventilation is required. Deciding what is "sufficient access and working space," especially for equipment that will not be serviced while energized, is the responsibility of the authority having jurisdiction.

Section 110.26(A)(1)(a) reads as follows:

Dead-Front Assemblies. Working space shall not be required in the back or sides of assemblies, such as dead-front switchboards or motor control centers, where all connections and all renewable or adjustable parts, such as fuses or switches, are accessible from locations other than the back or sides. Where rear access is required to work on non-electrical parts on the back of enclosed equipment, a minimum horizontal working space of 762 mm (30 in.) shall be provided.

The first sentence of this rule clearly states that working space is not required around the backs and sides of equipment where all equipment that requires servicing while energized is accessible from other than the rear or sides. This part of the rule applies to virtually all installations. The second sentence provides a reduction in working clearance to 30 inches where there are only non-electrical (never energized) parts that will be serviced, examined, or maintained. In other words, where equipment can be serviced from the front while energized, and there is no equipment that will be serviced at the sides or rear, the *NEC* does not establish specific minimum rear and side working space. For example, in common practice, most panelboards are mounted on or in a wall with no rear access and little if any space at the sides because all parts (whether energized or not) are accessible from the front of the enclosure.

Although the spaces for industrial machinery were at one time modified by Section 670.5, that section was deleted from the 2005 *NEC,* so the working space for industrial

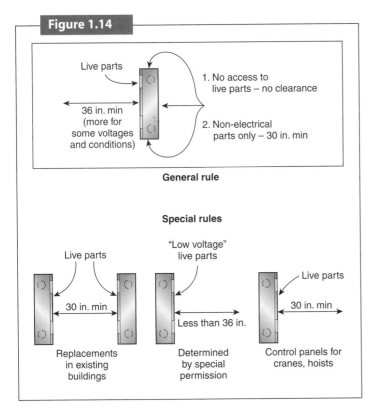

General and special rules for clearances from enclosures housing live parts that are likely to be subject to energized work.

machinery is now the same as for other electrical equipment. (See 670.1, FPN.) A reduction in the required clearance to 2.5 feet applies to controls for cranes and hoists according to 610.57. Section 110.26(A)(1)(b) also permits reduced but unspecified clearances where the exposed parts operate at no more than 30 volts rms, 42 volts peak, or 60 volts DC. This rule requires permission in writing from the authority having jurisdiction. All reductions in clearance must also be acceptable to other authorities having jurisdiction. For example, the special rule for low voltages is also in NFPA 70E, *Standard for Electrical Safety in the Workplace,* as is the rule for crane and hoist controls.

Figure 1.14 illustrates and summarizes the general and special rules discussed here.

Chapter 2

Wiring and Protection

Question 2.1: Is gray permitted for any grounded conductor or just certain voltages?

Keywords

White
Gray
Natural gray
Grounded conductor
Conductor identification

NEC References

200.6

Related Questions

- Is gray permitted as the identifier for the grounded conductor of any system or just for 277-volt systems?
- Is the color gray intended for higher-voltage systems only?
- What happened to the color "natural gray"?
- May gray and white be used to distinguish grounded conductors of different systems in the same enclosure?

Answer

The old term "natural gray" is no longer used in the *NEC*. Natural gray was really just treated as another version of "white." There are many legends and myths that surround this term. Whether originally intended or not, the term "natural gray" did recognize the off-white colors that appear in some insulation types when white is the intended color. Now, gray and white are recognized as different colors according to Section 200.6.

Although the color scheme described in the question (gray used for the system with the higher voltage to ground) is common in many areas, and was common even before the *NEC* changed, there is no requirement in the code that white be used for certain systems and gray for others. Gray and white may be used to distinguish the grounded conductors of different systems, but neither color is required or preferred under code rules. You may use white or gray for all systems as long as you have a way to distinguish them if they ever come together in the same enclosure, as required in 200.6(D). Because gray and white are now recognized as different colors, these two colors can be used to distinguish between grounded conductors (usually neutrals) of different systems. Other methods, such as a colored stripe (not green) on white insulation, may also be used.

Question 2.2: Where is GFCI protection required?

Keywords

GFCI
Ground-fault circuit interrupter

Related Questions

- Are GFCI-protected receptacles required in all unfinished basement areas?
- Do all receptacles in basements and garages require GFCI protection?

- What is the definition of a kitchen for the purposes of requiring GFCI protection for receptacles?
- When did GFCI requirements first appear in the *NEC*?

Answer

The most commonly applied requirements for ground-fault circuit-interrupter (GFCI) protection are found in Section 210.8. Section 210.8(A) applies only to dwelling units as defined in Article 100, Section 210.8(B) applies to all other occupancies, and Section 210.8(C) applies to boat hoists in dwelling unit locations. These are not the only GFCI requirements (see the table and discussion at the end of this section). The requirements of 210.8 apply only to 125-volt, 15- and 20-ampere receptacles, and those are the types of receptacles covered by this discussion. Some other code sections extend the requirements to other equipment or receptacle ratings. Examples of GFCI requirements for other circuit and receptacle ratings can be found in Article 527 for temporary installations and in Article 680 for storable pools, spas, hot tubs, and the like.

The requirements in 210.8 correspond to many of the requirements in 210.52. The locations where receptacles are required in dwellings are discussed in Section 210.52, and the receptacles that require GFCI protection are specified in 210.8(A). Most of the receptacles covered by 210.8(A) are required to be installed in those locations by 210.52. The obvious exception is wet bar sinks, for which there are no requirements in 210.52. Starting with the 2005 *NEC*, 210.8(A) also covers laundry or utility sinks.

There are very few receptacles that are specifically required in commercial and other nonresidential occupancies, but if 125-volt, 15- or 20-ampere receptacles are located in any of the areas listed in 210.8(B), GFCI protection must be provided. For example, if heating equipment is located on a rooftop of a commercial building, a receptacle must be provided for servicing that equipment according to 210.63, and that receptacle must be GFCI protected according to 210.8(B)(3). Receptacles may be located in commercial and institutional kitchens as the user and designer see fit because 210.52(B) and (C) do not apply, but those receptacles that are located in such a kitchen must be GFCI protected even if they do not serve countertops. A definition of "kitchen" was added in the 2005 *NEC*, which says that for the purposes of Section 210.8(B)(3), a kitchen is "an area with a sink and permanent facilities for food preparation and cooking" that is part of a commercial or institutional installation. This definition does not necessarily include lunch rooms with vending machines and microwave ovens. Many designers and installers had been putting GFCIs in these locations anyway, just because they felt it was prudent, but the requirement in the *NEC* covering commercial and institutional kitchens was new in 2002.

Most of the requirements in 210.8(A) are quite clear. There are no exceptions for the locations described in bathrooms, crawl spaces, kitchen countertops, boathouses, or near wet bar sinks. If the receptacles are located in these locations, they must have GFCI protection. Similarly, a receptacle is not required for servicing evaporative coolers on one- and two-family dwellings, but since this equipment is universally located outdoors, GFCI protection must be provided if such receptacles are installed. This does not include receptacles that may be used for disconnecting means within the evaporative coolers.

Receptacles installed outdoors that are not readily accessible and supplied from circuits dedicated and installed for fixed deicing and snow melting are not required to have GFCI protection, as shown in Figure 2.2(a), but must have ground-fault protection of equipment, sometimes called GFEP or GFPE. This exception also applies to rooftop receptacles in other than dwelling units. GFPE devices trip at about 30 mA and provide primarily fire protection rather than shock protection.

The definition of "readily accessible" in Article 100 does not include a height limit. Because of language in the 1993 *NEC* and some previous editions and the current language

Keywords

Receptacle
Basement
Garage
Kitchen
Unfinished
Crawl space
Rooftop
Outdoors
Class A GFCI
Class B GFCI
Ground-fault protection of equipment

NEC References

210.8, 210.52, 210.63, 760.21, 760.41

Four outdoor receptacles on a dwelling unit: three required to be GFCI protected, and one that requires only ground-fault protection for equipment.

Source: Based on *NEC Handbook*, 2005, Exhibit 210.11.

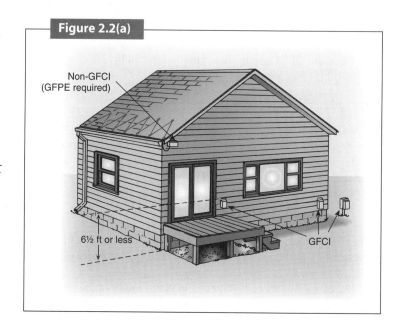

Figure 2.2(a)

for readily accessible switches in 404.8(A), the height limit is often interpreted to be 6.5 feet, as shown in Figure 2.2(a). However, under the current language of the *NEC*, the height limit for things that are to be "readily accessible" is established by the authority having jurisdiction. The issue in this case is, of course, ready access for inserting cord-and-plug-connected portable appliances, tools, and extension cords. The 6.5-foot dimension in Figure 2.2(a) is part of the rule for location of required outdoor receptacles from 210.52(E), but is not part of the GFCI requirement. For example, if the receptacle shown for deicing and snow-melting equipment were used and installed instead for holiday lighting, the GFCI requirement of 210.8 would apply, but the receptacle would not count as one of the required outdoor receptacles that must be readily accessible.

Similar exceptions to those for outdoor receptacles also apply in garages, accessory buildings with floors below grade, and unfinished basements, but in these locations, the receptacles may be exempted simply by being not readily accessible. A common example is a receptacle installed in the rafters or on the ceiling of a garage for a garage door opener, as shown in Figure 2.2(b). Receptacles dedicated to specific appliances, such as a freezer in the garage or a sump pump or washing machine in the basement, are also exempted if all such receptacles are used for those appliances and located in the spaces dedicated to those appliances. These exceptions are illustrated in Figures 2.2(b) and 2.2(c). A reason for this type of exception is that many appliances, especially older refrigeration appliances and appliances with immersed heating elements or motors, involve small inherent leakage currents that may cause nuisance tripping. This could create a serious problem with a freezer full of food or a sump pump.

Another exemption applies to receptacles in unfinished basements that are dedicated to supplying power to a fire alarm or burglar alarm system. Sections 760.21 and 760.41 prohibit a fire alarm system from being supplied through a GFCI. This exception does not apply to individual or interconnected smoke alarms, but only to fire alarm systems supplied by a fire alarm panel.

At least one of the GFCI-protected receptacles shown in Figure 2.2(c) would have to be within 25 feet of the furnace to comply with 210.63.

According to 210.8(A)(5), unfinished basements are portions or areas of a basement "not intended as habitable rooms and limited to storage areas, work areas, and the like." This rule is really more about the use of the area than its state of completion. Unfinished areas of a

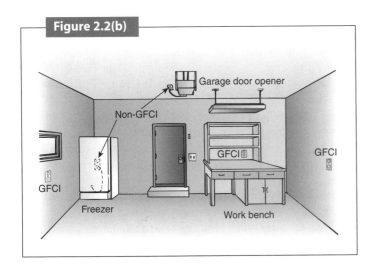

Figure 2.2(b)

Examples of receptacles in a garage that are and are not required to have GFCI protection.

Source: Based on *NEC Handbook*, 2005, Exhibit 210.10.

basement, like the areas of accessory buildings described in 210.8(A)(4), are more likely to be damp than areas that are finished. More significantly, these unfinished areas are also likely locations for the use of cords and portable tools, especially hand tools. Again, this requirement works in concert with Section 210.52. To protect persons using portable tools in areas where shock hazards are somewhat increased, 210.52(G) requires that receptacles be provided in each unfinished basement area, and 210.8(A)(5) requires that those receptacles have GFCI protection. According to the definition, an area in a basement that is "finished" into a work room or a shop, perhaps by building walls and finishing the ceiling, is still an unfinished basement for the purposes of requiring shock protection for the receptacles used in those areas. See the work area in Figure 2.2(c). We cannot tell from the drawing if this work area is "finished" or not.

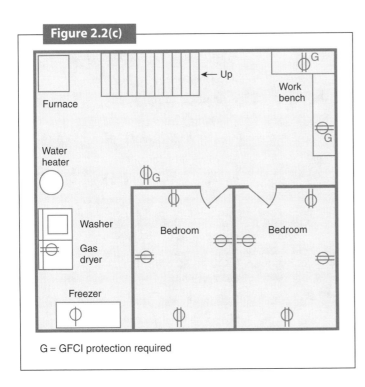

Figure 2.2(c)

G = GFCI protection required

Work-area receptacles in a dwelling unit basement are required to have GFCI protection. Receptacles in finished areas and those supplied only for specific appliances are exempt.

Source: Based on *NEC Handbook*, 2005, Exhibit 210.12.

The first requirements for GFCI protection under the *NEC* appeared in the 1968 code and applied only to underwater lighting in swimming pools. The rule was found in 680-4(g) and was one of three options for providing shock protection in such installations. Similar protection was also an option in earlier editions, but it was called "differential type circuit protection" and was essentially the same as what is now called a Class B GFCI. GFCIs are now defined in the *NEC* as Class A devices. Class B devices are intended to trip and Class A devices trip in the range of 4 mA to 6 mA ground-fault current.

The first GFCI requirements for receptacles were added to the 1971 *NEC* in 210-22(d) for "120-volt, 15- and 20-ampere receptacles installed outdoors" in residential occupancies. This rule had an effective date of January 1, 1973. The GFCI requirements were moved to their present location in 210-8 in the 1975 *NEC,* and the requirements were extended to residential bathrooms and to construction sites. As the technology was perfected, GFCIs became more reliable, and experience showed that they did prevent many electrical injuries, so the requirements were gradually expanded to many other locations where similar shock hazards exist. Now there are many applications where GFCI protection is required. Table 2.2 lists the numerous sections of the *NEC* that discuss locations that require GFCI protection. In most cases, these locations are places where portable tools, hand lamps, or appliances are used or

Table 2.2 Additional Requirements (Other Than Section 210.8) for the Application of Ground-Fault Circuit-Interrupter Protection

Applicable Section(s)	Location
215.9	Feeders
406.3(D)(2)	Replacement receptacles
422.49	High-pressure spray washers
511.12	Commercial garages
517.20(A), 517.21	Health care facilities
525.23	Carnivals, circuses, fairs, and similar events
590.6	Temporary installations
550.13(B) and (E), 550.32(E)	Mobile and manufactured homes
551.40(C), 551.41(C)	Recreational vehicles
551.71	Recreational vehicle parks
552.41(C)	Park trailers
555.19(B)(1)	Marinas and boathouses
600.10(C)(2)	Signs, mobile or portable
620.85	Elevators, escalators, and moving walkways
625.22	Electric vehicle charging systems
640.10(A)	Audio system equipment
647.7(A)	Sensitive electronic equipment
680.22(A)(1), (A)(5), and (B)(4); 680.23(A)(3)	Pools, permanently installed
680.32	Pools, storable
680.51(A)	Fountains
680.57(B) and 680.58	Signs with fountains and adjacent receptacles
680.71	Hydromassage bathtubs

are areas that may be wet or where there is an increased hazard of electrical shock. In some cases, as in 550.13(E), the GFCI protection also provides for a measure of protection from fire ignition and is used instead of ground-fault protection of equipment (GFPE).

Question 2.3

What types of devices are permitted for providing AFCI protection?

Related Questions

- Where should an arc-fault detector be installed?
- How many AFCIs are needed in a house?
- Why does the *NEC* require AFCI protection only in bedrooms?
- Are AFCI devices available to residential builders, electricians, and homeowners?
- Can GFCI receptacles be used on AFCI-protected circuits?
- Are smoke alarms in bedrooms required to be on AFCIs?
- Do AFCI requirements apply to guest rooms of hotels and motels?
- Is AFCI protection required in rooms associated with bedrooms, such as sitting rooms or a "parent's retreat"?

Keywords

AFCI
Arc-fault circuit interrupter
Feeder/branch type
Combination type
Bedroom
Branch circuit
Outlets

NEC References

210.12

Answer

Section 210.12 of the *NEC* requires listed combination-type arc-fault circuit interrupters (AFCIs) on all 120-volt, single-phase, 15- and 20-ampere branch circuit outlets in dwelling unit bedrooms. Combination-type devices "combine" the modes of protection in both branch/feeder and outlet AFCI devices. There are a couple of combination-type AFCIs that may be used. One is a circuit breaker that also includes overcurrent protection for the entire circuit and is installed in a panelboard. The other is a device that is installed like a ground-fault circuit-interrupter (GFCI) receptacle in a box. Either is acceptable, and both should be available from electrical equipment distributors and some home improvement stores by the effective date of the rule. Because the requirement for combination-type protection was new in the 2005 *NEC*, an effective date was included that permits the use of a branch/feeder device of the type originally required in the 1999 *NEC* until January 1, 2008. Figure 2.3(a) shows a branch/feeder-type AFCI device of the type required in the 1999 and 2002 editions of the *NEC*. Combination types are similar in appearance or may take the form of an outlet device.

Generally, an AFCI device should be at the origination of the branch circuit. Since the branch circuit is defined by its overcurrent device, this seems to imply that a circuit breaker type must be used. However, an exception permits the other type to be used if it is installed within the first 6 feet of branch circuit conductors that are installed in metal raceway or a cable with a metallic sheath from the branch circuit overcurrent device to the AFCI device. Figure 2.3(b) shows the three basic ways that AFCI protection can be provided as required in the 2005 *NEC*.

The number of AFCIs that must be used depends on the number of branch circuits and bedrooms in the dwelling and the size of the spaces, because the load calculation from 220.12 and the resulting number of required circuits from 210.11 are based on floor area. Also, the

An example of a branch/feeder type of AFCI device with overcurrent protection.

Figure 2.3(a)

circuit layout may affect the number of devices required. The *NEC* does not require separate circuits for receptacles and lighting in bedrooms. Both are considered to be part of the general lighting load. Where installers, designers, or owners prefer more circuits or separated circuits, additional AFCI devices may be required.

The requirement for combination-type AFCI devices was added in the 2005 *NEC* and is based on a desire to have a device that also protects the cords plugged into the branch circuit. A combination-type device is designed to provide protection for branch circuit (or feeder) wiring as well as cords. Branch/feeder devices are intended to protect fixed wiring but are not subjected to tests that establish their ability to protect portable appliance cords.

Code-Making Panel 2 did not have any specific statistics that led them to choose bedrooms over other areas within a dwelling. The original proposal that led to 210.12 was for all 125-volt lighting and appliance branch circuits to have AFCI protection, but after the public comment period it was decided to scale back the initial requirement and mandate AFCI protection for bedroom receptacle circuits only. In most dwellings, the requirement for bedrooms will generally pick up two or three rooms that do not have special protection such as GFCIs, and the panel felt it was a good compromise and a good place to start. Many arguments have since been advanced to justify this particular area, some of which make good sense in light of what is known about bedrooms. For example, bedrooms often include many cord-connected appliances, such as clocks, televisions, radios, and electric blankets, some of which are left connected for years at a time and may be subject to damage, and the occupants are likely to be sleeping and unaware of their surroundings.

In regard to the operational characteristics of AFCI devices, part of the product standard protocol requires proper operation where connected in series with GFCI-type receptacles. This application cannot result in adverse operation of either the AFCI or the GFCI. At least one circuit breaker manufacturer makes a combination AFCI/GFCI-type device. In fact, a variety of ground-fault protection is included in most AFCI devices to detect leakage currents

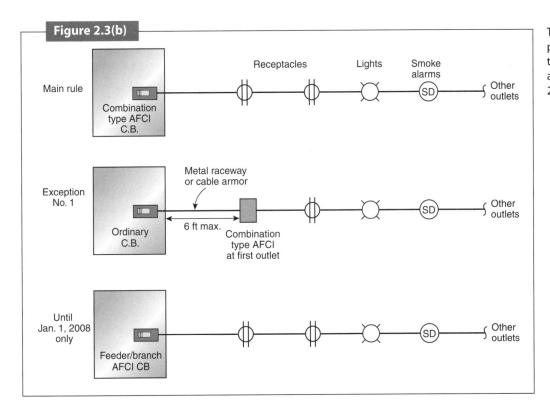

Three methods of providing combination-type AFCI protection in accordance with Section 210.12 in the 2005 *NEC*.

to ground, although the trip level is equivalent to that of GFPE devices like those used for protecting deicing and snow-melting equipment or pipeline tracing and is not designed for personnel (shock) protection.

When the AFCI requirement first appeared in the 1999 *NEC*, it applied only to receptacle outlets in bedrooms. This was expanded to all 125-volt, 15- and 20-ampere "outlets" in bedrooms in the 2002 *NEC*. Since an outlet is defined in Article 100 as "a point on a wiring system at which current is taken to supply utilization equipment," the typical smoke alarm that is supplied from a nominal 120-volt system is connected at an outlet, and the smoke detector is subject to the AFCI requirements. Note that the purpose of the AFCI is to interrupt power before a fire starts, and building and fire alarm code requirements mandate dual power sources, so a smoke alarm that is properly installed and maintained as required by codes and manufacturer's instructions will not be disabled by the tripping of an AFCI device.

AFCI requirements apply to guest rooms of hotels and motels only if the rooms are also dwelling units as defined in Article 100. This definition requires that a guest room have "permanent provisions for living, sleeping, cooking and sanitation" before the AFCI requirements of 210.12 would apply. The guest rooms in some suite hotels and short-stay residential hotels and motels do meet this definition, but most guest rooms do not have permanent provisions for cooking. (Provisions for cooking in this sense do not include portable microwave ovens.)

The *NEC* does not define a bedroom. A connected bathroom or closet would not be considered a bedroom, because these areas are separately defined. A bathroom is defined in the *NEC*. The definition of a bedroom and the determination of what associated or connected areas are part of a bedroom are left to building codes, dictionaries, and the authority having jurisdiction. A room that is connected to a bedroom but separated by doors and having a separate function, such as a reading room or a "parent's retreat," would probably not be considered a bedroom, but a sleeping porch or similar area that is used as a bedroom probably would be considered part of a bedroom or a separate bedroom.

Question 2.4

Why must single receptacles match the rating of the circuit but not duplex receptacles?

Keywords

Receptacle
Single receptacle
Multiple receptacle
Individual branch circuit

NEC References

210.21, 210.23, 240.4

Related Questions

- If a 120-volt, 20-ampere individual branch circuit runs to the location of a single item of utilization equipment such as a refrigerator, and the circuit has only a single receptacle outlet, may a 15-ampere duplex receptacle (less than the branch circuit rating) be used?
- If an outlet is obviously intended for the refrigerator, why is the rating for the receptacle allowed to be only 15 amperes if duplex, but 20 amperes is required for a single receptacle?

Answer

Sections 210.21(B)(1) and (B)(3) read as follows (exceptions omitted):

(B)(1): A single receptacle installed on an individual branch circuit shall have an ampere rating not less than that of the branch circuit.

(B)(3): Where connected to a branch circuit supplying two or more receptacles or outlets, receptacle ratings shall conform to the values listed in Table 210.21(B)(3)...

According to the referenced table, a circuit rated 15 amperes may supply receptacles rated not over 15 amperes, and a circuit rated 20 amperes can supply receptacles rated at 15 or 20 amperes.

A receptacle is defined in Article 100 as "a contact device installed at the outlet for the connection of an attachment plug. A single receptacle is a single contact device with no other contact device on the same yoke. A multiple receptacle is two or more contact devices on the same yoke." Article 100 also defines an individual branch circuit as one that serves only one piece of utilization equipment. Since one duplex receptacle on a circuit could serve more than one piece of utilization equipment, such a circuit is not an individual branch circuit, even though there is only one "receptacle outlet."

The solution to the questions can be found in Sections 210.21(B)(2) and 210.23(A) of the *NEC*. According to 210.21(B)(2), where a 15-ampere duplex receptacle is installed on a 20-ampere circuit, the maximum load allowed to be connected to either receptacle is 80 percent of 15 amperes, or 12 amperes. Then, according to 210.23 and 210.23(A), the maximum load on a 20-ampere circuit cannot exceed 20 amperes and the maximum load for *any one* cord-and-plug-connected load is 80 percent of the branch circuit rating, or 16 amperes. However, the 16-ampere load cannot be connected through a 15-ampere receptacle. In order to connect a 16-ampere load, the receptacle would have to be a 20-ampere single or multiple receptacle. If you have determined that the load is greater than the 12 amperes permitted to be connected to a 15-ampere receptacle, then a 20-ampere receptacle is an obvious choice.

Suppose that you choose instead to provide an individual 20-ampere branch circuit for the load. Now there is only one receptacle at only one receptacle outlet, as shown in Figure 2.4. Certainly the overcurrent protection for that single receptacle should be appropriate for the receptacle and the load, but 210.21(B)(2) does not apply to individual branch circuits. However, Section 210.23(A)(1) still limits the single cord-and-plug-connected load to 16 amperes, and this load can only be supplied through the one receptacle. In this case, the single receptacle must be rated for the maximum load that can be supplied; that is, it needs to be a

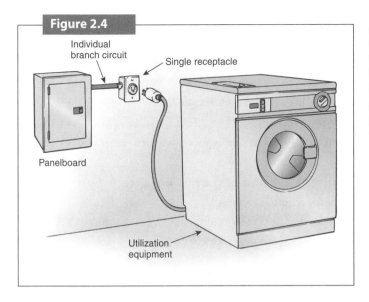

Figure 2.4

An individual branch circuit supplying one item of utilization equipment through a single receptacle.

Source: Based on *NEC Handbook*, 2005, Exhibit 100.7.

20-ampere receptacle. Also, a listed appliance that draws 16 amperes (or over 12 amperes) should come with a 20-ampere attachment plug, and that plug can only be connected to a 20-ampere receptacle. A receptacle rated 20 amperes can accept either a 15- or a 20-ampere attachment plug, but only 15-ampere attachment plugs will fit a 15-ampere receptacle.

The choice of multiple 15- or 20-ampere receptacles or a single receptacle is a design decision that should be made based on anticipated use.

There is another issue in the questions, however. That is, a single 15-ampere duplex receptacle could have two appliances plugged into it, and the total rating of those two appliances could exceed the branch circuit rating. When a user plugs in loads that are in excess of the circuit overcurrent device rating, such as two 12-ampere loads on a 15- or 20-ampere duplex receptacle, the overcurrent device will open the circuit and prevent damage to the receptacle. Listed receptacles are evaluated by testing agencies at 150 percent of their rating.

Loads on individual branch circuits are easy to predict because there is only provision for a single load. Loads on multiple-outlet receptacle circuits are much less predictable. For this reason, Section 240.4(B) does not permit rounding up to the next standard size overcurrent device on multi-receptacle circuits. Rounding up or otherwise increasing the overcurrent device above the conductor or receptacle rating is permitted for some single receptacles supplying motor loads, for receptacles supplying only welders, and for some receptacles used with electric discharge lighting.

Question 2.5

Where are receptacle outlets required in dwelling units?

Related Questions

- How is a railing counted in determining wall space?
- How should receptacles be located in finished basement areas?

Keywords

Receptacle
Outlet

Keywords

Location
Dwelling unit
Wall space
Habitable area
Finished
Unfinished
Kitchen
Countertop
Island
Peninsula
Bathroom
Garage
Basement
Outdoors
Hallways
HACR equipment

NEC References

210.8, 210.11, 210.23, 210.50, 210.52, 210.60, 210.63, 210.70, 400.7, 400.8, 406.4, 406.8, 410.4, 680.43, 680.71

- Is more than one receptacle required for long hallways?
- Are outdoor receptacles required for dwelling units in multifamily buildings?
- How many receptacles are required on kitchen peninsulas and islands?
- How is countertop space measured?
- What is the smallest countertop that requires a receptacle?
- Is a separate circuit required for a refrigerator?
- Is a laundry circuit permitted or required to supply more than one receptacle outlet?
- How far must receptacles be from a tub or shower?
- Is a cord pendant with a connector permitted as a receptacle outlet?
- Is there a height requirement for required receptacles?

Answer

Part III of Article 210 provides the general requirements for locations of outlets. These requirements cover both receptacle and lighting outlets. The more specific requirements, and probably the most discussed rules, apply to receptacle outlets in dwelling units. These rules are found in Section 210.52. The most general requirements apply to all occupancies and are found in 210.50. Other rules for receptacle outlet locations are found in Sections 210.60 for guest rooms and guest suites, in 210.62 for show windows, and in 210.63 for receptacles for servicing heating, air-conditioning, and refrigeration (HACR) equipment in all occupancies.

Section 210.70 covers lighting outlets that are required. Again, the requirements for lighting outlets are more comprehensive and specific for dwellings than for other occupancies. However, other rules in Chapters 5, 6, and 7 of the *NEC* may include additional requirements for outlets in specific occupancies or for special equipment or systems. For example, Article 620 requires receptacles in elevator machine rooms and hoistways, regardless of occupancy, and Article 517 requires receptacles in patient bed locations in health care facilities.

Section 210.50 contains some rules that apply to all occupancies. The first of these, 210.50(A), says that a pendant cord with a connector that accepts a male attachment plug is considered to be a receptacle outlet. This type of installation is common in certain commercial and industrial facilities and could also be used in dwellings, though it is seldom applied in dwellings. The second rule requires a receptacle outlet where any flexible cords with attachment plugs are used in any occupancy, but allows a cord connection to an outlet without a receptacle where such cord use is permitted. (Uses of cords that are permitted and not permitted are covered in Sections 400.7 and 400.8, respectively.) This rule, in 210.50(B), does not include any dimension for the location of the outlet, but in dwelling units, 210.50(C) requires a receptacle outlet within 6 feet of the intended location of an appliance.

The requirements for locating receptacle outlets in 210.52 apply only to dwelling units. For the purposes of this section, and as the term "receptacle" is used in this discussion, the required receptacles are of the convenience type, that is, receptacles rated 125 volts, 15 and 20 amperes, that will accept the common 15-ampere attachment plugs found on most small appliances and tools. As noted, a cord pendant may be counted as a required receptacle if the connector is located at or below 5.5 feet above the floor, and receptacles in baseboard heaters may be counted. However, receptacles that are part of a luminaire or appliance or located in cabinets or cupboards are not counted.

Other than the 5.5-foot maximum height, the *NEC* does not have a height requirement for general-use receptacles. In fact, according to 210.52(A)(3), receptacles located in floors may be counted if they are not more than 18 inches from the wall. The rules in the Americans with Disabilities Act (ADA) do specify higher (than the floor) minimums and lower (than 5.5 feet) maximums, but those rules do not apply to every individual dwelling unit. ADA does apply to those dwellings constructed to meet ADA requirements, which may be a percentage of the homes in a development, the apartments in multifamily housing, or the guest rooms in a hotel or motel.

Section 210.52 is divided into a number of subsections that apply to various parts of a dwelling unit. The general provisions for locating receptacles are found in 210.52(A). This section applies to the wall spaces in habitable areas not specifically covered elsewhere. The list of rooms covered includes kitchens, family rooms, dining rooms, living rooms, parlors, dens, sunrooms, bedrooms, recreation rooms, and similar rooms or areas. Many houses include rooms, sometimes called "great rooms," that combine the functions of one or more of the rooms listed; such rooms are covered under "similar rooms or areas." It may be easier to list the rooms or areas that are *not* covered by these general provisions for wall spaces—that is, hallways, kitchen countertops, bathrooms, closets, stairways, garages, laundry rooms, mechanical or utility rooms, and unfinished spaces. Basements with finished areas are covered if the areas are finished as bedrooms, family rooms, or the like. The distinction is that receptacles are required at a certain spacing around the walls of rooms used for habitation where appliances and portable lamps are likely to be used as opposed to rooms used for special purposes (bathrooms, laundry, and equipment spaces) or used primarily for passage or storage (halls, foyers, stairways, closets). Many of the special-purpose areas have separate rules. There are no requirements for receptacles in the floors or in the center of large rooms unless fixed room dividers are present.

Section 210.52(A) provides the spacing requirements for receptacles in walls in habitable areas, and 210.52(B) explains how and what wall areas are counted. The spacing rules require that there be a receptacle within 6 feet of any point on a wall in a habitable room (every 12 feet around an unbroken wall). Very short sections of walls, less than 2 feet wide, such as the space between two adjacent doors, are not counted because such small spaces are not likely to be used for equipment connected by a cord and plug.

Each unbroken wall section is counted separately. A wall is considered to be broken by openings such as doors or fireplaces, but not by windows. "Walls" are created by fixed panels in exterior walls, such as full-height windows, but not by movable panels such as sliding doors. Railings and other fixed room dividers are also treated as wall spaces. These spacing requirements are summarized in Figure 2.5(a). An example of a railing that

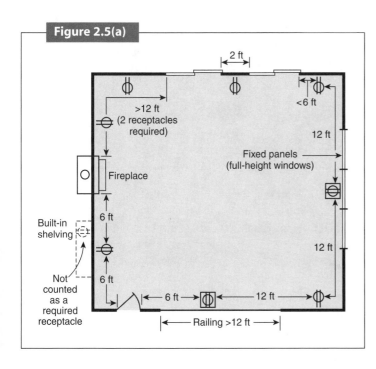

Figure 2.5(a)

Typical receptacle layout, showing minimum number of receptacles for a room with fireplace, sliding doors and door opening, railing, fixed panels, and short wall sections where receptacles are required.

is treated as a wall, with a floor receptacle used to satisfy the requirement for an outlet, is shown in Figure 2.5(b).

Sections 210.52(B) and (C) together cover the requirements for small appliance and kitchen countertop receptacles. Section 210.52(B) does not specify placements, but does explain which receptacles are to be supplied from the small appliance circuits required by 210.11(C)(1). The rooms where small appliance receptacles are to be provided are listed in 210.52(B)(1). These rooms include kitchens, pantries, breakfast rooms, dining rooms, and similar areas. They are similar because they are areas where small appliances used for cooking and food preparation, such as toasters and coffee makers, are most likely to be used. All the outlets in these rooms—and only those outlets, whether required for countertops or wall spaces—are to be supplied by small appliance branch circuits, although some exceptions apply. For example, a switched receptacle outlet in a dining room, used as a lighting outlet to comply with 210.70, may be on a general lighting circuit. Also, a separate circuit dedicated to a refrigerator is permitted but not required. (The general lighting circuit or a separate circuit for a refrigerator could have a rating other than the 20-ampere rating required for small appliance circuits.) Also, the small appliance circuits may be used to supply the small load of an electric clock or the supplemental equipment, such as clocks, timers, igniters, and controls, on a gas-fired cooking appliance. Small appliance circuits are not intended for fixed appliances such as dishwashers, waste disposers, electric ranges, ovens, and cooktops, or for built-in or other microwave ovens that require individual branch circuits. Individual branch

Figure 2.5(b)

Floor receptacle outlet used to comply with requirements for a receptacle at a railing that exceeds 12 feet in length where the first receptacle is installed in the column.

circuits for such appliances may be required either by nameplate information or because of the load restrictions of 210.23.

Receptacle spacing on countertops in kitchens and dining areas is the subject of 210.52(C). For countertops that sit against a wall, the width of a countertop is the dimension parallel to the wall, and is measured in the same manner as the cabinets that usually support a countertop. Unless a countertop turns a corner and continues down another wall, the wall space at the end of a countertop is not counted. Receptacles must be located in such a way that there is a receptacle within 24 inches horizontally from any point on the wall behind a countertop that is 12 inches or more in width.

"Width" is not so clear for countertops that form an island or peninsula, so requirements are based on short dimensions (usually depth) and long dimensions (usually width). One receptacle is required for an island or peninsula countertop that is at least 12 inches or more in the short dimension and 24 inches or more in the long dimension. (These dimensions correspond to the 12-inch width of a wall countertop that is typically 24 inches in depth.) Islands stand apart, so dimensions are fairly obvious, but for peninsulas, the dimension is taken from the edge of the connecting countertop that is against a wall. Each separate countertop space is considered individually, and, according to 210.52(C)(4), if any countertop is separated by a sink, range, or refrigerator, the countertop on each side is considered as a separate space. In general, the wall space behind a sink or range is not counted as a separate space, but where the space is larger than shown in Figure 210.52 in the *NEC*, it may be counted as a usable countertop space, and a receptacle may be required. Note that if an island or peninsula countertop is unbroken, no more than one receptacle is required, although additional receptacles are permitted. The requirements for countertop receptacles are summarized in Figure 2.5(c).

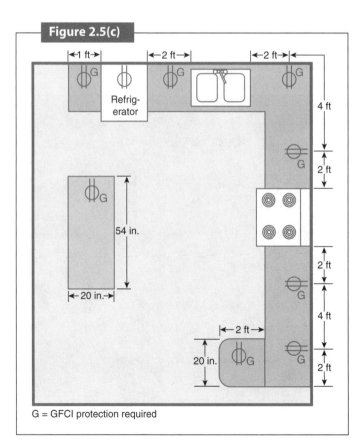

Locations for receptacles serving countertop spaces in a dwelling unit kitchen.

Source: Based on *NEC Handbook*, 2005, Exhibit 210.26.

The requirements for receptacle height at countertops are found in 210.52(C)(5). In general, the receptacle must be located above, but not more than 20 inches above, the countertop. Receptacles must be readily accessible, so receptacles that are located behind appliances fastened in place or occupying dedicated spaces and receptacles concealed in appliance garages cannot be counted as required receptacles. There is no minimum height above the countertop, although there may be practical limits due to the form and size of a backsplash.

Often, countertops that form peninsulas or islands do not have any backsplashes or cabinets above the countertop to provide a mounting location for receptacles. In such cases, 210.52(C)(5), Exception, allows for receptacles to be located below the countertop where the countertop does not extend more than 6 inches beyond the support base. This exception may also be used to accommodate access to receptacles for the physically impaired. Section 406.4(E) prohibits a receptacle from being mounted face-up in a countertop.

Because of the variety of kitchen designs, in some cases there may be no acceptable location for a receptacle above or below an island or peninsula countertop, and some jurisdictions do not recognize the exception that allows receptacles below countertops because they believe that a cord extending over the edge of a countertop creates an increased hazard. In such cases, other arrangements or designs may be mandated, or the receptacles may be omitted in accordance with the policy and interpretation of the authority having jurisdiction. For example, many island and peninsula "countertops" are intended as areas for eating rather than areas for food preparation or cooking, and the authority having jurisdiction may decide the countertop is really a table, like the bars or the permanent table tops used in restaurants. This is a decision that must be made by the authority having jurisdiction. Some jurisdictions may require receptacles in all cases because they may believe that the intended use will not preclude the space being used for cooking appliances such as food warmers or fondue pots, and the lack of a receptacle will encourage the use of extension cords. Figure 2.5(d) shows an example where the receptacle was omitted because the peninsula countertop is designed and intended as an eating area or table rather than an area for food preparation.

Peninsula "countertop" constructed for use as a table with no provision for a receptacle outlet.

Figure 2.5(d)

Receptacles are also required in some nonhabitable areas in dwellings. Sections 210.52(D), (E), (F), (G), and (H) cover bathrooms, outdoor areas, laundry areas, basements and garages, and hallways, respectively. At least one receptacle is required in each of these areas, and two are required outdoors: one each at the front and back of the dwelling. However, some of the rules apply only to one- or two-family dwellings.

All bathrooms must have at least one receptacle outlet located on an adjacent wall or partition and within 3 feet of the outside edge of each basin. Where there is more than one basin, one receptacle may suffice if it can be located within 3 feet of each basin as shown in location B in Figure 2.5(e). Section 210.11(C)(3) requires that the bathroom receptacle(s) be supplied from a 20-ampere circuit that is dedicated either to receptacles and other equipment within that one bathroom or to receptacles only in more than one bathroom. There is no specific spacing required from a toilet, tub, or shower, but 406.8(C) does not allow a receptacle in a bathtub or shower space. (The space defined in 410.4(D) for bathtub and shower spaces applies to luminaires, not receptacles.) If the bathroom includes a spa or hot tub, 680.43(A)(1) requires the receptacle to be at least 5 feet from the spa or hot tub. Hydromassage bathtubs are treated differently. Section 680.71 permits a receptacle within 5 feet of a hydromassage bathtub if the receptacle is GFCI protected. Of course, 210.8(A)(1) requires all convenience-type receptacles in dwelling unit bathrooms to be GFCI protected.

Note the interplay between 210.52 and 210.8 in this case. Section 210.52(D) requires the receptacle in a bathroom, and 210.8(A) requires that receptacle to have GFCI protection.

Figure 2.5(e) Two arrangements of receptacle outlets that comply with the location requirements for receptacle outlets in bathrooms.

Ⓐ Two receptacle outlets, one for each basin
OR
Ⓑ One receptacle outlet located to serve both basins.

There is no requirement for a receptacle in a bathroom that is not part of a dwelling unit or guest room, but 210.8(B)(1) requires GFCI protection for any such receptacles if they are located in bathrooms. Section 210.11(A)(3) also specifies dedicated circuits for receptacles in bathrooms in dwelling units, but there are no similar requirements for bathroom receptacles in other occupancies.

Outdoor receptacles are required at the front and back of single-family dwelling units and at the front and back of each unit of a two-family dwelling unit if the unit is at grade level. In other words, two outdoor receptacles are required for each unit of a two-family dwelling if both are at grade level, as in a typical side-by-side duplex or two-unit townhouse. If a duplex is arranged so that one unit is at grade level and the other is above it, outdoor receptacles are required only for the grade-level unit. Receptacles are not required at balconies or decks above grade so long as the "front-and-back" requirements are met. Receptacles mounted more than 6.5 feet above grade are not considered readily accessible from grade and do not meet the requirement, although they may be at the front or back of a dwelling. Such receptacles are permitted as additional receptacles. Outdoor receptacles are not required in multifamily dwellings, except at least one is required in those units with grade-level access and individual exterior entrances.

An individual unit of a multifamily dwelling is not considered a single-family dwelling for the purposes of 210.52. This is not to say that outdoor receptacles should not be installed in other applications, but that 210.52(E) only requires them in one- and two-family dwellings and certain grade-level units of multifamily dwellings. However, where a multifamily dwelling unit is built with firewalls in such a way that each unit is considered to be a separate building under the applicable building code, a multifamily dwelling may be considered to be a group of single-family dwellings, as shown in Figure 2.5(f). Outdoor receptacles that are installed at any dwelling unit must have GFCI protection in accordance with 210.8(A)(3), whether or not the receptacles are required. (Section 210.8(A)(3) provides an exception for receptacles that are not readily accessible and are used for deicing and snow melting, but ground-fault protection of equipment is required for those circuits according to 426.28.)

Unless a dwelling unit is part of a two-family or multifamily dwelling where laundry facilities are provided on the premises or laundry facilities are not permitted, at least one receptacle must be installed for the laundry. More than one receptacle or receptacle outlet may be installed in a laundry area, and the laundry circuit required by 210.11(C)(2) may supply more than one outlet, but only the laundry area may be supplied by this circuit. For example, a receptacle must be installed for a washing machine, but another receptacle may be installed on the same circuit for a gas dryer or for other laundry equipment, such as a clothes iron. The GFCI requirements in 210.8 do not apply to these receptacles unless the laundry equipment is in an unfinished area of a basement or the receptacle is located within 6 feet of a laundry or utility sink.

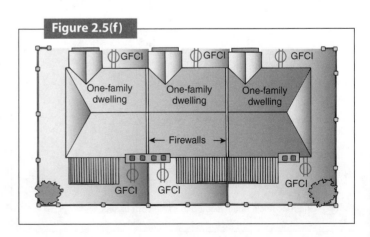

Multifamily housing separated by firewalls, treated as separate single-family units, with receptacles at the front and back of each unit.

Source: Based on *NEC Handbook*, 2005, Exhibit 210.27.

Requirements for receptacles in garages and unfinished basements apply only to one-family dwellings. Only attached garages and garages with electric power are covered. Power is not required at a detached garage, but if power is supplied, the use of tools becomes more likely and a receptacle must be supplied as well. Again, 210.8 requires GFCI protection for most such receptacles, unless the receptacles are not readily accessible or are located in spaces dedicated to specific appliances and used for those appliances. Also, receptacles in basements used to supply power to fire alarm systems (systems with central fire alarm panels) cannot be GFCI protected. Where basements are partially or totally finished, 210.52(G) applies only to the unfinished portions, but at least one GFCI-protected receptacle must be installed in each separate unfinished space. Receptacles in finished spaces must be provided in accordance with the requirements in 210.52 for the type of finished room. Figure 2.5(g) illustrates the requirements for a partially finished basement with laundry facilities, HACR equipment, and separate unfinished spaces.

Hallways are covered in 210.52(H). This section applies to hallways in any dwelling unit. One receptacle is required unless the hallway is less than 10 feet long. Most hallways in dwelling units are not very long, so in most cases, no receptacles or only one receptacle is required. If the hall is separated by doorways, each section is considered separately. This may result in more than one receptacle, one for each section, or no required receptacles if each separate section is less than 10 feet long, as shown in Figure 2.5(h).

Receptacles for servicing heating, air-conditioning, and refrigeration (HACR) equipment are required in most dwelling units because one or more of those types of equipment are likely to be present. In fact, such equipment is often located both inside and outside. The receptacle must be within 25 feet and on the same level as the HACR equipment. Usually, one of the other requirements of 210.52 will result in a receptacle being located to meet this requirement. However, if the outside HACR equipment is more than 25 feet from the recepta-

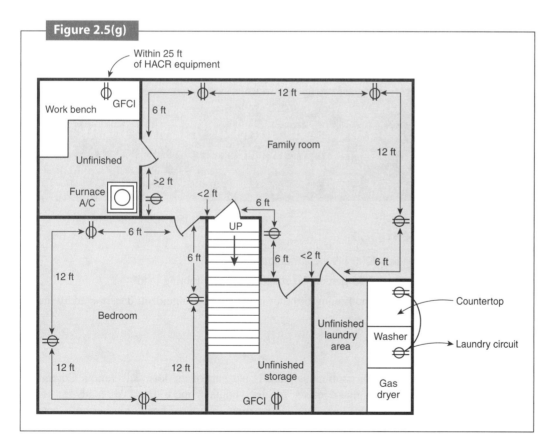

Typical requirements for receptacles in partially finished basement, showing one required receptacle in each unfinished area.

Receptacle requirements for hallways.

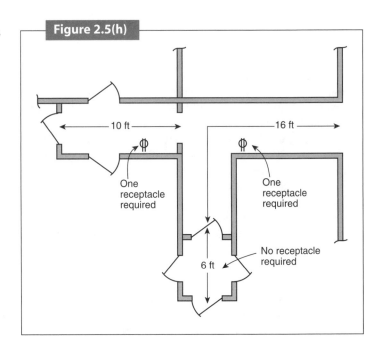
Figure 2.5(h)

cles located at the front or back, or if the equipment is on a roof, or in an attic or crawl space where receptacles are not required by other rules, additional receptacles will be required. Again, 210.8 may require GFCI protection for these receptacles, especially if the HACR equipment is located outdoors, including on roofs, or in crawl spaces or unfinished basements, as shown in Figure 2.5(g). The requirements for receptacles for servicing HACR equipment are found in 210.63. Receptacles are not required for servicing of evaporative coolers for one- and two-family dwellings.

Question 2.6 Do unit load values for general lighting include factors for continuous loads?

Keywords

Load calculation
Continuous load
125 percent
Unit load

NEC References

210.19, 210.20, 215.2, 215.3, 220.12, 220.43, 220.50, 220.60, 230.42, 310.15, Table 310.16, 410.101, 430.22, 430.24

Related Questions

- Is the sign load of 1200 VA a continuous load?
- Should an additional 25 percent be added to a track lighting load?
- If heating and air conditioning or other loads are noncoincident, does the additional 25 percent apply to both?

Answer

The 125 percent factor for continuous loads is not part of the load calculation. Continuous loads are defined as loads that continue for three hours or more. However, loads are not actually increased just because they continue for more than three hours. Furthermore, conductor ampacity, is, by definition, the amount of current that a conductor can carry *continuously* under specific conditions of use.

Article 220 is the basis for load calculations in the *NEC*. The load calculations of Article 220 are intended as the basis for determining the ampacity of a conductor according to 315.15(B), FPN. The allowable ampacity tables, such as Table 310.16, provide values that, when corrected for temperature and adjusted for adjacent load-carrying conductors (the conditions of use) may be used for continuous loads without additional corrections. This considers only the conductor, however. The things to which the conductor connects must also be considered.

An adjustment must be made for continuous loads to accommodate the test methods used for overcurrent devices and their operating characteristics. Since conductors must typically be provided with overcurrent protection based on their ratings, a rule that requires an increase in the rating of an overcurrent device will usually also require a corresponding increase in the ampacity of the protected conductor. Although in some cases the higher-rated overcurrent device could still comply with 240.4 for protection of the conductor, larger conductors must be used with continuous loads to help ensure that the overcurrent device will not overheat and will operate as tested.

Overcurrent devices are tested with conductors of an ampacity that corresponds to the rating of the overcurrent device. A conductor connected to an overcurrent device tends to act as a heat sink for the overcurrent device, so an undersized conductor that does not properly conduct heat away from an overcurrent device may cause the overcurrent device to act differently than intended. This operating characteristic does not change the actual load to which a conductor is exposed, so it does not change the selection of a conductor ampacity when applying the rules of 310.15 based on the load from Article 220.

The rules that sometimes require an increase in the rating of an overcurrent device and the conductor to which it is connected are found in 210.19 and 210.20 for branch circuits, in 215.2 and 215.3 for feeders, and in 230.42 for service conductors. These rules establish a minimum size of overcurrent device and conductor that is separate from the size required to carry the load under the actual conditions of use. This minimum conductor size may be selected from Table 310.16 without consideration (or knowledge) of the conditions of use. The conductor sized based on branch circuit, feeder, or service rules may be bigger or smaller than the conductor sized based on Article 220 and the conditions of use. Conditions of use are taken into account by application of the correction and adjustment factors included in the ampacity tables and 310.15(B)(2)(a), but this is a separate issue. Obviously, if the *NEC* has two or three rules that provide minimum sizes or ratings for a given component, the largest of those minimums must be chosen in order to meet the requirements of all the rules. (See the separate discussion on selection of ampacity in Chapter 3.)

With regard to sign loads and unit load values from 220.12 for general lighting, track lighting, or noncoincident loads, the 125 percent factor will apply to sizing the overcurrent protection and conductors for branch circuits, feeders, or services if the loads are continuous, but the 125 percent factor does not change the load that is calculated in Article 220. Signs and general lighting in commercial occupancies are likely to be continuous loads. Track lighting loads on *feeders* are calculated based on the length of the track in most occupancies, according to 220.43(B), but track lighting loads on *branch circuits* are based on connected load or the rating of the track, according to 220.12 and 410.101(B). Again, track lighting loads may well be continuous, such as in merchandising displays or galleries, but this does not change the load calculation. It does limit the load that may be installed on any one circuit.

Noncoincident loads are an issue only for *feeder load calculations*, according to 220.60. Branch circuits must usually be sized for the entire load they supply. So, for example, where two continuous loads on separate branch circuits are noncoincident, the 125 percent factor will apply in determining a minimum rating and conductor size for each branch circuit, and the load calculation on each branch circuit will be based on 100 percent of the circuit load. (Remember, a load is continuous because it continues for three hours or more when it operates, but it may not operate all of the time. Thus, two continuous loads may not operate at the same time, but will still be continuous loads when they do operate.) The load calculation for the feeder will only have to include 100 percent of the larger of the noncoincident loads. The

Load calculations and minimum conductor sizes for continuous noncoincident branch circuit and feeder loads.

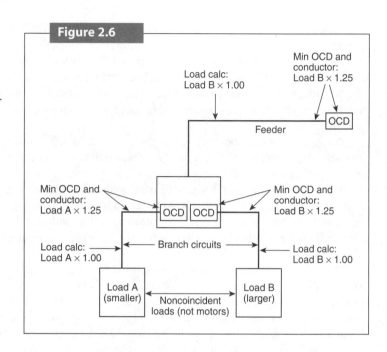

smaller noncoincident load will be omitted from the load calculation by 220.60. The minimum rating and conductor size for the feeder will also be based on 125 percent of the larger continuous load.

This application is illustrated by Figure 2.6. Load A and Load B represent two loads that are not applied to the feeder at the same time, but when they do operate, they may operate for more than three hours. The "Load calc" is the load calculated according to Article 220. This value is used in sizing conductors and determining ampacities under Article 310. The "Min OCD and conductor" is the minimum overcurrent device rating and minimum conductor ampacity determined from 210.19 and 210.20 for branch circuits or from 215.2 and 215.3 for feeders.

Motor loads are somewhat different. According to 210.1, motor branch circuits are covered by Article 430 rather than Article 210. However, most motors are considered to be continuous duty, and the minimum size for most motor circuit conductors includes 125 percent of the largest motor on a conductor per Sections 430.22 and 430.24. Because Sections 220.14(C) and 220.50 refer to the conductor size rules to determine motor loads, 25 percent of the largest motor *is* added in the load calculation for motor loads on a given conductor. (The conductor size from 430.22 may be based on more or less than 125 percent if the motor is time-rated or used for intermittent loads, but in any case, the result from 430.22 is still used for load calculations.)

Question 2.7 May a building be supplied by sources in addition to a service?

Keywords
Service
Outside

Related Questions
- Do the provisions of Article 225, Part II, apply when buildings are provided with their own individual services complying with Article 230?

- Can a multibuilding complex under single management have both a service per Article 230 and also be served by an additional branch circuit or feeder from another building or service where none of the conditions of 230.2(A) through (D) apply?

Keywords

Feeder
Separate building
Number of supplies

Answer

Article 230 applies to services, and Article 225 applies to outside feeders and branch circuits. Part II of Article 225 does apply to any feeders supplying power to a building from outside the first building. Nevertheless, a building supplied by a service is not prohibited from being additionally supplied by a feeder from another building that is itself supplied by a service. Perhaps the most common example of one service supply and one feeder supply is a building with a service and a separate emergency source supplied by a feeder, as shown in Figure 2.7.

Section 230.2 generally requires that a building be supplied by only one service, but more than one service is permitted for a variety of special conditions, special occupancies, large power capacity requirements, or for supplies with different characteristics. Similar rules are found for buildings supplied by feeders or branch circuits. Certainly, one supply is more easily and safely disconnected in an emergency than multiple supplies, but in some cases, the separate supply is purposely arranged so that it will not be easily or inadvertently disconnected in an emergency. Any time a building has multiple supplies, all supplies must be readily identified and located, which is the purpose of the requirements in 225.37, and 230.2(E). Where the

NEC References

225.30, 225.37, 230.2

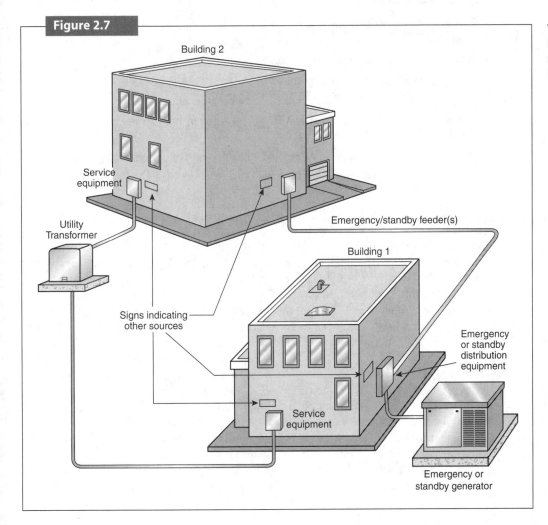

Two buildings served by both feeders and services as permitted in 230.2 and 225.30.

other sources are used for emergency or standby power, signs indicating the locations of other sources are also required by 700.8, 701.9, and 702.8, as shown in Figure 2.7.

Although neither Article 225 nor Article 230 absolutely or specifically restricts the overall number of supplies, the general rules in 225.30 and 230.2 would restrict the number of supplies to one of each type. If an application does not comply with any of the special provisions of 230.2(A) through (D) or 225.30(A) through (E), an enforcement authority could reasonably restrict the use of multiple supplies. As noted, the reason for restricting multiple supplies is a legitimate safety issue (ease of isolating the building from power sources), and the legitimate reasons for allowing multiple supplies are well covered in 230.2 and 225.30. If there is no legitimate reason for decreasing safety in this way, it probably should not be done.

Question 2.8 How many services are permitted for a building?

Keywords

Service
Supply
Disconnecting means
Grouped
Location
Service conductors
Feeders

NEC References

Article 100, 225.30, 230.2, 230.40, 230.70, 230.71, 230.72

Related Questions

- Are multiple service laterals to grouped main disconnects considered to be multiple services?
- How many mains are permitted for one service?
- When are more than six mains permitted?
- How close together must disconnects be to be considered "grouped"?
- When and how must I get permission for more than one service or more than one location?
- May I supply multiple service-entrance conductors with one service drop or lateral?
- How do I recognize the service-entrance conductors?

Answer

The general requirement of 230.2 is that a building be supplied by only one service. A similar rule applies to buildings supplied by feeders according to 225.30. Both rules have similar special rules that permit more than one service or supply. In some cases, such as supplies for fire pumps or some emergency power, additional separate services or feeder supplies are required, and these requirements for "special conditions" are accommodated by 230.2(A) and 225.30(A). Other reasons for having more than one service or supply are covered by 230.2(B) and 225.30(B) for "special occupancies," by 230.2(C) and 225.30(C) for "capacity requirements," and by 230.2(D) and 225.30(D) for supplies with "different characteristics." Other rules require identification of multiple sources and allow multiple feeders where documented safe switching procedures are established.

For the most part, if one of the conditions covered in 230.2 or 225.30 is met, more than one service or supply is permitted, and special permission is not required. This does not mean that the authority having jurisdiction (AHJ) has no say in the matter. Some of the conditions, such as "a building or structure that is sufficiently large" are subject to the interpretations of the AHJ. Nevertheless, the only rule that specifically requires special permission is 230.2(C)(3), where a separate service may be permitted when the conditions of 230.2(C)(1) or (2) are not met and the AHJ decides that more than one service is justified. "Special permission" is defined in Article 100 as written permission from the AHJ.

Section 230.2 covers only services, and Section 225.30 covers only branch circuits and feeders, so the rules do not restrict each other. Typically, one supply of each type is permitted under each of these sections. For example, a common application uses a normal source of power from a utility (one service) and a separate source for emergency or legally required standby power (one feeder supply). This situation falls under the basic rules of 230.2 and 225.30.

The next question is, What constitutes a service, or when does a service become two services? The answer to this question also applies to feeders, except that there will be no service-entrance conductors, service laterals, or service drops, so the rules about those conductors will not apply to feeder or branch circuit supplies.

A "service" is defined in Article 100 as "the conductors and equipment for delivering electrical energy from the serving utility to the wiring system of the premises served." If we also examine the definitions of "premises wiring," "feeder," "service conductors," and "service equipment," we see that a service ends at the service disconnecting means. Thus, we must examine the rules for service equipment and service disconnecting means to determine what constitutes a service and how service disconnecting means are to be provided.

Part VI of Article 230 covers "Service Equipment—Disconnecting Means." For the purposes of this discussion, Sections 230.70, 230.71, and 230.72 are most applicable. Section 230.70 requires the service disconnecting means to be located in a readily accessible location, either outside or inside, that is nearest the point of entrance of the service conductors. This location requirement makes the "main disconnect" easily accessed when needed and limits the amount of service conductor that will be inside a building. Service conductors present a greater risk to a building than feeder or branch circuit conductors because service conductors are only required to have overload protection, and feeders and branch circuit conductors must generally have overload, ground-fault, and short-circuit protection.

Section 230.71 permits up to six disconnects and limits the number of disconnects to six in any one location. According to 230.71(A), certain disconnects for power monitoring, ground-fault protection control circuits, transient voltage surge suppression, or power-operable disconnects are not counted. Generally, additional disconnects in another location constitute an additional service. As we have seen, 230.2 restricts the number of services. These requirements are supported and refined by 230.72, which applies when there is more than one disconnect. Section 230.72 requires the two to six disconnects permitted by 230.71 to be grouped. Additional disconnects for fire pumps or standby systems are treated as separate services, as permitted in 230.2, and are required to be located remote "to minimize the possibility of simultaneous interruption of supply." (Even though mentioned in 230.2, Section 230.72(B) does not mention emergency systems. Typically, separate disconnects from the same utility source are not recognized as suitable secondary sources for emergency systems, but that judgment is left to the AHJ.)

Nothing in Part VI of Article 230 requires multiple services to be in separate locations, nor does it restrict the number of disconnects to six in one location if two services (perhaps for different voltages) are permitted. The restriction on six disconnects in 230.71 is no more than six "per service" in any one location. Where more than one service is permitted under 230.2(A)(1) through 230.2(A)(4) or 230.2(B), the services can be expected to be in separate locations, but multiple services may be located in one place in other cases.

The *NEC* does not say how big an area may be to be considered all one location. Nor does it say how far apart disconnects must be to be in different locations or how close they must be to be considered "grouped in any one location." These judgments are left to enforcement authorities for good reason. Disconnects vary greatly in size, as do the panelboards or switchboards that contain disconnects. For example, two large adjacent panelboards that contain main disconnects may be arranged so that the mains are also adjacent or widely separated, and thus may be considered to be in one location or two, and may constitute one service

Two examples of a service consisting of two grouped disconnects with more than one set of service-entrance conductors supplied by a single service drop.

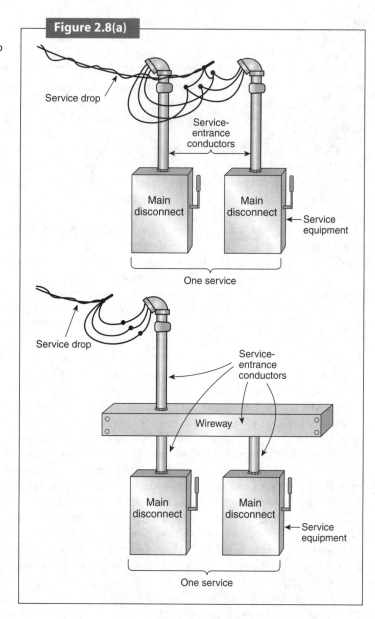

Figure 2.8(a)

or two. Two mains on either side of an aisle may be considered grouped, but that depends on how clearly they are identified and how readily they could be operated. An AHJ must make this decision based on the intent of the *NEC*, that is, based on safety considerations. For that matter, the limitation of six disconnects or six operations is not a magic number. It is simply a reasonable requirement and limitation for readily and quickly disconnecting power when necessary.

The *NEC* does help to clarify some other common interpretation issues. For example, the diagrams illustrating this question show three arrangements that are all considered to be one service. Figure 2.8(a) shows two examples of services with two disconnects grouped together and more than one set of service-entrance conductors supplied by a common service drop. Although Section 230.40 states that "each service drop or service lateral shall supply only one set of service-entrance conductors," Exception No. 2 specifically permits applications like

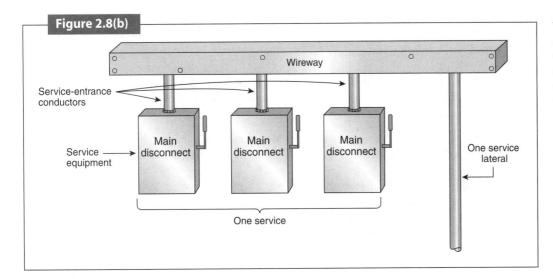

Three disconnects constituting the service equipment for one service with three sets of service-entrance conductors supplied by one service lateral.

this one. Figure 2.8(b) illustrates multiple sets of service-entrance conductors supplied by a common service lateral, another case that is covered by the same rule and exception.

Figure 2.8(c) shows two disconnects grouped in one location supplied by what appear to be two separate sets of service lateral conductors. Section 230.40, Exception No. 2, applies here as well, because 230.2 says, for the purposes of this example, that the two sets of conductors "connected together at the supply end but not connected together at their load end shall be considered to be supplying one service." In other words, the two conductor sets in Figure 2.8(c) are treated as one service lateral even though they are not truly a parallel set because they are not connected together at both ends. (In effect, Section 310.4 permits paralleled conductors to be treated as a single conductor in some respects, primarily with regard to establishing ampacity, although this permission is not explicitly stated.)

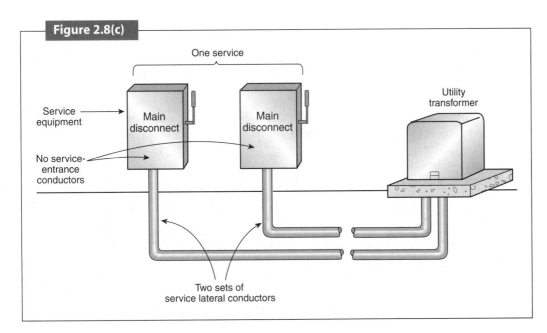

Two sets of service lateral conductors supplying one service with no service-entrance conductors.

Notice that there are no service-entrance conductors in Figure 2.8(c). This is consistent with the definition of service-entrance conductors in Article 100 and the Fine Print Note, which reads as follows:

Service-Entrance Conductors, Underground System. The service conductors between the terminals of the service equipment and the point of connection to the service lateral.

FPN: Where service equipment is located outside the building walls, there may be no service-entrance conductors, or they may be entirely outside the building.

If the service lateral conductors in Figure 2.8(c) were interrupted by metering equipment before they were terminated on the service equipment, the laterals would end at the metering equipment, and service-entrance conductors would extend from the metering point to the service disconnects. (See the definition of "service lateral conductors" in Article 100.)

Question 2.9 Are goosenecks permitted to be located below service drop conductors?

Keywords
Gooseneck
Service head
Weatherhead
Service drop
Raintight
Point of attachment

NEC References
90.5, 230.54

Related Questions
- Are goosenecks in service cables permitted by the *NEC*?
- Are goosenecks prohibited when they cannot be located above the point of attachment of the service drop conductors?

Answer

As with many interpretation issues, the rule in question must be taken in its entirety. Section 230.54(A) requires service raceways to be equipped with a raintight service head at the point of connection of the service drop conductors. Section 230.54(B) (with the Exception) requires service cables to be equipped with a service head *or* gooseneck for the purpose of being made raintight. The exception to 230.54(B) clearly permits the use of a Type SE (service-entrance) cable with a formed and taped gooseneck as a method of making the cable raintight.

A common misinterpretation likely arises from a misunderstanding of the terminology. The term "service head" applies to the fittings used both on raceways and on cable. However, where a service head, often called a "weatherhead," is not used, a service cable (but not raceway) may be made raintight by forming the cable in a gooseneck, as shown in Figure 2.9.

Section 230.54(C) requires the service heads and goosenecks to be located above the point of connection of the service drop conductors where they are attached to the building or structure, as shown in Figure 2.9. This is intended to keep rain or other moisture that may come down the service drop from being transmitted into the cable or raceway. The exception to 230.54(C) applies both to the service heads used on a raceway or cable and to the gooseneck. The exception is a "permissive rule" as described in 90.5(B). It allows but does not require something different from what the main rule requires. Where the point of attachment for

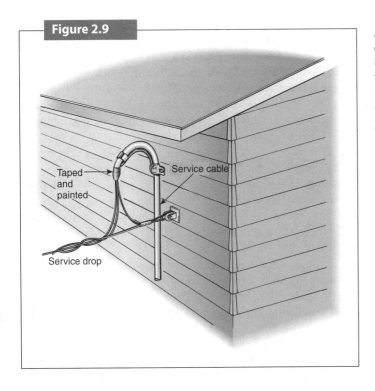

Figure 2.9

An example of a service-entrance cable formed into a gooseneck and terminated without a raintight service head.

Source: Based on *NEC Handbook*, 2005, Exhibit 230.25.

the service drop is necessarily above the top of the service cable, either a service head or a formed gooseneck is acceptable.

Prior to the 2005 *NEC*, the exception to 230.54(C) did not prevent the use of a service cable formed into a gooseneck, but goosenecks had to be located above the point of attachment of service drop conductors, as shown in Figure 2.9, whereas cables or raceways with service heads could be located below but within 24 inches of a service drop attachment. Now the exception applies to formed goosenecks or service heads. In any case, the conductors emerging from the raceway or cable service head must still be formed in a "drip loop" to prevent water from entering the cable or raceway, as required by Sections 230.54(F) and (G).

Question 2.10

What size service is required for temporary power for construction?

Related Question

- How is the minimum rating for a service supplying temporary power calculated?

Answer

The required minimum rating for temporary power depends on the number and rating of the circuits supplied. According to Section 590.4(A), temporary services must comply with

Keywords

Temporary power
Service
Service conductors
Feeders
Circuits

NEC References

Article 220, 230.23, 230.31, 230.42, 230.79, 590.4

Article 230. Article 590 is primarily about alternative wiring methods and ground-fault protection for temporary power—whether for construction or for other purposes. Article 230 addresses the issue of minimum sizes in Sections 230.23, 230.31, 230.42, and 230.79. Section 230.23 covers overhead service drop conductors, 230.31 covers underground service lateral conductors, 230.42 covers service-entrance conductors, and 230.79 covers the service disconnecting means. Typically, either 230.23 (overhead conductors) or 230.31 (underground conductors) must be applied along with 230.79 (service disconnect rating) in determining a temporary service size.

A temporary service may or may not have service-entrance conductors. For example, many temporary services are installed before there is a building, perhaps using an underground-fed meter-main-panelboard combination device for service equipment supplied directly from a pad-mounted transformer. In other cases, the overhead conductors are installed from the utility transformer to the service equipment without a splice. In these cases there may be no service-entrance conductors, and Section 230.42 (service-entrance conductor rating) may not apply.

There are two sets of rules for minimum ratings. First, Sections 230.23, 230.31, 230.42, and 230.79 each provide some minimum sizes. Second, each of the sections mentioned also refers to Article 220 for load calculations. Based on the references provided, the smallest rating permitted for a single-circuit installation is 15 amperes, but this size usually cannot be used for temporary power for construction because at least two circuits are required to provide separate circuits for tools and lighting according to 590.4(D). For two circuits, the minimum size for service-entrance conductors and disconnecting means is 30 amperes, based on 230.79(B), and the minimum service drop or lateral size is 8 AWG copper, based on 230.23(B) and 230.31(B). For more than two circuits, the minimum rating of the disconnect

Wiring for temporary power for construction is covered by Article 590, but service ratings are still covered by Articles 220 and 230.

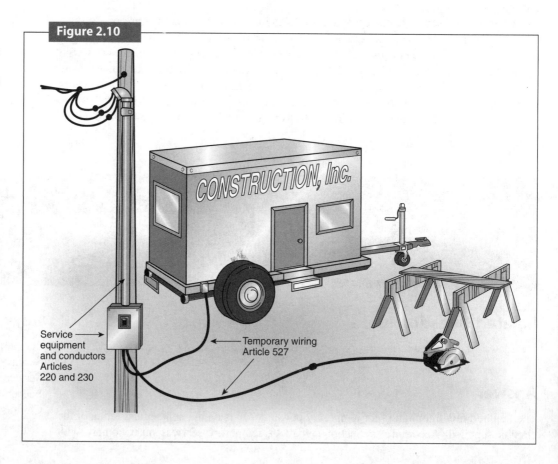

Figure 2.10

and service-entrance conductors for temporary power is 60 amperes, and service drop or lateral conductors must have an equivalent rating.

The rules summarized so far are only minimums that do not necessarily consider the actual load. However, as noted, service disconnects and conductors must always be adequate for the load calculated according to Article 220, as illustrated by Figure 2.10. Where temporary heating, welders, cranes, lifts, construction trailers, brick saws, slip forms, or the like are used, the required service may be much larger. Certain types of equipment used in high-rise and industrial construction, such as stud welders and tower cranes, may require three-phase, 480-volt temporary services that are rated in hundreds (even thousands) of amperes, with separate feeders supplying separately derived systems for portable tools and temporary lighting. In short, the rating required depends on the nature of the project and the equipment that will require power. For single-family residential construction sites, the smallest size that is adequate and code compliant is usually 60 amperes.

Question 2.11

Are "gang-operated" isolating switches required ahead of primary metering equipment?

Related Questions

- Does the *NEC* require a gang-operated switch or switches ahead of high-voltage primary metering?
- What code rules apply to a 13.5 KV service and metering?

Answer

This answer assumes the question is about placing switches so as to disconnect *service conductors* as defined in Article 100. Because a service is defined as coming from a utility, most equipment ahead of the service is owned by a utility. Therefore, the *NEC* does not apply to most utility power distribution and metering installations, according to 90.2(B)(5). Typically, the utility owns the metering equipment once the installation is complete and energized, even though part of it may be supplied by the customer or a contractor working under the *NEC*.

Article 230, Part VIII, and Article 490, Part II, cover services and service equipment over 600 volts. Section 490.21(B)(7) requires a gang-operated disconnecting switch for metal-enclosed switchgear and substations where high-voltage (over 600 volt) fuses are used. However, this would usually only apply to the service equipment and feeder devices that are downstream from the utility metering.

Generally, according to Section 230.204, isolating switches are required where the blades of the service disconnecting means are not visible, such as in oil switches or in nonviewing circuit breakers (such as air, oil, vacuum, and sulfur hexafluoride types). The use of a gang-operated or isolating switch with visible contacts as shown in Figure 2.11 provides a ready indication of the status of the switch. However, the code *does not require* isolating switches on service conductors to isolate metering. Also, the code does not specifically address "cold" or "hot" sequence metering. In fact, the *NEC* does not require meters or meter sockets at all, but it does accommodate their use. For the most part, the code permits the utilities (or other standards or jurisdictions) to set their own metering requirements.

Keywords

Isolating switches
Primary metering
Service disconnects

NEC References

Article 100, 230.204, 490.21

Isolating switch on a pole, shown in the closed position.

Figure 2.11

Question 2.12

Where is an overcurrent device required to be located in a circuit?

Keywords

Tap rules
Overcurrent protection
Tap conductor
Transformer
Secondary conductor
Feeder tap
Ampacity
Overcurrent device
Device

Related Questions

- Is an overcurrent device permitted at a location other than the beginning of the circuit?
- Do tap rules apply to service conductors?
- How is overcurrent protection provided for tap conductors?
- Are transformer secondary conductors considered to be tap conductors?
- Must all feeder taps terminate in an overcurrent device?
- Is the rating of an overcurrent device at the end of a tap permitted to be rounded up?
- Is a tap conductor permitted to terminate on an MLO panelboard?
- Is a tap conductor permitted to terminate on a panelboard main circuit breaker?

- Is a transformer permitted to supply more than one set of secondary conductors?
- How is overcurrent protection provided for conductors connected to a transformer secondary?
- Are separate overcurrent devices or disconnects required for transformer protection?
- Is a contactor permitted in transformer secondary conductors? Is a transformer permitted to supply more than one set of secondary conductors?
- May the overcurrent device rating on a transformer primary be rounded up?
- How is panelboard or transformer overcurrent protection provided under the 25-foot industrial rule?

Keywords

Service conductor
Overload
Short circuit
Ground fault

NEC References

215.2, 240.2, 240.4, 240.20, 240.21, 240.22, 240.24, 240.92, 450.3

Answer

The rules found in Part I of Article 240, specifically Sections 240.4, 240.5, and 240.6, cover only the *ratings* of overcurrent devices. However, the basic requirement for providing overcurrent protection based on the ampacity of the conductors implies that the overcurrent devices will be located at the beginning of the conductor. Although overload protection can be provided anywhere in a circuit as long as it is in series with the entire load, short-circuit and ground-fault protection is only provided for conductors on the load side (downstream) of an overcurrent device. That is, short-circuit and ground-fault protective devices can only sense and interrupt currents from such faults if they are on the load side of the device. Section 240.4(E) recognizes this and permits tap conductors to be protected in accordance with other rules that are more specific with regard to the location of overcurrent devices.

A tap conductor is defined in 240.2 as "a conductor, other than a service conductor, that has overcurrent protection ahead of its point of supply that exceeds the value for similar conductors that are protected as described elsewhere in 240.4." This definition applies only in the context of Article 240, since the term "tap" is used in other ways in other places in the *NEC*. Figure 2.12(a) shows two examples of possible tap conductors as defined, along with four examples of conductors that are not tap conductors as the term is used in Article 240. The examples that meet the definition are (1) not service conductors, (2) have an overcurrent device ahead of their point of supply, and (3) the overcurrent device does not comply with 240.4 for the protection of the conductor in question. The illustration does not show how the conductors are terminated or the nature of the load. Termination is a primary factor in whether these applications are permitted, and is addressed by the "tap rules," which will be discussed later. The nature of the load may also have a bearing on the permissible ratings of overcurrent devices according to 240.4(G).

Referring to Figure 2.12(a), Example A represents a feeder tap, that is, a feeder tapped from a feeder. Example B is a tap conductor as defined, but might not be a tap conductor if the conductor were instead a branch circuit conductor and the load consisted of a welder or motor(s) or the conductors were otherwise protected in accordance with 240.4(G). The 130 A conductor in Example C may be considered to be protected by the 150 A overcurrent device according to 240.4(B), so it does not meet the definition of a tap conductor. Example D shows a change in conductor size, which is often thought of as a tap, but the 200 A conductor is protected by the 200 A overcurrent device, so no tap conductor is shown in this example. Similarly, in Example E, the connection between conductors might be called a tap connection, but the conductor is not a tap conductor. Finally, Example F shows taps in service-entrance conductors, but the conductors are not tap conductors because they are service conductors and they do not have overcurrent protection ahead of their point of supply.

Part II of Article 240 covers the location of overcurrent devices, both in circuits and on the premises. For example, Section 240.20 requires overcurrent protection in each ungrounded conductor, Section 240.22 says that overcurrent protection is generally not permitted in grounded conductors, and Section 240.24 requires most overcurrent devices to be readily accessible to occupants.

Illustration of the definition of a tap conductor, with two conductors that meet the definition in 240.2 and four that do not. Numbers refer to ratings of overcurrent devices and ampacities of conductors. Only A and B represent tap conductors.

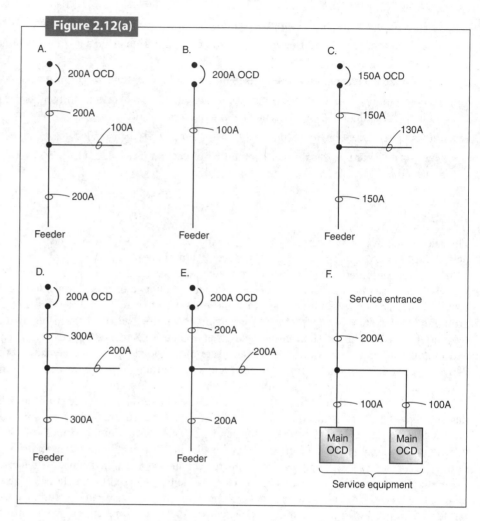

Figure 2.12(a)

Section 240.21 provides requirements for the location of an overcurrent device in a circuit. The general rule is that the overcurrent protection should be located at the point where a conductor receives its supply. Typically, branch circuit or feeder conductors that are connected to overcurrent devices in a panelboard meet this general rule. However, in many cases, conductors must be run some distance from their supply to an overcurrent device, or must be reduced in size to connect to utilization equipment, and the special rules in 240.21(A) through (G) accommodate these needs.

The special rules in 240.21 are often called the "tap rules," but not all the rules are about taps. For example, the service conductors covered by 240.21(D) and the conductors that connect directly to generator terminals covered by 240.21(G) must run some distance before they can be connected to an overcurrent device. The distance may be relatively long in the case of service conductors, or may be quite short if, for example, a generator overcurrent device is mounted on the generator. These conductors are not tap conductors by definition because there is no overcurrent protection for the conductors ahead of their point of supply. Service conductors and taps from service conductors are only required to be provided with overload protection (i.e., short-circuit and ground-fault protection is not required) in accordance with 230.91, and the other tap rules of Article 240 do not apply. In other words, there are no 10-foot or 25-foot tap rules or other length limitations for service conductors.

The tap rules of Section 240.21 are essentially broken into three parts, covered by 240.21(A), (B), and (C). (A) covers taps from branch circuits, (B) covers taps from feeders, and (C) covers conductors connected to transformer secondaries. For the most part, transformer secondary conductors do not meet the definition of a tap conductor unless, perhaps, they are supplied by a two-wire secondary or from a delta-delta connection so that the primary overcurrent device is considered to protect the secondary conductors as permitted by 240.4(F). Since most transformer secondary conductors are not really tap conductors as defined, they have been covered under their own rules since the 1999 NEC. Separate rules also apply to motor circuit taps and busway taps. The motor circuit feeder tap rules in 430.28 are essentially the same as the 10-foot, 25-foot, and 100-foot rules of 240.21(B). Other special rules that apply only to supervised industrial facilities as defined in 240.2 can be found in 240.92.

One difficulty that many people have with tap rules is that the conductors appear to be unprotected. However, by definition, tap conductors are not really unprotected, because they have an overcurrent ahead of their point of supply that will provide some short-circuit and ground-fault protection, and they are terminated in a device that will limit the load on the conductor and that therefore usually provides overload protection. The tap conductors are unprotected only in the sense that all of the overcurrent protection is not located at the point where they receive their supply, and the short-circuit and ground-fault protection is not sized directly from the tap conductor ampacity.

The tap rules of 240.21 differ in the length of the "unprotected conductor," the termination, the minimum size in relation to the overcurrent device ahead of the tap or on the primary side of the transformer, the location, and, in some cases, the occupancy. The rules are similar in that the conductors are always required to have physical protection, they must always be big enough for the load supplied, and they must terminate in devices that will tend to limit the load on the conductors.

In some ways, the tap rules are more restrictive than the overcurrent sizing rules of 240.4. Sections 240.21(B) and (C) both prohibit the use of 240.4(B) with feeder taps or transformer secondary conductors. For example, according to 240.21(B)(3), primary conductors that are protected at their ampacity are not counted in the 25-foot total length of the primary and secondary conductors, but this does not allow the overcurrent device on the primary to be rounded to the next larger standard size as might otherwise be permitted. This means that if a tap is made with a primary conductor that is rated at, say, 130 amperes and protected at 150 amperes as permitted by 240.4(B), that conductor is not "protected at its ampacity" and must be counted in the total 25 feet. As another example, the conductor in Figure 2.12(b) that extends from the 110 A fuse to the 8 AWG feeder tap conductor must have an ampacity of at least 110 amperes or its length would have to be included in the 25-foot allowance. (This in-

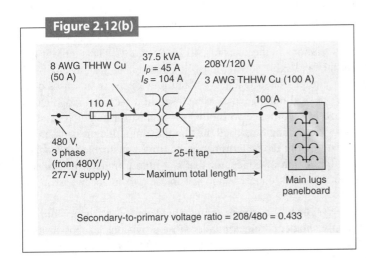

An example in which the transformer feeder tap conductors (primary plus secondary) are not over 25 feet long in accordance with 240.21(B)(3).

Source: Based on *NEC Handbook*, 2005, Exhibit 240.9.

terpretation only applies to feeder tap conductors supplying transformers. If no tap is made, 240.21(C)(6) allows 25 feet of secondary conductor and does not govern primary conductors that are protected in accordance with 240.4.) Similarly, the rating of the secondary overcurrent device (or, in the case of 240.21(C)(3), the sum of the ratings) may not be rounded up. Except under 240.21(C)(3), tap conductors or secondary conductors must be terminated in a *single device* (usually a single overcurrent device) that limits the load to the ampacity of the tap or secondary conductors. This language does not permit rounding up to the next larger rating as permitted for conductors in 240.4(B).

Before reviewing the details and differences in the tap rules, we should recognize one important point: The rules in 240.21 are about protection of conductors, not about protection of panelboards or transformers or other equipment. Other rules for protection of equipment are referenced in Section 240.3.

Compliance with the rules of 240.4 and 240.21 will provide protection for conductors but will not ensure that other equipment is properly protected. Similarly, compliance with the rules for protecting specific equipment will not necessarily provide protection for conductors. For example, the "10-foot tap rules" of 240.21(B)(1) and 240.21(C)(2) allow a conductor to be terminated in a device—not necessarily an overcurrent device—if the rating of the device is not greater than the rating of the conductors. This would permit a tap or a transformer secondary conductor to be terminated in a panelboard without a main overcurrent device. However, many panelboards are required to have overcurrent protection that complies with 408.36. Where protection of the panelboard is required, the overcurrent device may be ahead of or within the panelboard enclosure, but must protect the panelboard bus assembly.

Consider another example of the difference between rules for overcurrent protection of conductors and rules for overcurrent protection of equipment. According to Tables 450.3(A) and (B), the secondary protection of a transformer may consist of up to six overcurrent devices (breakers or sets of fuses). However, as we have seen, single overcurrent devices are required for secondary conductor protection. Up to six sets of conductors, each of which terminates in a single overcurrent device, could be used if the rating of each overcurrent device did not exceed the rating of its supply conductors as required by 240.21, and the sum of the ratings of the overcurrent devices did not exceed the value permitted for a single overcurrent device for the transformer as required by 450.3. In short, a given installation may require overcurrent protection for more than one purpose, and all applicable rules must be observed and coordinated to comply with the code.

Note that an overcurrent device provided for transformer secondary conductors could also be used to provide secondary protection of the transformer if properly selected to meet both requirements, but only 240.21(C)(3) *and* 240.92(B), *both* of which apply only to industrial occupancies, permit more than one overcurrent device for a single set of secondary conductors, and 450.3 allows no more than six overcurrent devices to protect a transformer. Also, although transformers are often protected by overcurrent devices on the primary only, primary overcurrent protection is usually not considered to protect secondary conductors. (See 450.3, FPN No. 1.)

Table 2.12 summarizes the variations in the rules of 240.21(B) and (C). This table is not comprehensive, and all rules of the applicable section must be observed, as well as any rules that apply to the transformers or equipment supplied. As noted previously, in all cases conductors must be protected from physical damage and/or enclosed in a raceway. Conductors must also be adequate for the load in accordance with other provisions for conductors, such as 215.2 or 310.10. Also, with regard to length restrictions, the maximum length is the maximum length of the conductor itself, from tap to termination or from transformer secondary to termination. This means that the length of conductor that is within a transformer, panelboard, junction box, or other enclosure is counted, not just the length in raceways or the scaled distance on a drawing.

The main rule of 240.21 says that "No conductor supplied under the provisions of 240.21(A) through (G) shall supply another conductor under those provisions, except through an overcurrent protective device meeting the requirements of 240.4." This rule does not

Table 2.12 Summary of Tap Rule Provisions and Applications

Section Number	Tap or Secondary Conductor	Maximum Length	Rating Ratio*	Location	Occupancy	Termination	Other Restrictions or Notes
240.21(B)(1)	Tap	10 ft	1/10	Any	Any	Single device	1/10 ratio applies to field taps; termination may be in device other than overcurrent device.
240.21(B)(2)	Tap	25 ft	1/3	Any	Any	Single OCD†	
240.21(B)(3)	Tap and Secondary	25 ft total	1/3	Any	Any	Single OCD†	Portion of primary that is protected at its ampacity is not counted in total length, but applies only where primary conductors are tap conductors. Secondary ampacity must be corrected for transformer.
240.21(B)(4)	Tap	100 ft	1/3	Inside	High-bay manufacturing buildings only	Single OCD†	Conductors not smaller than 6 AWG cu or 4 AWG al. No penetration of walls, floors, or ceilings permitted.
240.21(B)(5)	Tap	No limit	N/A	Outside	Any	Single OCD†	Must remain outside except at OCD termination.
240.21(C)(1)	Secondary	No limit	N/A	Any	Any	Not specified	Ampacity must be corrected for transformer. Limited to two-wire secondary and three-wire delta-delta transformers where secondary conductor may be protected by primary overcurrent device. See 240.4(F).
240.21(C)(2)	Secondary	10 ft	1/10	Any	Any	Single device	Termination may be in device other than overcurrent device.
240.21(C)(3)	Secondary	25 ft	N/A	Any	Industrial only	Multiple OCD grouped together	Conductor ampacity must be at least equal to transformer secondary rating. Sum of overcurrent device ratings cannot exceed conductor ampacity.
240.21(C)(4)	Secondary	No limit	N/A	Any	Any	Single OCD†	Must remain outside except at termination.
240.21(C)(5)	Secondary	25 ft	1/3	Any	Any	Single OCD†	Same as 240.21(B)(3)—applies only where primary conductors are tap conductors. Secondary ampacity must be corrected for transformer.
240.21(C)(6)	Secondary	25 ft	1/3	Any	Any	Single OCD†	Applies where primary conductors comply with 240.4. Secondary ampacity must be corrected for transformer.

*Rating ratio is the least permitted ratio between the ampacity of the conductor and the rating of the overcurrent device, corrected for transformers where applicable. N/A, not applicable, means that no ratio applies in this rule.
†Single OCD = Single overcurrent device with rating not exceeding ampacity of conductors.

Multiple separate tap conductors or multiple separate secondary conductor sets are permitted if other requirements for protection of conductors and transformers are met.

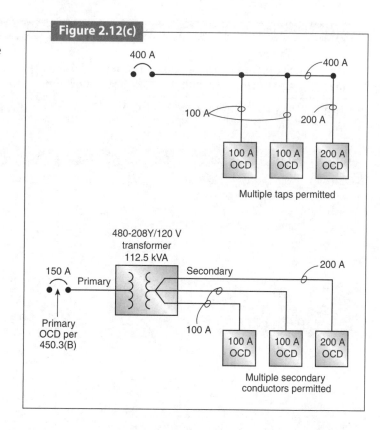

prohibit a conductor that is protected in accordance with 240.4 from supplying more than one tap conductor, nor does it prohibit a transformer from supplying more than one secondary conductor. These multiple taps or secondary conductors could each comply with different rules, for that matter. These permitted arrangements are illustrated in Figure 2.12(c). What is prohibited is a tap conductor being supplied directly from another tap conductor or a tap conductor being supplied from a transformer secondary conductor, as shown in Figure 2.12(d). This restriction is commonly stated as "You can't tap a tap."

Additional tap conductors are prohibited except through an overcurrent device meeting the requirements of Section 240.4.

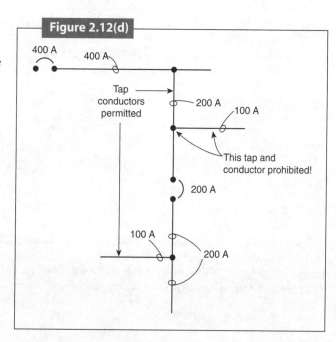

Question 2.13

May I terminate a grounding electrode conductor in a meter base?

Related Questions

- Is a bonding jumper required to bond a meter base ahead of the service equipment?
- Is a bonding jumper required for a conduit nipple installed in a threaded hub?

Answer

The location of system grounding connections is covered by Section 250.24(A) in the *NEC*. Section 250.24(A)(1) says the connection of the grounding electrode conductor to a grounded service conductor "shall be made at any accessible point from the load end of the service drop or service lateral to and including the terminal or bus to which the grounded service conductor is connected at the service disconnecting means." A meter base is typically located between the end of the service lateral or drop and the service disconnect, so it is a permissible location for the grounding electrode connection, as shown in Figure 2.13(a). However, this must be an accessible point. Many utilities assume ownership of everything up to and including the meter base, and access to other than utility personnel may be denied. Therefore, the meter base is often considered inaccessible to inspection and is not a permitted termination point for grounding electrode conductors. In most cases, the service disconnect is the preferred location (shown as a solid line in Figure 2.13(a)) because it is accessible and, if listed for use as service equipment, includes provisions for terminating a grounding electrode conductor.

Most meter bases also have a terminal for a neutral conductor that is connected directly to the meter enclosure. In effect, when a grounded conductor is landed on such a terminal, the direct connection to the meter base also serves to bond the meter base in accordance with Section 250.92(B)(1). Since the meter base is almost automatically bonded, no additional bonding jumper is usually needed. This bonding connection is shown in Figure 2.13(a) as a solid line because the connection is usually inherent to the construction of the meter base, even if the grounding electrode conductor is not terminated in the meter base.

When metering is done on the load side of the service disconnect, the inherent bonding feature in meter bases may cause problems. Generally, any such connection to a grounded conductor on the load side of the service is prohibited by 250.24(A)(5), but a special rule in 250.142(B), Exception No. 2, allows ordinary meter bases to be used if there is no ground-fault protection on the service, the meter enclosures are "near" the service disconnect, and the grounded service conductor is no smaller than would be permitted for an equipment grounding conductor. Meter bases may be ordered with isolated neutrals, but they are usually not readily available.

Section 250.92(A) requires that all non-current-carrying metal parts of service raceways and enclosures (meter fittings are specifically mentioned) be bonded using techniques specified in 250.92(B). As we have seen, a direct connection to the grounded service conductor can be used and typically is used to bond the meter enclosure. But what about the raceways coming into and out of the meter base?

The service equipment (disconnect) enclosure is usually bonded through the main bonding jumper. In some service equipment, the grounded terminal is directly connected to the service enclosure, which would also bond the service enclosure. If a service raceway or conduit nipple enters the enclosure through a wrenchtight connection to a threaded hub (threaded "boss," in code terminology), the nipple is bonded by that connection. The nipple does not

Keywords

Grounding electrode conductor
Grounded service conductor
Bonding at services
Meter base
Meter fitting
Meter enclosure
Service equipment
Service disconnect

NEC References

250.24, 250.92, 250.142

An illustration of three permitted points of connection between a grounding electrode conductor and the grounded service conductor.

Source: Based on *NEC Handbook*, 2005, Exhibit 250.8.

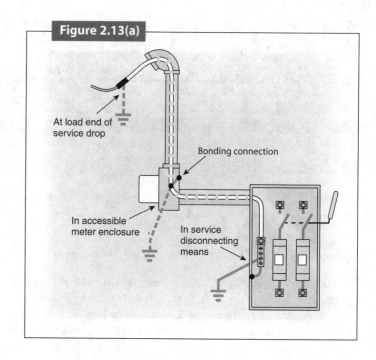

Bonding of components of service equipment using hubs and other inherent features.

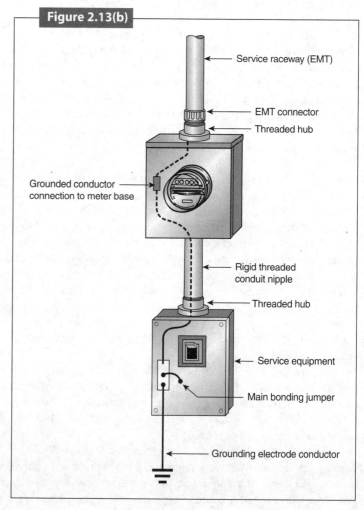

also have to be bonded to the meter base, because both the nipple and the meter base are already bonded. If another hub is used on the meter base and the service raceway is similarly connected, then the service raceway is also bonded and no other special bonding devices or jumpers are required.

Section 250.92(A) simply provides a list of items that must be bonded at or ahead of service equipment, and 250.92(B) provides a list of acceptable methods. These rules do not impose any requirement that the items be bonded in a daisy-chain, a loop, a star, or any other specific pattern, nor do they require that each item be individually bonded with a jumper. They simply require that all the items listed in 250.92(A) be bonded by one of the methods listed in 250.92(B). Thus, the items in Figure 2.13(b) are considered to be bonded to "service grade" requirements even though no bonding or grounding bushings or bonding jumpers are used.

Consider the enclosure bonding illustrated in Figure 2.13(b). The service equipment enclosure is bonded to the grounded conductor and the grounding electrode conductor by the main bonding jumper. The conduit nipple is bonded by connection to the threaded hub and the service enclosure. The meter base is bonded by its direct connection to the grounded service conductor. The EMT service raceway is bonded by the threadless connector, which is bonded by its threaded connection to the hub, which in turn is bonded by its bolted, welded, or sometimes riveted connection to the meter base. In this example, all of the elements that are required to be bonded are bonded by inherent connections. There is no requirement for an additional bonding connection between the threaded nipple and the meter base, so a bonding or grounding bushing is not required at that connection point, and an ordinary insulated bushing could be used.

In many other configurations, bonding jumpers may be required. In fact, many service laterals are connected to service enclosures through knockouts at the bottom of the enclosures rather than with threaded hubs. In such cases, bonding bushings or grounding bushings with jumpers or an equivalent method from 250.92(B) must be used, as shown in Figure 2.13(c).

Service equipment with service raceway entering through concentric knockouts. The service raceway is bonded to the service enclosure with a grounding bushing and jumper.

Question 2.14

What is meant by a "parallel path"?

Keywords

Parallel path
Objectionable current
Separately derived system
Buildings supplied by feeders or branch circuits
Separate building
Grounded conductor
Equipment grounding conductor
Grounding electrode
Bonding jumper

NEC References

220.61, 250.6, 250.24, 250.30, 250.32, 250.102, 250.142, 250.166

Related Questions

- Why are parallel paths prohibited in grounding arrangements for separately derived systems and separate buildings?
- What is an "objectionable current over grounding conductors"?
- Are bonding jumpers and electrodes always required at separate buildings?
- Where is the bonding jumper to be installed on a separate building or separately derived system?

Answer

When Article 250 was extensively reorganized and rewritten for the 1999 edition of the *NEC*, a number of issues with regard to grounding of separately derived systems and at separate buildings were clarified. One of these had to do with the issue of parallel paths for neutral current. Section 250.30, covering grounding of separately derived systems, and Section 250.32, covering grounding at separate buildings supplied by the same service, were both revised to specifically mention (and prohibit) these parallel paths. Neither of these were really new issues in the *NEC*, because the issue of "objectionable current over grounding conductors" was already in Article 250. This old rule is now found in 250.6, but for many people, applications of the rule are still not obvious.

Neutral currents are intended to be carried by grounded conductors, not grounding conductors, so any portion of the normal neutral current that is carried on a grounding conductor, especially an equipment grounding conductor, is an objectionable current. Grounding conductors are intended to establish an equal potential between various non-current-carrying metal parts and the earth so that there will be no shock hazard from contact with those parts. Current on a conductor will create some voltage drop across that conductor. Ideally, there will be no current on the grounding paths, so that no voltage difference can develop between parts that are intended to be grounded. Some grounding conductors (equipment grounding conductors and bonding conductors) are expected to carry ground-fault currents when such faults occur, but those temporary currents are not "objectionable" according to 250.6(C). Equipment required to be grounded must still be grounded, but grounded conductors are provided for normal neutral currents, and grounding or bonding conductors are required for establishing equal potential between metal parts and for carrying fault currents. Grounded conductors are usually not permitted to be used for grounding of equipment on the load side of service equipment. (There are a few special cases in 250.142(A).) Maintaining separate isolated paths will keep unintended currents off of grounding conductors.

Section 250.30(A) was revised to require the bonding jumper and grounding electrode conductor connections to the grounded conductor (usually a neutral) to be in the same location. That is, where a transformer is the source, both the bonding jumper and the grounding electrode conductor could be connected to the grounded conductor in the transformer, or those connections could be made in the first disconnecting means. An exception was added (250.30(A)(1), Exception No. 1) that permits a bonding jumper in both locations "where doing so does not establish a parallel path for the grounded circuit conductor." This exception accommodates a few necessary special rules and situations, but also requires a judgment to be made as to what parallel paths might be possible. Many of the possible metallic paths could be created by other (non-electrical) systems, by metal conduits, or by interconnected grounding electrodes.

Consider the example in Figure 2.14(a), in which both the bonding connection and the grounding electrode conductor connection to the grounded (neutral) conductor are made in the transformer. From this point on, the functions of the grounded and grounding conductors are separated, with the grounded conductor providing a path for normal currents and the grounding conductors a path only for fault currents.

These separate paths are important for the reasons explained previously and because the two conductors are sized by different rules. The grounded conductor is sized according to 220.61 for the unbalanced current that may be imposed—a current that could exist for a relatively long time, even continuously. The equipment bonding jumper between the transformer and the disconnect is sized in the same manner as the system bonding jumper within the transformer, that is, according to 250.30(A)(2) and (A)(1), respectively. These rules refer to 250.102(C) and 250.28(D), each of which in turn refers to 250.66 (or 12-1/2 percent of the area of the derived phase conductors for larger conductors). The references to 250.66 and the "12-1/2 percent rule" are intended to make the conductor big enough for fault current when the fault is ahead of the overcurrent device. The equipment grounding conductors on the load side of the disconnecting means (none are shown in Figure 2.14(a)) are sized based on the overcurrent device rating and according to 250.122 for the short-time fault current that could flow until the overcurrent device operates.

Figure 2.14(b) shows a similar installation, but in this example the connections to the grounded conductor are made in the first disconnect. In this case, the grounded conductor must carry both normal current and fault current between the source and the first disconnect. From the first disconnect on, the functions are separated as before. This figure also shows a bonding jumper between the transformer and the first disconnect for grounding of and providing a fault current path from the transformer frame and enclosure to the grounding point in the disconnect and back to the secondary of the transformer. The proper sizing of bonding jumpers and grounded conductors between the source and the first disconnect is covered by

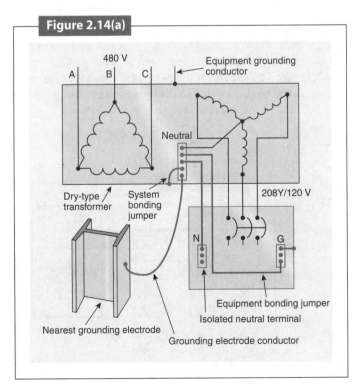

Figure 2.14(a)

A grounding arrangement for a separately derived system in which the grounding electrode conductor connection is made in the transformer.

Source: Based on *NEC Handbook*, 2005, Exhibit 250.13.

A grounding arrangement for a separately derived system in which the grounding electrode conductor connection is made in the first disconnect.

Source: Based on *NEC Handbook*, 2005, Exhibit 250.14.

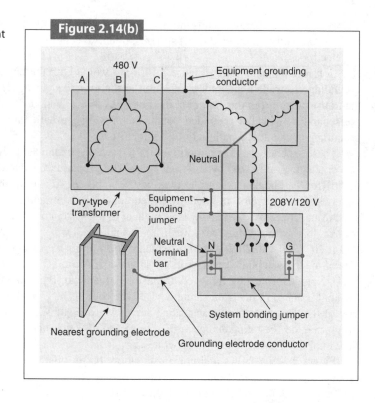

250.30(A)(2) and (8) to be certain that these conductors will be adequate for the currents they are expected to carry.

Figure 2.14(c) shows an example similar to Figure 2.14(b). However, in this example, a parallel path is created by the metal raceway and the addition of another bonding jumper in the transformer. In this case, the second bonding jumper (in the transformer) is not needed, because the metal raceway could be used as a fault current path from the transformer to the bonding point in the disconnecting means. The additional bonding jumper in the transformer would be permitted if the raceway were nonmetallic and there were no other parallel paths, in which case Section 250.142(A)(3) specifically permits the use of the grounded conductor for grounding the transformer frame in the same way that it is permitted to use the grounded conductor for grounding equipment ahead of the disconnecting means at a service.

A *prohibited* grounding arrangement for a separately derived system in which the grounding electrode conductor connection is made in the first disconnect and a parallel path is created.

Source: *NEC* Seminar.

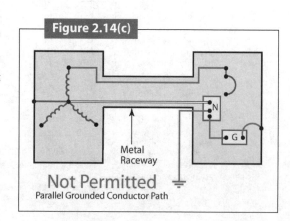

Section 250.32 provides somewhat different rules that address similar issues where a building is supplied by a feeder or branch circuit (not by a service) from a service or other supply in another location. First, 250.32(A) addresses the issue of a grounding electrode at a building supplied by a feeder or branch circuit (also called a separate structure in this discussion). Then 250.32(B) explains how that electrode will be connected to the wiring system.

A grounding electrode is not required at a separate structure where (1) there is only one branch circuit (including one multiwire branch circuit) *and* (2) an equipment grounding conductor is run with the circuit conductors. The equipment grounding conductor will provide for grounding of non-current-carrying metal parts and will also provide a fault current path. Where there is only one branch circuit but no equipment grounding conductor, a new electrode must be established for grounding of equipment in the second structure.

Where an electrode is required, 250.32(B) provides two ways for connecting that electrode to the electrical system and equipment. The first option applies only where an equipment grounding conductor has been supplied with the supply conductors (usually a feeder). Section 250.32(B)(1) requires the grounding electrode conductor at the separate structure to be connected to the equipment grounding conductor of the supply as well as to the equipment grounding conductors for the circuits and equipment at the second structure. Grounding connections to the grounded conductor are not permitted. This is in accord with the rule of 250.24(A)(5) that prohibits such connections on the load side of the service in most cases. In this scenario, as illustrated by Figure 2.14(d), the grounded conductor carries only normal neutral currents, and the equipment grounding conductors carry only fault currents. The grounding electrode provides for a local reference to earth and a local path to earth, but not the required low-impedance path for fault current. Isolating the neutral or grounded conductor from the grounding conductors and grounding electrode conductor prevents the formation of a parallel path for neutral currents.

The second option, described in 250.32(B)(2), is illustrated by Figure 2.14(e). In this case, there is no equipment grounding conductor run with the supply conductors, so the grounded conductor must carry both normal and fault currents back to the service at the first building. The grounded conductor is treated like a grounded conductor at a service and is connected by a bonding jumper to the grounding electrode and the equipment grounding conductors in the separate building. This is permitted by 250.24(A)(5) because it is permitted here and in 250.142(A)(2). In effect, the grounded conductor in the supply is being used to ground equipment in the second building with respect to the first building. The connection to earth through the ground rod does not provide the required low-impedance fault path. Although the earth does form a parallel path, it is not a metallic parallel

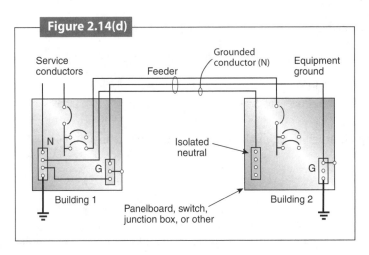

Grounding connections at a separate building when an equipment grounding conductor is run with the circuit conductors.

Source: *NEC* Seminar.

Grounding connections at a separate building when an equipment grounding conductor is not run with the circuit conductors and there are no parallel metallic paths.

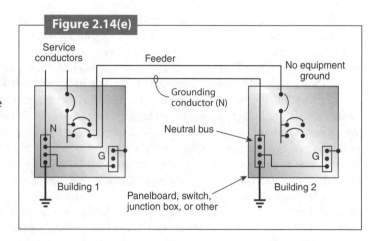

Figure 2.14(e)

path like those prohibited by 250.32(B)(2). Any metal piping system common to the two structures, whether water, gas, electrical, or part of the grounding electrode system, would probably create a parallel path, and, if so, this method could not be used. A parallel path might also be created by metal structures such as a pipe rack, a walkway canopy, or even continuous reinforced concrete. Where a connection to the grounded conductor at the second building creates a parallel path for the grounded conductor, the first method must be used and an equipment grounding conductor must be run from the service to the second building or structure to eliminate the parallel path.

Question 2.15 How are grounding electrode conductors sized?

Keywords

Grounding
Grounding electrode
Grounding electrode conductor
Ground-fault current
Grounded system

NEC References

250.2, 250.4, 250.30, 250.32, 250.52, 250.64, 250.66

Related Questions

- Does the size of a grounding electrode conductor depend only on the size of the ungrounded conductors or also on the type of electrode?
- Is an electrode required to clear fault currents in the wiring system?
- How should a grounding electrode conductor be sized if it runs to a system of electrodes?
- May multiple grounding electrode conductors be different sizes for the same service rating?
- What size grounding electrode is required when the only electrode is a ground rod or multiple ground rods?

Answer

Article 100 defines a grounding electrode conductor (GEC) by describing what it is supposed to do. A GEC is used to connect a grounding electrode to the equipment grounding conductors and to the grounded conductor in grounded systems. In ungrounded systems, the GEC connects a grounding electrode only to the equipment grounding conductors and, indirectly, to conductive enclosures and other non-current-carrying metal parts.

Sections 250.4(A)(1) and (2) and 250.4(B)(1) explain the purpose of grounding systems and equipment. The connection to earth through an electrode is made for three general reasons: (1) to limit voltages to ground during events such as lightning strikes and contact with higher-voltage lines, (2) to stabilize the voltage to ground on grounded systems, and (3) to limit the voltage to ground on equipment within the system. In the first case, the GEC provides a path to ground for abnormal currents that use the earth as a part of the circuit and is expected to carry those types of currents. In the second and third cases, the function is primarily to provide a reference to ground so we know what ground is in relation to our system and equipment; the GEC is not expected to carry much current in these cases. In simple terms, the equipment grounding system ties all metal parts together so that everything we can touch is the same, and the "reference" connection to a grounding electrode makes everything we can touch the same as what we are standing on (at least at the point where the connection to earth is made). The quality of this reference may vary according to the type and extent of the electrode that was used.

On the one hand, when the GEC is called on to carry current, the magnitude of that current is likely to be directly related to the size of the normal circuit conductors because the size of those conductors is a very general indication of their impedance. However, the nature of the electrode itself and the quality of its connection to earth will also affect the impedance of the circuit. On the other hand, when the GEC is used for reference, little if any current is expected to flow, so the primary consideration is not the size of the GEC but its reliability and permanence. Of course, we must install a GEC for both purposes, and it must be adequate for both purposes.

The GEC is not intended to carry fault currents from within a premises wiring system. Equipment grounding conductors and grounded system conductors complete the fault current path to the source. Sections 250.4(A)(5) and 250.4(B)(4) explain that the earth to which the GEC connects is not considered to be a path for ground-fault currents. The definition of "effective ground-fault current path" in 250.2 and the performance requirements of 250.4(A) also explain that this path is a path back to the electrical supply source, not to the earth, and not through the earth. A connection to earth does not provide the low-impedance path required.

When we understand the purposes of a grounding electrode system and a GEC, we can see that the type of electrode will obviously affect the current that will flow in a GEC and therefore will be a factor in sizing a GEC. This is precisely how Section 250.66 works when we apply it to specific applications. In general, Section 250.66 tells us to use Table 250.66 for sizing a GEC, and Table 250.66 is based on the size of the service-entrance conductors. According to the table notes and 250.30, this table also applies to separately derived systems, where the derived conductors are used as the basis for sizing the GEC. Table 250.66 is also used at separate buildings, as referenced in 250.32(E), where the feeder or branch circuit conductors provide the basis for sizing the GEC.

The GEC may be larger than required by Section 250.66, but it may not be smaller. However, as we have noted, the type of electrode should also be considered. Therefore, Section 250.66 requires the use of Table 250.66 unless the GEC connects to (A) a rod, pipe, or plate electrode, (B) a concrete-encased electrode, or (C) a ground ring.

When connected to a rod, pipe, or plate electrode, as permitted in 250.52(A)(5) or (6) (not the water pipe covered in 250.52(A)(1)), the GEC need not be larger than 6 AWG copper. An aluminum size is also given, but, according to 250.64(A), aluminum may not be suitable for this type of electrode because it may not be terminated within 18 inches of the ground. The portion of a GEC that connects to a concrete-encased electrode need not be bigger than 4 AWG copper. Where the "re-bar" itself is not used, 4 AWG is the minimum size copper wire that may be used as a concrete-encased electrode, according to 250.52(A)(3). The minimum size that may be used to form a ground ring is 2 AWG. Larger sizes are often used to form ground rings, and 250.66(C) says that the part of a GEC that connects to a ground ring need not be any bigger than the ground ring. (Where the ground ring is larger than the GEC size required by Table 250.66, the GEC may be smaller than the ground ring.) Figure 2.15 illustrates these sizing requirements. The "full-size" label in the diagram refers to the GEC size determined from Table 250.66. The other GEC sizes apply regardless of the size of the ungrounded supply conductors.

The collection of electrodes and bonding jumpers shown in Figure 2.15 could be rearranged in any way the designer or installer desires. A separate GEC could be run to each grounding electrode, or the electrodes could be grouped in a number of ways. For example, the ground ring might be connected only to the ground rods and concrete-encased electrodes, with separate conductors run to the building steel and water pipe. In such cases, 250.64(F) applies, which both allows multiple GECs and requires that a GEC run to a group of electrodes "be sized for the largest grounding electrode conductor required among all the electrodes connected to it." In this example, one GEC would have to be sized for the ground ring—the largest required for a ground rod, ground ring, or concrete-encased electrode. The other GEC(s) run to the water pipe and building steel would be full-sized. Using this same rule, we see that a GEC run to a group of ground rods need never be bigger than 6 AWG, although larger sizes would be permitted. Figure 2.15 shows a similar situation, where the main GEC is full-sized for the water pipe and building steel, but the jumpers that go to the ground rod, concrete-encased electrode, and ground ring can be smaller in many cases.

In effect, Section 250.66 and Table 250.66 provide the minimum sizes required for various sizes of ungrounded supply conductors and the maximum sizes required for certain specific types of electrode. Table 250.66 never requires any GEC to be larger than 3/0 AWG copper. Conductors larger than minimum sizes are permitted, and may be desired in some cases, such as for considerations of mechanical strength. When Table 250.66 is used for other purposes, such as sizing the main bonding jumper or other bonding jumpers ahead of the service overcurrent device, larger sizes may be required because these elements are actually expected to carry ground-fault current back to the electrical supply source. Remember, the GEC

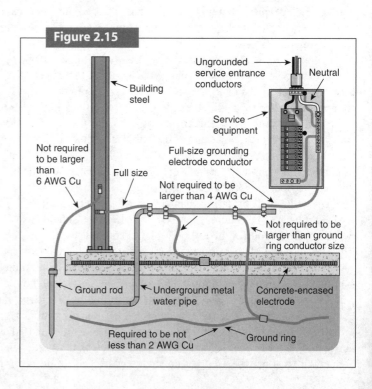

Sizing grounding electrode conductors.

Source: Based on *NEC Handbook*, 2005, Exhibit 250.29.

itself is not expected to be a significant path for ground-fault current, but the same table may be used for elements that do carry fault current because the sizes in the table are based on short-time current-carrying capacities of wires.

A few examples may be useful to illustrate the proper use of Table 250.66 for the full-sized GECs that might be needed for connections to water pipe or building steel.

Example 1

The service-entrance conductors consist of three parallel sets of aluminum conductors, each of which is 500 kcmil. The equivalent area of these three sets is 1500 kcmil. From Table 250.66, over 900 kcmil through 1750 kcmil aluminum requires a 2/0 AWG copper or 4/0 AWG aluminum (or larger) grounding electrode conductor.

Example 2

Size 2/0 AWG copper conductors are used to supply a 200 A service on a dwelling unit in accordance with 310.15(B)(6). From Section 250.66 and Table 250.66, for 2/0 ungrounded copper service-entrance conductors, the minimum size of a copper GEC to a metal underground water pipe or metal structure used as an electrode or to a concrete-encased electrode is 4 AWG. Where the electrode used is a driven ground rod or pipe or a plate, the GEC need not be larger than 6 AWG copper.

Example 3

The service-entrance conductors consist of six sets of 350 kcmil copper conductors per phase. The equivalent area is 6 × 350 = 2100 kcmil. For conductors larger than 1100 kcmil, the GEC must be at least 3/0 AWG copper.

Example 4

The service consists of two disconnecting means with separate sets of service-entrance conductors. One set of service-entrance conductors consists of 3 AWG copper conductors and the other of 3/0 AWG copper conductors. The equivalent area is determined by adding the circular mil area of the two sets. The circular mil area of the 3 AWG conductors is 52,620 cm, and the area of the 3/0 AWG conductors is 167,800 cm. (The circular mil areas for AWG sizes are found in Table 8, Chapter 9 of the *NEC*.) The sum of the areas is 52,620 + 167,800 = 220,420 cm. From Table 250.66, the minimum-size GEC for 3/0 AWG through 350 kcmil is 2 AWG copper. The taps run from this common GEC to each enclosure are permitted to be sized on the basis of the service-entrance conductors supplying each enclosure, so the taps could be 8 AWG copper for the 3 AWG conductors and 2 AWG copper for the 3/0 AWG conductors in accordance with 250.64(D).

Example 5

The service supplies a sign that requires only two 20-ampere circuits. According to 230.42 and 230.79, the minimum rating for the service-entrance conductors is 30 amperes, so 10 AWG conductors are chosen. Even though the service-entrance conductors are only 10 AWG, Table 250.66 requires at least an 8 AWG GEC conductor.

Example 6

A small, 500 VA transformer is used to create a grounded, separately derived system for a Class 1 control circuit. The derived phase conductors for this transformer are 14 AWG. In this case, Sections 250.30(A)(3), Exception No. 3, and 250.20(A)(1), Exception No. 3, permit the bonding jumper to the enclosure to be used in place of a grounding electrode conductor, and the bonding jumper is not required to be larger than 14 AWG.

Question 2.16

Must water piping and steel be bonded where there is no building?

Keywords

Water piping
Building
Structure
Steel frame
Bonding

NEC References

250.104, 250.136

Related Questions

- Does Section 250.104 apply to outdoor equipment yards?
- Where a service serves an open area and not a "building," are the metal water piping and structures such as fences required to be bonded or to be used as electrodes?

Answer

Sections 250.104(A) and 250.104(C) of the *NEC* require bonding of "metal water piping system(s) installed in or attached to a building or structure" and "exposed structural steel that is interconnected to form a steel building frame." A metal column or fence used for supporting electrical and mechanical equipment is a "structure," but would not itself qualify as a "steel building frame" even if it were considered to be a building, and is not required to be bonded unless it "is not intentionally grounded and is likely to become energized." However, according to 250.136(A), the column (or fence) is likely to be effectively (and intentionally) grounded if it is used to support electrical equipment that itself is likely to require grounding. If a fence column supports only mechanical equipment that includes no electrical equipment, such as a hose connection and reel, the column is not likely to become energized and is not required to be bonded. The metal water pipe in the yard is not required to be grounded unless it is "in or attached to a building or structure," so if it is not attached to the structure (or it is not metal), it is not required to be bonded. The yard itself would usually not be considered a structure even though it could, in some sense, have been constructed.

A structure is defined in Article 100 as "that which is built or constructed." However, for certain purposes, certain types of "structures" are not treated as such. For example, separate structures are supposed to have their own disconnecting means for electrical equipment in or on that structure, but the exceptions to 225.32 allow disconnecting means for light poles and poles for signs to be located elsewhere. Section 250.32(A), Exception, also exempts these separate "structures" from a requirement for a grounding electrode if they are served by only one branch circuit that includes an equipment grounding conductor.

The *NEC* does state (through a Fine Print Note that follows 250.104(B)) that bonding of all metal piping will provide additional safety. That is certainly a true statement, but the FPN is not a requirement of the code. However, this "recommendation" in the FPN might also be considered in selecting an electrode.

Section 250.50 requires certain items to be used as an electrode "if present at each building or structure served." If there is an electrical service, most likely there is some sort of "structure" that includes or supports electrical equipment. If metal underground water pipe is available at the service equipment in a yard, it must be used as an electrode, along with any of the other electrodes listed in 250.52. At the least, one or more ground rods or plates must be used in accordance with 250.50 and 250.52. Figure 2.16 shows a service for a well pump

Figure 2.16 Well pump for irrigation in an open field where the underground piping (if metal) or well casing could be used as an electrode according to 250.52(A)(1) and (A)(7).

for a center-pivot irrigation system, where the well casing and underground piping could be used as electrodes and may be supplemented by a ground rod or other electrode. Fences are not included in the list of items that would be used for electrodes. Typically, a fence is not required to be grounded unless it supports electrical equipment. (This is not true for substation fences, but the grounding of public utility substations to eliminate touch or step potentials is not covered by the *NEC*.)

Question 2.17

Is a lighting pole considered grounded by a direct connection to an electrode?

Related Questions

- Are equipment grounding conductors required for grounding outside metal lighting poles or similar structures?
- May a separate ground rod be used for grounding outside metal lighting poles or similar structures?

Keywords

Grounding
Equipment grounding
Ungrounded system
Supplemental electrodes

Keywords

Luminaire
Pole
Effectively grounded
Effective ground-fault current path

NEC References

250.4, 250.20, 250.21, 250.50, 250.54, 250.110, 250.112, 250.118, 250.122, 410.15, 410.20, 410.21

Answer

Section 250.110 requires exposed non-current-carrying metal parts of fixed equipment likely to become energized to be grounded under any of the six conditions listed in that section. In addition, Section 250.112 refers to Article 410, Part V, for grounding of luminaires. In turn, Section 410.21 says luminaires and equipment (the pole, for example) will be considered grounded when connected to an equipment grounding conductor as specified in 250.118 and sized according to 250.122. Section 250.118 requires the equipment grounding conductor to be either a conductor run with the circuit conductor, one of several metal raceways, or, in some cases, a metallic cable sheath.

If the conductors are installed in a nonmetallic raceway as shown in Figure 2.17, an equipment grounding conductor must be installed within the rigid nonmetallic (PVC or fiber) conduit, routed with the circuit conductors, and attached to the pole. This equipment grounding conductor must be sized in accordance with Table 250.122, which is based on the size of the overcurrent protection device protecting the circuit conductors. If metal conduit (appropriate for the conditions) were used, the conduit could be the required equipment grounding conductor according to 250.118.

If the electrical supply *system* is not grounded and is not required to be grounded according to 250.20 or 250.21, all metal electrical *equipment* must still be grounded by means of a grounding electrode conductor run to a grounding electrode system as described in Section 250.50. The equipment grounding conductor for the lighting stanchions should originate at a distribution panelboard equipment grounding terminal bar and be run to each lighting fixture pole, where it is connected to a grounding terminal at the pole and to the luminaire. According to Sections 410.15(B)(3) and 410.20, both the pole and luminaire must include a means for connecting the equipment grounding conductor to the pole and luminaire.

Section 250.4(A)(5) covers the performance of a ground-fault current path. This section covers only grounded systems. The requirements for ungrounded systems are similar and are

Typical detail for a lighting pole and mounting base.

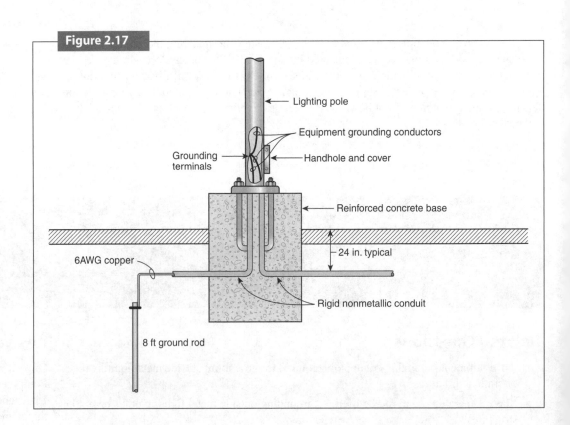

Figure 2.17

found in 250.4(B)(4). The last sentence of each of these two sections states: "The earth shall not be considered as an effective ground-fault current path." If the pole is grounded by a connection to a local electrode, this is only a "supplemental" connection. Such supplemental electrodes are permitted by 250.54 and may be desired for lightning or to reduce or control a local voltage difference between the pole and the earth. However, such a connection to ground does not provide the required *effective* ground-fault current path. An effective ground-fault path must be a low-impedance, highly reliable path that will carry a fault back to the source, not just to earth. In the case of an ungrounded system, it is a second fault that is likely to use this fault path. In cases like that shown in Figure 2.17, the installer may want to use the earth as the grounding means and eliminate the equipment grounding conductor, which would eliminate the effective ground-fault return path. This is not allowed.

Question 2.18

Is metal roofing, siding, and metal veneer required to be grounded or bonded?

Related Questions

- What size bonding jumper is required to ensure bonding of the metal roof and siding?
- How is the size of a bonding jumper for metal roofs or siding determined?

Answer

Section 250.116 of the *NEC* requires three specific types of nonelectric equipment to be grounded. It does not specify metal roof, metal siding, or metal veneers like those shown in Figure 2.18. However, the Fine Print Note (FPN) following Section 250-116 states: "Where extensive metal in or on buildings may become energized and is subject to personal contact, adequate bonding and grounding will provide additional safety." The FPN is not mandatory language, is not enforceable, and is for additional information only, but that does not necessarily mean it should be disregarded. Generally, FPN information of this type is information that should be considered by designers but not enforced by jurisdictions.

Does the *NEC* say how to size a bonding jumper to make these metal parts "electrically continuous"? Local rules may require bonding of the metal siding and roofing even though it is not required by the *NEC*. In that case, the local law should also say how this should be done. Since the *NEC* does not contain a rule about bonding siding and roofing, it cannot be expected to answer this question directly. However, Section 250.90 states: "Bonding shall be provided where necessary to ensure electrical continuity and the capacity to conduct safely any fault current likely to be imposed." The equipment that may energize the metal siding is the electrical equipment that is mounted on the metal parts of the structure. Therefore, the amount of current that could be imposed on the metal skin of the building in question is determined by calculating the fault current that may be imposed by the electrical equipment. The bonding conductor need not have an ampacity (continuous current rating) adequate for the fault current, though. Sections 250.66 and 250.122 are about sizing conductors based on their short-time current capacity, and that should be the nature of the fault current contemplated in this question.

Keywords

Metal building
Bonding
Grounding
Bonding jumper

NEC References

250.66, 250.102, 250.104, 250.116, 250.122

Metal-frame building with metal roof and siding.

Figure 2.18

Under Sections 250.102(C) and (D) of the *NEC*, the size of the bonding jumper is based on either the size of the conductors supplying the service or the rating of the overcurrent device supplying the equipment. Unless this is service equipment, the overcurrent device supplying the equipment that is determined to be likely to energize the metal wall or roof will determine the size of the bonding conductor. Therefore, the bonding jumper could be based on the metal "skin" of the building being energized by the service or being energized by a feeder or branch circuit. Let's say the service rating is 2000 amperes. If a 2000-ampere overcurrent rating were used, because we decided that the steel frame and siding could be energized by a fault in a 2000-ampere main feeder (assuming there is one), the bonding jumper from 250.122 would be 250 kcmil copper. However, for bonding of a building steel frame, Section 250.104(C) makes it clear that the size of the bonding conductor must be based on Table 250.66, which is based on the size of the service-entrance conductors, and this bonding conductor is never required to be larger than 3/0 AWG.

The siding and roofing is all connected to the metal building frame, and the building steel must be bonded according to 250.104(C). Therefore, the rule used for bonding the metal building frame is probably the best rule to use for bonding the siding and roofing. Since the metal siding and roofing are all screwed to the building frame, and since we rely on ordinary fasteners to bond the various parts of the metal frame together, the fasteners that connect the siding and roofing to the metal frame should be adequate to heed the advice of 250.116, FPN, or to meet the requirements of a local rule, unless the local rule specifies otherwise.

Question 2.19

Where can I find flexible metal conduit that is listed for grounding?

Related Questions

- What stipulations or criteria must be met to allow listed conduit and fittings to be used as the grounding conductor?
- Is flexible metal conduit listed for grounding available?

Answer

The most commonly available type of flexible metal conduit (Type FMC) is the type referred to in Section 250.118(5) and illustrated in Figure 2.19(a). This type is not *listed* for grounding, but with listed fittings it is *permitted* for grounding with significant restrictions. Section 250.118(5) limits the total length of FMC in the ground-fault return path to 6 feet, requires fittings that are listed for grounding, and limits the overcurrent device in the circuit to 20 amperes. Also, where used for grounding, the conduit may not be installed to provide for flexibility after installation. This last provision is meant to require a redundant equipment grounding conductor if the flexible conduit is subject to flexing while in use. (Obviously, all

Keywords

Flexible metal conduit
FMC
Listed
Grounding
Fittings
Special permission

NEC References

90.4, 110.2, 250.118

Figure 2.19(a)

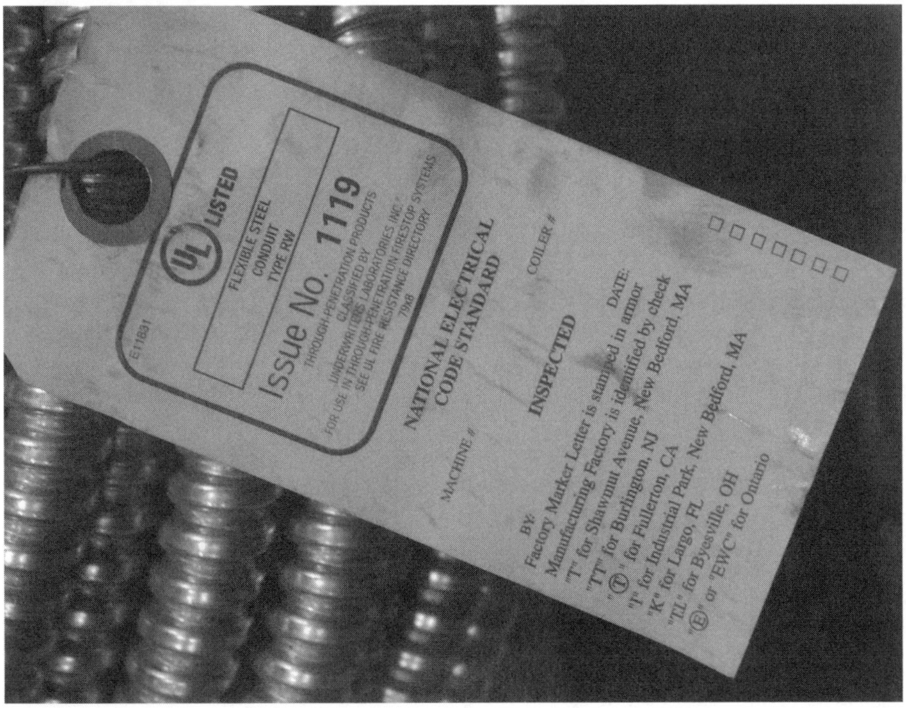

Listed flexible metal conduit, not listed for grounding.

of it will be "flexed" during manufacture, packaging, and installation.) Flexing in use may transmit strain to the connectors and cause the connector to become loose or separated from the conduit, which would impair or eliminate the grounding path.

When an equipment grounding (or bonding) conductor is installed, either for flexibility or because the conduit or overcurrent device is too large, it may be installed on the outside of the conduit, as shown in Figure 2.19(b). This use is generally permitted by 250.102(E) when the length is not more than 6 feet. Section 250.102(E), Exception, permits longer lengths for applications at outside pole locations.

According to the Underwriters Laboratories (UL) *Electrical Construction Equipment* directory, fittings are considered to be suitable for grounding if they are of the type that clamps around the conduit. In sizes 3/8 through 3/4, listed fittings of other types are considered suitable for grounding if the requirements of the *NEC* that were mentioned previously are met.

The UL directory also covers the suitability of the flexible conduit as a grounding means. In this regard, the UL directory says: "Flexible metal conduit no longer than six ft. and containing circuit conductors protected by overcurrent devices rated at 20 amps or less is suitable as a grounding means." It goes on to say that "flexible metal conduit longer than six ft. has not been judged to be suitable as a grounding means." Therefore, based only on the UL information, the type of flexible conduit in the question, and previously mentioned in Section 250.118(5) in the 2002 *NEC*, that has no limits on length, size, or overcurrent device ratings, does not exist. However, some jurisdictions recognize other product standards organizations that do test and list flexible metal conduit for grounding. The acceptability and availability of this product is mostly dependent on local jurisdictions' acceptance of the listing agency, product standard, and marking means. For that reason, the production, testing, and listing of flexible metal conduit that is acceptable for grounding may only apply in a specific local market.

In short, the main criterion for FMC that is suitable for grounding is listing. This product is available in some areas, but many jurisdictions do not accept it, or do not accept it as a grounding means. In other words, the other criterion is approval, because, according to Section 110.2, all products and equipment mentioned in the *NEC* are acceptable only if approved. "Approved" is defined in Article 100 as "acceptable to the authority having jurisdiction." In effect, although previous versions of the *NEC* recognized the product described, the product is not listed under UL standards, is no longer recognized by the *NEC*, and is not

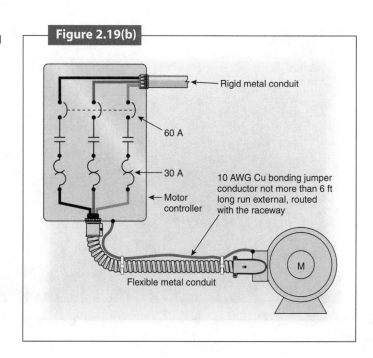

A bonding jumper around the outside of a flexible metal conduit.

Source: Based on *NEC Handbook*, 2005, Exhibit 250.43.

widely available or widely approved. Therefore, both the product listing and the use of the product are subject to approval and, from the standpoint of the *NEC*, special permission in writing from the authority having jurisdiction is required according to 90.4.

Question 2.20

Should a service lateral include a grounding or bonding conductor?

Related Questions
- What is meant by "objectionable currents" on grounding conductors?
- May a grounded conductor be used as a grounding or bonding conductor?
- Where can I find a rule that requires isolated neutrals in "subpanels"?

Answer

Like other portions of the electrical system, non-current-carrying metal parts ahead of the service equipment must be bonded to ensure an adequate path for fault current, and grounded to eliminate any difference of potential between such parts. However, the methods of grounding and bonding on the supply side of the service are very different from the methods used on the load side of the service.

A grounded conductor is required to be connected to equipment grounding conductors and grounding electrode conductors at the service, and is generally prohibited from having any such connections on the load side. These requirements are found in 250.24(A) and (B). In effect, equipment grounding conductors are "created" at the service equipment. Ahead of the service equipment, the grounded conductor (usually a neutral) is intended to carry both normal currents (usually unbalanced neutral currents) and any fault currents. Therefore, a grounded service conductor is permitted and expected to carry any fault current back to the utility source, and also serves to ground equipment on the supply side of the service. Making the grounded conductor big enough for fault current is the objective of 250.24(C). Separating the grounding and bonding functions at the service and using separate pathways exclusively on the load side of the service is the objective of 250.24(A)(5). The permission to use a grounded service conductor for grounding and bonding supply-side equipment is given in 250.92(B)(1) and 250.142(A). Section 250.142(B) describes the very limited cases where a grounded conductor is permitted for grounding or permitted to have additional grounding connections on the load side of the service.

Sizing rules are also different for grounded conductors and grounding (or bonding) conductors, and the rules are different on the line side and the load side of the service. On the load side of service equipment, grounding and bonding conductors and jumpers are based on the rating of the overcurrent device for the circuit under consideration. On the supply side, grounding and bonding functions are covered by the sizing rules of 250.24(C), 250.102(B), and 250.28(D), which are based on the size of the supply conductors rather than the rating of an overcurrent device.

Since grounded service conductors must be sized to handle all types of current likely to be imposed, no separate grounding or bonding conductors are required to run between the

Keywords
Grounding
Grounding conductor
Bonding
Grounded conductor
Neutral
Objectionable current
Subpanel

NEC References
250.6, 250.24, 250.28, 250.30, 250.32, 250.92, 250.102, 250.142

service and the utility supply. Separate grounding and bonding conductors should *not* be supplied if they form a parallel path for normal currents (such as neutral currents) that are intended to be carried by the grounded service conductor. The use of a grounded conductor for other than normal current (that is, for fault current) is also permitted on the supply side of the disconnecting means (and bonding point) of separately derived systems and certain separate buildings, as covered in 250.30 and 250.32. In 250.30 and 250.32, parallel paths for grounded conductor currents are specifically prohibited. The same prohibition applies to service conductors, but it is not so explicitly stated.

Section 250.6 covers objectionable current over grounding conductors. This section requires that connections be arranged to "prevent objectionable current over the grounding conductors or grounding paths." Subsection (B) requires alterations in connections where such objectionable currents occur. Keep in mind that ground*ing* conductors and ground*ing* paths are not intended to carry the normal or off-balanced currents that should be imposed on the ground*ed* or neutral conductor. If a bonding or grounding conductor is run with service conductors, it must be connected to the grounded conductor at the service equipment through the main bonding jumper according to 250.24(B) in order to complete the required "effective ground-fault current path." This grounding or bonding conductor has nowhere to terminate in a utility transformer except on the transformer frame, grounding electrode, or grounded bus. (Section 250.24(A)(2) requires an additional connection to a grounding electrode at outdoor transformers.) Since the transformer frame, grounding electrode, and grounded bus are connected together at utility transformers, a grounding or bonding conductor in a service raceway or cable becomes a parallel path for currents that are intended to be carried only by the grounded conductor, as shown in Part A of Figure 2.20. Then, in order to comply with the *NEC*, one connection to the grounded conductor must be eliminated, as shown in Part B of Figure 2.20, and the grounded conductor becomes the fault path again, as intended. Rather than run such a grounding conductor that must be disconnected, equipment on the supply side of the service should be grounded and bonded by connection directly to or through a jumper to the grounded service conductor, as shown in Part C of Figure 2.20.

Figure 2.20 is based on underground installations, but overhead installations are essentially the same because most service drop cables include a bare messenger that also serves as the grounded conductor, with no separate grounding or bonding conductor. The path through the earth between the electrodes is unavoidable, but there is usually very little current in this circuit because of the high impedance of the soil in comparison to the grounded service conductor. If the two electrodes are connected together by a grounding grid or ground ring system, objectionable currents may be present, and alterations are permitted as outlined in 250.6(B), but the alterations may not eliminate the effective ground-fault path required by 250.4(A)(5).

On the load side of a service, 250.24(A)(5) specifically prohibits parallel paths for grounded conductor currents by prohibiting load-side grounding connections to grounded conductors. This rule preserves the intended functions of the grounded conductors, equipment grounding conductors, and bonding conductors and jumpers by keeping the grounding and bonding conductors separated from the grounded conductors. The idea is that grounded conductors in feeders and branch circuits should only carry normal currents, and grounding and bonding conductors should carry no current except fault current. This is why grounded and grounding conductor terminals are separated in "subpanels."

Although "subpanel" is a widely used term among electricians and electrical engineers, designers, and inspectors, the *NEC* does not use this term. The term is normally used to refer to a panel that is supplied by a feeder, protected by an overcurrent device, and has separated buses that isolate neutral conductors from enclosures and equipment grounds. Service panels or main panels are supplied by service conductors, may have no overcurrent protection or may have overload protection only for the service conductors, and have neutral or grounded buses that are connected to the enclosure, to grounding electrode conductors, and to equipment grounding conductors.

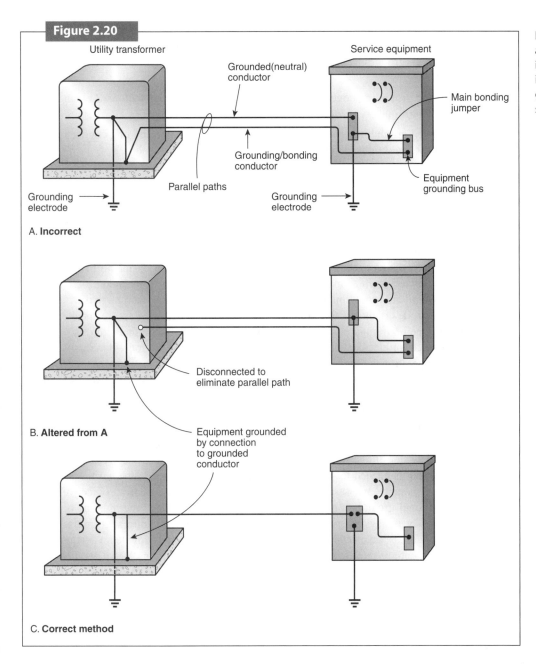

Figure 2.20

A. Incorrect
B. Altered from A
C. Correct method

Elimination of parallel paths and objectionable currents in underground service installations, and correct grounding/bonding of line-side equipment.

Is a separate electrode system permitted to establish a "clean ground"?

Question 2.21

Related Questions

- Are grounding connections permitted to be made to additional local electrodes?
- May a single isolated grounding conductor be used for multiple outlets or receptacles?

Keywords

Isolated ground
Isolated equipment grounding conductor

Keywords

Equipment grounding conductor
Grounding electrode
Isolated grounding receptacle
Grounding terminal
Objectionable currents
Subpanel

NEC References

250.4, 250.6, 250.96, 250.118, 250.122, 250.134, 250.146, 300.3, 406.2, 517.16, 640.6, 647.7

- Are isolated grounding conductors required to be connected to boxes and panelboards through which they pass?
- Are isolated grounding conductors required to pass through a panelboard without termination?
- Are isolated grounding conductors permitted to terminate in a panelboard on an isolated grounding terminal?
- If an isolated grounding receptacle is used, may the isolated grounding conductor terminate at a subpanel rather than at the service or derived system source?

Answer

The permission and requirements for using isolated equipment grounding conductors are based in Section 250.146(D). All references to isolated grounding in the NEC refer back to this section. This section is about receptacles with terminals that are insulated from the mounting means, which also isolates the grounding terminal of the receptacle from the box and normal equipment grounding conductors such as metal conduits. The usual objective of isolated grounding is to connect the equipment grounding conductor of a piece of utilization equipment directly to the grounding electrode conductor to eliminate other grounding pathways to the specific equipment. This may help to reduce electrical noise on a grounding conductor that often also serves as a signal reference for electronic equipment. Whether an isolated ground will actually solve a noise or other operational issue is a design problem that is not within the scope of the *NEC*. The *NEC* simply permits the practice as long as the safety objectives of equipment grounding are not significantly compromised. The first sentence of 250.146(D) provides the permission, and the rest of the section provides the requirements. Furthermore, 250.6(D) explains that alterations in required grounding connections may not be made to eliminate currents that cause "noise or data errors," and that such currents are not the "objectionable currents" addressed by 250.6.

Often, a designer of a system may believe that a separated electrode, such as a separate driven ground rod, is needed to improve the operation of equipment. Whether this would actually help is debatable, but such a solution definitely violates the safety objectives of the *NEC*. An additional grounding electrode may be used, but the electrode must be bonded to other electrodes of the system in accordance with 250.58, and an equipment grounding and bonding pathway must be provided to eliminate shock hazards and provide an adequate, low-impedance fault path as required by 250.4(A)(2) through (5). Section 250.54 permits a separate "supplementary" grounding electrode to be bonded to the equipment grounding conductor of a circuit; such a connection meets the requirements of 250.58 by making the supplementary electrode part of the main grounding electrode system. A separate isolated ground rod is not permitted.

The requirements of the second and third sentences of 250.146(D) are intended to provide for the low-impedance fault path and to meet the basic safety requirements of 250.4. The requirement to run the insulated equipment grounding conductor with the other circuit conductors helps to reduce the impedance of the equipment grounding conductor under fault conditions. This is simply a reiteration of the requirements found in 250.134 and 300.3. The third sentence requires that the insulated/isolated grounding conductor terminate on an equipment grounding terminal of the derived system or service that supplies power to the equipment. It also requires that this connection be made in the same building, so as to be connected to the reference grounding point and electrode system of that building and complete the fault path back to the source.

Section 250.146(D) does not say how far back to go within the same building except to require a connection to the grounding point of the applicable power source. A designer may conclude that termination in the first panelboard (the panelboard nearest the load) is sufficient for equipment needs, and termination on a grounding terminal of a panelboard that is part of the supply to the equipment does meet the requirements of 250.146(D). More commonly,

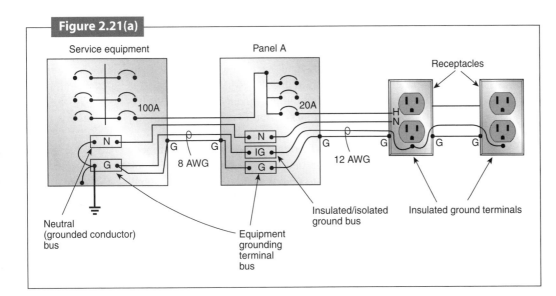

Grounding connections in an isolated grounding system.

however, a designer or equipment specifier wants an isolated conductor to remain isolated back to the common point of connection between the grounded conductor, the electrodes, and the grounding conductors of the system, so as to have a direct route to earth ground and to eliminate ground loops in the grounding path. To accomplish this, an isolated grounding conductor is run with the feeder conductors to panelboards from which the equipment branch circuits are supplied. An example of this type of arrangement is shown in Figure 2.21(a). However, the isolated grounding conductors from the receptacles would still be in compliance with the *NEC* if they were terminated on the regular (not isolated) equipment grounding bus of Panel A. (In a building supplied by a feeder rather than a service, the common point of connection may not include a connection to the grounded conductor, but will include a connection to the equipment grounding conductors within the building and with the supply to the building, along with a connection to a local electrode.)

Figure 2.21(a) also shows more than one isolated grounding receptacle on the same circuit and sharing the same isolated grounding conductor. Nothing in the *NEC* prohibits this practice. However, the designer may recognize that electronic equipment may impose noise on other equipment in some cases and may desire separate isolated grounding conductors for each receptacle. Again, this is a design issue.

The sizing of one or more equipment grounding conductors and the equipment grounding conductors for more than one circuit is done according to 250.122, where the ordinary rules for sizing equipment grounding conductors are found. In Figure 2.21(a), the size of the equipment grounding conductor to the receptacles is based on the rating of the branch circuit, that is, the rating of the overcurrent device protecting the circuit as required by 250.122. Where an isolated grounding conductor is run with a feeder that supplies multiple circuits with isolated grounds, the isolated grounding conductor is sized based on the overcurrent device protecting the feeder in the same manner as an ordinary equipment grounding conductor. Thus, the 12 AWG size of the equipment grounding conductor(s) for the receptacles is based on the 20-ampere rating of the circuit overcurrent device. Similarly, the 8 AWG minimum size for the isolated grounding conductor run with the feeder is based on the 100-ampere rating of the feeder overcurrent device. All other requirements in 250.122 for sizing equipment grounding conductors also apply to isolated equipment grounding conductors.

A difference between isolated grounding conductors and ordinary equipment grounding conductors is in required terminations in intermediate boxes and panelboards. Ordinarily, 250.148 requires equipment grounding conductors that are spliced or terminated in a metal box to be also connected to the box and to other equipment grounding conductors in the box. Since this would defeat the purpose of an isolated equipment grounding conductor, isolated equipment grounding conductors are exempted from this requirement by 250.148, Exception. A similar rule and exception for isolated grounding conductors passing through panelboards is found in 408.40. Note that both of these rules and all other references to isolated equipment grounding conductors (250.96(B), 406.2(D), 517.16, 640.7(C), and 647.7(B)) also refer to 250.146(D) for the essential requirements for those installations.

The Fine Print Note to 250.146(D) reminds the code user that metal boxes must still be grounded. The fact that an isolated grounding conductor is not required to be connected to a box means that another equipment grounding conductor must be supplied to the box. This point is also illustrated in Figure 2.21(a), where the black dots labeled "G" symbolize required grounding/bonding connections, which are in addition to the isolated equipment grounding conductors and connections. In the diagram, metal raceways may be used as the other equipment grounding conductors, as permitted by 250.118 and 250.134. Where a nonmetallic raceway, excessive lengths or large sizes of flexible metal conduit, cables such as Type MC cable, or other wiring methods are used that don't supply an inherent equipment grounding conductor in the raceway or cable armor, more than one separate equipment grounding conductor must be provided: one for the boxes and panelboards, and one for the isolated grounding terminals.

Another feature that is illustrated in Figure 2.21(a) is an isolated grounding bus in Panel A. A grounding bus of this type is not mentioned in the *NEC*. It is not prohibited, however, and it is readily available from manufacturers of listed panelboards. Isolated grounding buses or terminal bars provide a good way to make connections to multiple isolated equipment grounding conductors.

Figure 2.21(b) shows an isolated grounding receptacle. The marking on this receptacle complies with the requirements of 406.2(D). The triangle on the face is orange to indicate that it includes an isolated ground. The dot is green to signify that this receptacle is "hospital grade," as required for patient bed location receptacles in health care facilities (see 517.18(B)). The insulator that separates the mounting means from the grounding terminal of the receptacle can also be seen. This receptacle has self-grounding mounting screws to ground the metal yoke and faceplate. Some isolated grounding receptacles are supplied with a separate green grounding screw on the yoke.

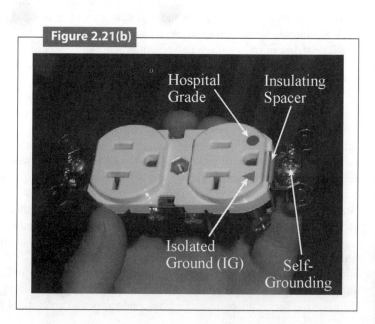

Isolated grounding receptacle, hospital grade, with self-grounding mounting screws.

Question 2.22

Where must isolated grounding conductors terminate?

Related Questions

- Must an isolated grounding conductor terminate only at the associated transformer or service?
- May I terminate the isolated ground wire to the ground terminal of a distribution panelboard on the load side of a transformer or the ground terminal of a panelboard supplied by the separately derived system?
- May the designer decide where to make the common point of connection between isolated ground and normal ground?

Answer

The critical language in this case is the last sentence of 250.146(D), which says:

> This grounding conductor shall be permitted to pass through one or more panelboards without connection to the panelboard grounding terminal as permitted in 408.40, Exception, so as to terminate within the same building or structure directly at an equipment grounding conductor terminal of the applicable derived system or service.

Figure 2.22 illustrates a typical termination point for an isolated grounding conductor.

Keywords

Isolated grounding
Equipment grounding
Grounding terminal
Separately derived system

NEC References

90.1, 250.4, 250.118, 250.134, 250.146

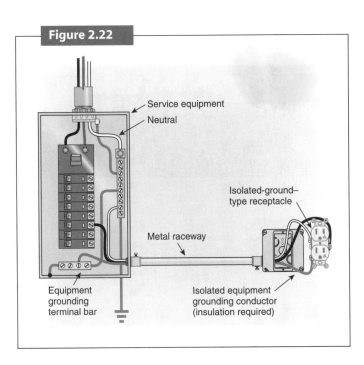

An isolated grounding receptacle with the isolated grounding conductor terminated at an equipment grounding terminal of the system.

Source: Based on *NEC Handbook*, 2005, Exhibit 250.55.

The entirety of the isolated grounding rule in 250.146(D) is permissive. (All other rules about isolated grounding also refer to 250.146(D).) The only time an isolated grounding conductor is required in the *NEC* is when an isolated grounding receptacle is used. This rule is not for reducing noise; it is to ensure that an adequate grounding path is provided. Passing through panelboards *without connections to the ordinary equipment grounding terminal* is also permitted (that is, it can be done), but it is not required. However, termination at an equipment grounding terminal of the system *is required* by this rule and many other rules of Article 250, as well as by the general performance requirements of 250.4.

The code requirement covering the termination of an isolated equipment grounding conductor does not specify that it has to terminate at the source of a separately derived system. The isolated equipment grounding conductor is permitted to terminate at an equipment grounding terminal in any panelboard supplied by a separately derived system, provided that termination does not interrupt the ground-fault current return path to the system source, and provided the equipment grounding conductor is "run with the circuit conductors" as required by Sections 250.118, 250.134, and 250.146(D), among others. In other words, an isolated equipment grounding conductor is permitted to terminate at the equipment grounding terminal of a panelboard (with other equipment grounds, both isolated and nonisolated) and use the equipment grounding conductor of the panelboard feeder circuit as the return path for fault current. The *NEC* concern is not how far the isolated ground extends, but rather that the fault current path is a low-impedance path and is continuous to the source, whether through isolated or nonisolated equipment grounding conductors.

Isolated grounding is permitted by the *NEC* as a possible method of improving the performance of some circuits by reducing noise on the grounding conductor. The ordinary functions of equipment grounding conductors must still be provided; that is, they must still eliminate hazardous voltage differences between exposed metal parts of electrical equipment and must still provide a reliable, low-impedance fault path. The suitability of isolated grounding for reducing noise in any given situation is a judgment left to the designer. If the system, equipment, or circuit designer believes that the "clean ground" can be achieved by connection to the equipment grounding terminal of an intermediate panelboard, that is the designer's prerogative. Acceptable levels of noise as well as possible sources of noise differ from one application to another. According to Section 90.1, the primary purpose of the *NEC* is providing safe installations, not optimal installations.

Question 2.23: Is a 250-volt DC system required to be grounded?

Keywords

Grounded system
Grounded conductor
DC system
Batteries
Grounding electrode conductor
Separately derived system

Related Questions

- Is a battery installation a separately derived system?
- How are DC systems grounded?

Answer

Battery installations are covered by Article 480, but Article 480 does not say anything specific about grounding of battery systems, so Article 250 must apply. In Article 250, the requirements for grounding of DC systems are covered in Part VIII. Section 250.162(A) requires grounding of two-wire DC systems that are greater than 50 volts but not greater than 300 volts, so most 250-volt DC systems must be grounded. There are exceptions, the broadest of which applies to rectifier-derived systems but not to batteries. The other exceptions are for industrial equipment in "limited areas" and certain fire alarm systems. Ground detectors must be installed for the industrial equipment exception to apply, and modern fire alarm systems are mostly less than 50 volts, so the fire alarm exemption does not apply to new installations.

Figure 2.23 illustrates the main rules for grounding of DC systems. DC systems up to 50 volts (most modern fire alarm circuits and most DC instrumentation, communication, and data circuits) are not required to be grounded, and the *NEC* has no rules for two-wire DC systems over 300 volts. The exceptions for systems from 50 through 300 volts were discussed earlier. All three-wire DC systems are required to be grounded, regardless of voltage.

The methods of grounding a DC system are also covered in Part VIII. Typically, for battery-supplied systems, one of the DC conductors must be grounded, usually between the source and the first disconnecting means, so the rules for DC systems in 250.164(B) are similar to the rules for separately derived systems in 250.30. For separately derived AC systems, grounding electrode conductors are sized according to 250.66 and Table 250.66. For DC systems, the grounding electrode conductor is usually required to be as big as the neutral conductor, where there is one, or as big as the largest conductor supplied in other cases, and in any case, not smaller than 8 AWG copper, according to 250.166(A) and (B). Thus, the general rule for the grounding electrode conductor of a DC system often requires a larger conductor than would be required for an AC system. However, the grounding electrode conductor size may be smaller when the electrode is a rod, pipe, or plate, a concrete-encased electrode, or a ground ring. These rules for specific electrodes in 250.166(C), (D), and (E) require the same sizes as those found in Section 250.66 for AC systems using the same types of electrodes.

Battery installations may or may not be separately derived. A separately derived system is defined in Article 100 as a system that has "no direct electrical connection, including a solidly connected grounded circuit conductor, to supply conductors originating in another system." If the DC source is used directly and not inverted to AC, the batteries are probably separately derived because the only possible connection to another system is through the battery charger, which is likely isolated. Otherwise, the way the inverter, UPS system, and battery chargers are connected, both internally and externally, will determine whether the battery installation is a separately derived system.

NEC References

Article 100, 250.30, 250.66, 250.162, 250.164, 250.166, 480.5

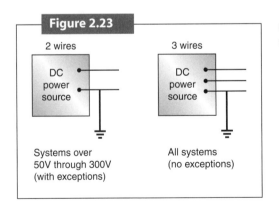

General rules for grounding of DC systems according to Section 250.162.

Chapter 3

Wiring Methods and Materials

Question 3.1

Is wiring restricted from sharing a penetration with plumbing?

Keywords

Penetrations
Physical damage
Plumbing
Ampacity corrections

NEC References

300.4, 300.5, 300.21, 310.10, 310.15

Related Questions

- What is the *NEC* reference that relates to electrical wires near plumbing?
- Does the *NEC* require some spacing between plumbing lines such as main drainage pipes or potable water lines in interior walls of the building?
- May a plumbing pipe and electrical wiring run through a single opening in the wall framing?

Answer

Refer to Article 300 of the *National Electrical Code* and Figure 3.1. This article covers general wiring methods, including protection of wiring from damage. However, nothing in this article prohibits cables or raceways from being routed near plumbing inside a wall. For that matter, 300.5 lists requirements for protection of underground installations, but the use of a common trench for electrical and other services is not prohibited. There are also requirements in Section 300.4 for running cables and raceways between framing members, but these rules are about physical protection of cables from common hazards such as nails or screws. In Figure 3.1, point A illustrates a location where nonmetallic (NM) cable is routed through a penetration with vent piping. There is no *NEC* rule that prohibits this practice as long as the cables are properly protected and supported. Point B shows nail plates installed for the protection of the plumbing piping. These nail plates would also protect the NM cable if needed (the plumbing piping is closer to the edge of the framing than the NM cable).

Although the plumbing lines are not likely to be a source of physical damage to the wiring, in some cases they might change the heating characteristics of the circuit. A drain line was mentioned in the question, and a drain line or plumbing vent like those shown in Figure 3.1 is likely to be about the same temperature as the ambient temperature near the shared penetration. In such cases, the plumbing pipe is not likely to interfere with heat dissipation from the electrical wiring or add any significant heat to the wiring. However, if the pipe is for hot water or steam, the close proximity of a heat source may adversely affect the conductors, so an ampacity correction would be necessary. Section 110.11 addresses the issue of excessive temperatures that might damage conductor insulation. See 310.15 and the associated tables and notes as well as 310.10, both of which address heat in conductors. Steam lines especially can produce enough heat to damage insulation and cause circuit failure if adequate spacing is not provided.

Some penetrations will require fire sealing in accordance with 300.21, especially in vertical runs in fire-resistive construction. Fire sealing a shared penetration is possible, but in some cases, it may be more difficult to fire seal the shared penetration than it would be to make separate penetrations for the electrical and plumbing systems.

When a Type NM cable is routed with one or more other NM cables through wood framing that is to be fire-stopped or draft-stopped with thermal insulation or sealing foam, the allowable ampacity must be adjusted. The adjustment is based on Table 310.15(B)(2)(a) and the number of current-carrying conductors in the cables that are bundled together and applies to the length of the bundled cables. However, this is a very specific rule from Section 334.80 that applies to only one cable type; it does not apply to similar installations where other cable types are used or to cables installed with other systems. This discussion has only covered the electrical code aspects of the question. Other issues with regard to plumbing support or penetrations may arise based on

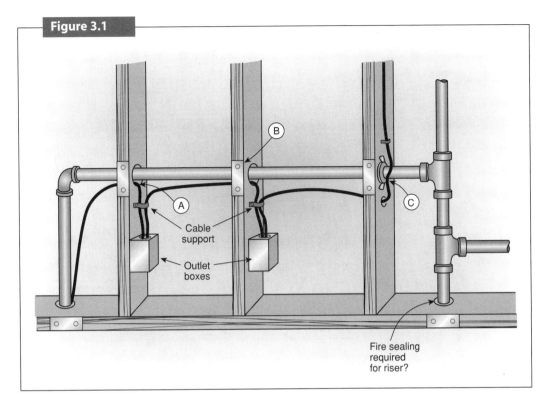

Figure 3.1 Type NM cable installed in a wall with plumbing drain, waste, and/or vent piping.

the applicable plumbing code. For example, many plumbing support fittings are made to fit into (and fill) a hole in framing, as shown at point C in Figure 3.1, or the hole may be required to be made to fit the piping in order to provide the support required for the plumbing piping.

Many people may prefer that electrical cables or raceways have their own penetrations. Although this could sometimes be a safety issue as noted, in many cases it is more of a workmanship issue; it is a method that is acceptable to some and not to others, but is not specifically addressed in the *NEC*.

Question 3.2

Is air tubing permitted in a raceway for conductors?

Related Question

- Does NFPA 70 (the *NEC*) or NFPA 79 permit air or other tubing inside electrical conduits, wireways, or other enclosures?

Answer

Section 300.8 of the *NEC* prohibits the installation of any pipe, tube, or equivalent for steam, air, water, gas, or drainage in a raceway or cable tray that contains electric conductors. NFPA 79, *Standard for Industrial Machinery,* follows the *NEC* on any issue not specifically covered in NFPA 79. There is no clear modification of this rule in NFPA 79. Therefore, the raceways

Keywords

Raceway
Air tubing
Enclosure
Pneumatic tubing

NEC References

300.8, NFPA 79

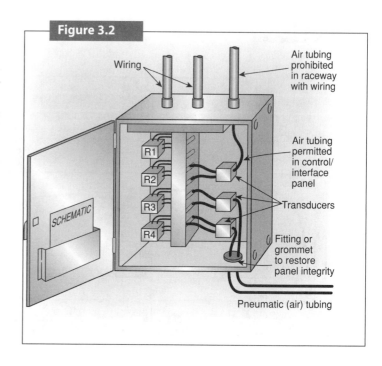

Wiring and pneumatic tubing sharing a common enclosure but not a common raceway.

and wireways ("ducts" and "cable trunking systems," as they are also called in NFPA 79) and support systems (like cable trays) installed under NFPA 79 must meet the same requirements as in the *NEC*. Section 14.5.7 of NFPA 79 allows machine compartments of industrial machinery like columns and bases to be used for wiring (essentially as raceways) if those spaces are entirely enclosed and isolated from coolants and oil. However, there is no prohibition in the NEC from having electric conductors in the same enclosure (such as a box, control panel, or cabinet) with piping or tubing for other systems. Many pneumatic control panels have the air/electric transducers or other interfaces within one enclosure along with the wiring and tubing connecting to these interfaces, as shown in Figure 3.2. However, Section 12.2.2.2 of NFPA 79 prohibits "pipelines, tubing, or devices (e.g. solenoid valves) for handling air, gases, or liquids" from sharing the same "enclosure or compartment" with electrical control equipment. Thus, unless a separate compartment for tubing is created within the enclosure in Figure 3.2, this installation may not be permitted by NFPA 79.

Note that NFPA 79 is a voluntary standard, not a code, that is intended to provide for consistency and safety in the construction and use of industrial machinery. NFPA 79 contains mandatory language, but, like a code, it must be adopted or be otherwise acceptable to the authority having jurisdiction in order to be applicable in any specific case.

Question 3.3 Are sleepers laid on a roof adequate for support of exposed raceways?

Keywords

Raceway
Securing

Related Questions

- Do wood sleepers provide adequate support for raceways?
- Should the sleepers themselves be somehow attached to the structure?

Answer

Raceways such as rigid metal conduit (RMC) or electrical metallic tubing (EMT) are often run across a roof to supply power to rooftop equipment. Similar installations are often used for feeders or other circuits. Often, the conduit is supported and kept off the roofing by laying short pieces of weather-resistant wood on the roof and attaching the conduit to the wood sleepers referred to in this question.

Section 300.11(A) and various sections in the raceway articles (344.30(A) for RMC and 358.30(A) for EMT are examples) require that raceways be supported and securely fastened in place. Sleepers that are not themselves securely fastened in place as shown in Figure 3.3 cannot provide the secure support that is required for the raceways. Many roofing systems are compatible with sleepers that are attached to the roof to provide the secure support required. In some cases, these types of supports are specifically made for use with certain types of roofing systems, such as built-up tar and gravel roofs or membrane-type roofs.

Keywords

Supporting
Sleepers

NEC References

300.11, 344.30, 358.30

Figure 3.3 Secure support required for rooftop raceway installations.

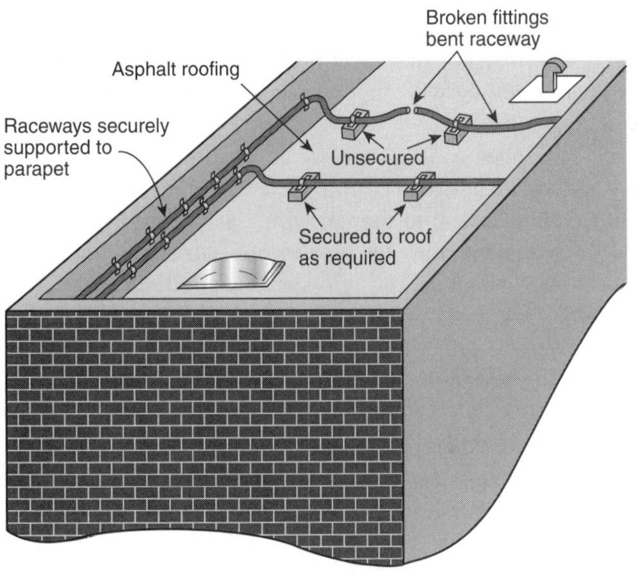

Question 3.4

When can raceways or cables be attached to ceiling grids or supports?

Related Questions

- What wiring is permitted to be attached to ceiling support wires?
- Are boxes or luminaires permitted to be supported by ceiling grids or ceiling support wires?

Keywords

Supporting
Securing
Suspended ceiling

Keywords

Grid ceiling
Support wires
Cables
Raceways
Luminaires

NEC References

300.11, 314.23, 410.16

- Are wires that are added and attached to ceiling grids permitted to support wiring?
- Are additional support wires permitted to be connected to ceiling grids?

Answer

Section 300.11 contains general requirements for securing and supporting wiring. According to its scope and title, Article 300 covers only wiring methods. Other rules apply to luminaires and boxes used to support luminaires.

Section 300.11 imposes some very specific requirements on cables and raceways and makes more general statements about "wiring," which includes cables and raceways, but also includes other equipment and fittings. For example, 300.11(A) says specifically, "Cables and raceways shall not be supported by ceiling grids." However, 300.11(A) also says that independent support wires (independent of and in addition to the ceiling grid support system) that are securely fastened at both ends can be used to provide sole and secure support for raceways, cables, boxes, and the like. In order for the added wires to provide secure support, they are permitted and generally required to be connected to the ceiling grid. Sections 300.11(A)(1) and (2) go on to say that wiring in the space created by a suspended ceiling space may not be "secured to, or supported by, the ceiling assembly, including ceiling support wires." The ceiling assembly includes the grid system or other framing system and the wires that support the ceiling system, but not the wires added specifically for supporting wiring. Figure 3.4 shows a suspended ceiling with support wires for the ceiling system and separate support wires for the wiring.

In general, then, a grid-type suspended ceiling may not be used to support any wiring, and raceways and cables may not be attached to a grid system. However, exceptions to 300.11(A)(1) and (2) do permit a ceiling system to support wiring and equipment if the ceiling system manufacturer provides documentation that support of electrical equipment is an intended and recognized use. Although many ceiling grid systems are designed to support some luminaires, as shown in Figure 3.4, grid system manufacturers have little incentive to encourage the use of the ceiling system for other purposes or to have the systems tested for those purposes, so they are unlikely to provide the needed documentation to allow the use of the exception.

Ceiling grid systems come in three standard ratings: light-, medium-, and heavy-duty. Medium-duty systems are the most commonly used type and are designed to support certain

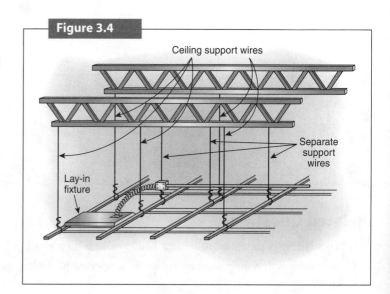

Separate support required for wiring methods in suspended ceilings.

Source: 1999 *NEC Changes*, p. 157.

items like luminaires up to a specified weight for a specified area. Light-duty systems are typically not intended to support anything but themselves, and heavy-duty systems may support greater weights, but not cables or raceways. The permitted uses are covered by listings, manufacturers' instructions, and building codes, and the *NEC* defers to those other documents and instructions for any permission to use a ceiling system for supporting electrical equipment.

As noted, many ceiling grid systems are intended to support "lay-in" luminaires and other luminaires, such as exit lights, up to a specified weight. Other equipment and heavier luminaires must be supported independently of the suspended ceiling system, although they may be directly or indirectly attached to the ceiling system. The *NEC* prohibition concerning support of raceways and cables does not extend to luminaires and boxes. The rules for luminaires and boxes are found in 410.16(C) and 314.23(D), respectively. According to these sections, a box may be attached to ceiling framing members or support wires (usually to support a luminaire or to connect to other equipment below the ceiling), and luminaires supported by ceiling framing members must be attached to the framing members. The method of attachment may be screws, bolts, rivets, or clips that are listed for the purpose. Many luminaires intended for such use come with clips that are listed as part of the listing of the luminaire. The luminaire in Figure 3.4 must be attached to the ceiling with support clips or screws. We will assume that the luminaire in Figure 3.4 is attached by screws or rivets but they cannot be seen in the diagram. Such connections should be verified in actual installations.

In addition to the *NEC* requirements, building codes usually require additional support wires to be attached to luminaires and other equipment to provide back-up support in case the ceiling framing system shifts, as may be likely during an earthquake. These additional "quake wires" are not required in all jurisdictions and are not shown in Figure 3.4.

Communications cables are also subject to the requirements of 300.11 because Section 300.11 is referenced in Chapter 8, specifically in Sections 800.24, 820.24, and 830.24. In addition, the *NEC* requires that communications installations be "neat and workmanlike" and prohibits accumulations of cables from preventing access to electrical equipment, which includes access to the cables themselves. (See 800.21, 820.21, and 830.21.) In addition, the limitations of the listing of the ceiling system still apply and are independent of the *NEC* requirements.

Question 3.5

Must neutrals be pigtailed at devices supplied by multiwire branch circuits?

Related Questions

- Does the restriction on the use of a device for splicing a neutral apply where there is only a two-wire circuit derived from a multiwire branch circuit?
- Why are pigtails required on neutrals to connect receptacles in multiwire circuits?

Answer

Section 300.13(B) reads as follows:

> **Device Removal.** In multiwire branch circuits, the continuity of a grounded conductor shall not be dependent upon device connections, such as lampholders,

Keywords

Multiwire
Branch circuit
Device
Continuity

NEC References

300.13

receptacles, and so forth, where the removal of such devices would interrupt the continuity.

The requirement for continuity of the "grounded conductor" pertains only up to what is shown in Figure 3.5 as the first receptacle outlet in the circuit. Beyond that point, a discontinuity in the neutral affects only a two-wire circuit. In other words, the portion of the circuit beyond the first receptacle outlet in Figure 3.5 is no longer a multiwire circuit because the neutral is not shared by other ungrounded conductors. In the two-wire extension, when the overcurrent device supplying the ungrounded conductor is turned off, there is no potential on a disconnected neutral from the other ungrounded conductor. Based on the *NEC* language in Section 300.13(B), the neutral splicing restriction only applies to that portion of the circuit "where the removal of such devices would interrupt the continuity" of the neutral in the multiwire circuit and not to portions where the neutral is only used with one ungrounded conductor. Opening the neutral in a two-wire portion beyond the point of divergence at the first receptacle does not affect continuity of the neutral in another two-wire portion or in the multiwire portions.

For example, Figure 3.5 illustrates a situation where two hot (H—ungrounded) conductors and one neutral (N—grounded, white) conductor run from the panel to a receptacle. We'll call these two circuits "1" and "3." At the first receptacle, which is connected to circuit 1, the receptacle connection for the white conductor is made via a pigtail jumper that is spliced to the other white conductors in the box, that is, to the one white conductor coming from the panel and the two going to other receptacles. (Conductor H1 is also shown connected to the receptacle by a pigtail—a good practice, but not a requirement of the *NEC*.) Beyond this point, the multiwire circuit diverges into two two-wire circuits, with one neutral conductor for circuit 1 and one for circuit 3. No more pigtails are required in the two-wire circuits beyond the point of divergence. (This is not to say that pigtails might not be preferred, depending on the type of connection at the device and the type of device.)

This rule is not only about hazards during servicing a portion of the circuit. It is also about the reliability of the neutral connections. A neutral connection that is opened in a multiwire circuit can cause unbalanced voltages in the rest of the circuit, and those unbalanced voltages can damage equipment or create other hazards. Furthermore, depending on the

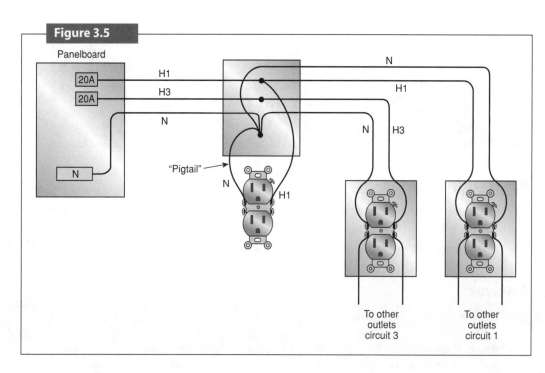

Continuity of neutral conductors in multiwire circuits (equipment grounding conductor not shown).

Figure 3.5

nature of the load on the circuit and the system from which the circuit is derived, the neutral of a multiwire circuit may carry more current than the ungrounded conductors, and heavy loading is more likely to cause heating and deterioration of the neutral connections at a device (or elsewhere). Therefore, the continuity of the neutral in a multiwire circuit is critical, and neutral connections in the multiwire portion must be highly reliable and not dependent on wiring devices such as switches and receptacles.

Question 3.6

Are receptacles permitted in plenum ceilings?

Related Questions

- Where are the requirements for installations in air-handling spaces?
- Are flexible cords permitted in suspended or dropped ceilings?

Answer

The requirements for electrical installations in air-handling spaces are found in Section 300.22. This section divides air-handling spaces into three types: (A) ducts for handling dust, loose stock or vapor, (B) ducts or plenums used for environmental air, and (C) "other space used for environmental air."

Ducts for transporting dust, loose stock, or vapor are covered in 300.22(A). No wiring is permitted in these spaces. Ducts or plenums used for environmental air are covered in 300.22(B). These ducts may contain wiring, but, in general, the wiring methods and enclosures must be metallic, and wiring is permitted only for equipment that acts on the air. Wiring that just passes through and wiring or equipment that does not need to be in such spaces is prohibited.

So-called plenum ceilings are referred to as "other space for environmental air" in the *NEC* and are covered by 300.22(C). Generally, wiring and equipment in these spaces must be metallic. Nonmetallic equipment and wiring must be listed for the use. However, wiring in these spaces may be there for any reason.

Section 300.22 does not mention receptacles specifically. A receptacle may be installed if made of appropriate (usually specifically listed) materials or if installed in suitable enclosures. The problem is that a receptacle is usually used with a flexible cord and an attachment plug, and the flexible cord is not permitted to be run into or through suspended ceiling spaces. Although a permitted use in Section 400.7 may seem to fit some desired use, Section 400.8 prohibits flexible cord "where run through holes in walls, structural ceilings, suspended ceilings, dropped ceilings or floors" or "where concealed by walls, floors, or ceilings or located above suspended or dropped ceilings." A use prohibited by 400.8 may be permitted if "specifically permitted in 400.7," but none of the permitted uses specifically mentions use in a ceiling space. Also, these rules do not consider whether the cord is suitable for an air-handling space or even if the ceiling is an air-handling space.

In short, the *NEC* does not prohibit the installation of a receptacle in a "space for environmental air" where the receptacle is used by service personnel for electrically powered tools, but does prohibit cords for connection of permanently installed equipment in such spaces. Figure 3.6 summarizes these restrictions.

Keywords

Plenum
Duct
Other space for environmental air
Receptacle
Flexible cord
Luminaire
Ceiling
Suspended
Dropped

NEC References

300.22, 400.7, 400.8

Flexible cord is not permitted in suspended or "plenum" ceilings.

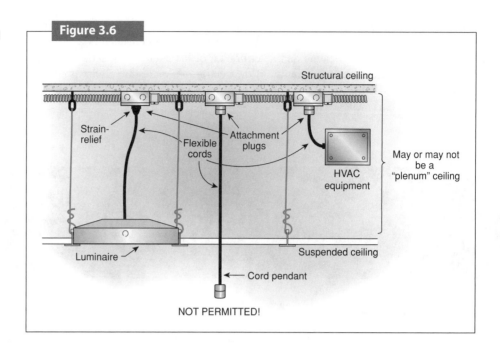

Often people want to use cord-and-plug connections for luminaires. However, 410.30(C)(1) permits this application for electric-discharge luminaires only where the cord "is visible for its entire length outside the luminaire," which effectively prohibits the use of cord for recessed fixtures or other fixtures installed on a suspended or dropped ceiling, without regard to whether the ceiling is an environmental air-handling space. However, a common application in some modern retail and restaurant spaces is to use a suspended grid system without any tiles to create a visual plane and a support system for luminaires or lighting track without creating an actual ceiling. This application could render a cord visible for its entire length and would not conceal the wiring above the grid system. The applications shown in Figure 3.6 would not be prohibited if the grid did not form a suspended ceiling and if other provisions for the use of cords (400.7 and 400.8) were met.

Question 3.7
What wiring methods and equipment are permitted in ducts, plenums, and other air-handling spaces?

Keywords
Transformer
Dry type
Duct
Plenum
Other space for environmental air
Combustible

Related Questions
- Must all equipment for use in "other space for environmental air" be listed or marked for the use?
- Are dry-type transformers permitted in plenum ceilings?
- Are any combustible wiring methods permitted in air-handling spaces?
- How do the articles in Chapter 7 work with Article 300 with regard to wiring in plenum ceilings and ducts?

Answer

Section 300.22 prohibits most combustible wiring methods and electrical equipment in ducts for dust, loose stock, or vapor or in ducts or plenums used for environmental air. For the most part, metallic wiring methods are required in these spaces if the wiring is permitted at all. No wiring is permitted in ducts for dust, loose stock, or vapor, according to 300.22(A). However, according to 300.22(B), certain wiring may be installed in ducts or plenums used for environmental air, but only if the "equipment or devices," including transformers, motors, instruments, or similar equipment, are "necessary for their direct action upon, or sensing of, the contained air." Thus, in general, dry-type transformers and other electrical equipment are not permitted in these spaces even if totally enclosed and constructed of noncombustible material because it's unlikely that they need to be in the airstream to serve their purpose. (Equipment such as damper motors and some types of sensors are types of equipment that could be installed in these ducts in some cases.) "Equipment" as defined in Article 100 includes all types of wiring methods and materials. Therefore, rigid nonmetallic conduit, electrical nonmetallic tubing, and similar products, as well as nonmetallic straps or nylon or plastic cable ties, are not permitted in the spaces described in 300.22(A) or (B).

Wiring in "other space for environmental air" is covered by the rules in 300.22(C). The most common example of this type of space is the so-called plenum ceiling or air-handling ceiling space above a suspended grid ceiling system. This space is not a "plenum" as the term is defined in the *NEC*, though it may be called that by other trades or codes or in trade jargon. Other codes and standards govern the installation and other uses of these spaces. For example, these spaces must normally be constructed of noncombustible materials such as steel or concrete structural elements with approved grid systems and tiles or "hard" drywall or plaster ceilings, and the contained mechanical systems, plumbing piping, and the like must be noncombustible. The primary difference between grid and hard ceilings from the standpoint of the *NEC* is simply that the wiring and equipment above a grid ceiling are considered to be exposed and are more likely to be accessible, but the rules are not different otherwise. The three types of ducts or spaces and the wiring that is permitted are summarized in Figure 3.7(a).

The rules of 300.22(C) do not apply to the combustible air-handling spaces in dwelling units that consist of spaces between joists or studs and are commonly used for handling return air. These spaces may contain nonmetallic or combustible wiring methods such as Type

Keywords

Metallic
Nonmetallic
Air handling

NEC References

300.21, 300.22, 450.13, 725.3, 725.61, 800.154

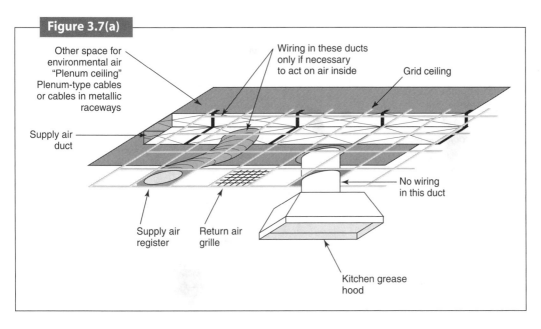

Wiring in ducts and plenums.

Source: *Limited Energy Systems*, 2002, Figure 3.23.

Cable passing through joist space of a dwelling unit.

Source: Based on *NEC Handbook*, 2005, Exhibit 300.21.

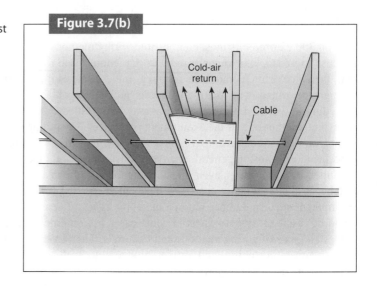

NM cable if the cable is run only perpendicular to the spaces, as shown in Figure 3.7(b). Figure 3.7(c) shows two types of spaces that are covered by 300.22. The fabricated duct is covered by 300.22(B). The return air space above the ceiling and in the fan room is covered by 300.22(C) because there is no return air duct connecting to the air-handler, so the ceiling and room are used for handling air. The habitable area shown in Figure 3.7(c) is not covered by 300.22 because air handling is not a primary function of that space.

The wiring and equipment that may be installed in other space for environmental air are restricted to metallic conduit types such as EMT, RMC, IMC, FMC, and metal cable types without nonmetallic jackets, such as Types MC and AC. Nonmetallic cables are permitted only if specifically listed for the use. Such cables are typically identified by a "P" in the type marking, such as CMP (communications plenum) or CL2P (Class 2 plenum). The permitted wiring methods are listed in 300.22(C)(1).

Although Sections 725.3(C) and 725.61(A) permit the use of CL2P and CL3P cables for Class 2 and Class 3 circuits in ducts, plenums, and other space for environmental air, and Article 725 has the authority to modify Article 300, Section 725.3(C) also references the terms of 300.22. Thus, the language in 725.61(A) is restricted by the reference to 300.22, and cable types permitted in Article 725 must also comply with 300.22. Therefore, CL2P and CL3P ca-

Ducts and plenums used for environmental air and other space for environmental air.

Source: Based on *NEC Handbook*, 2005, Exhibit 300.20.

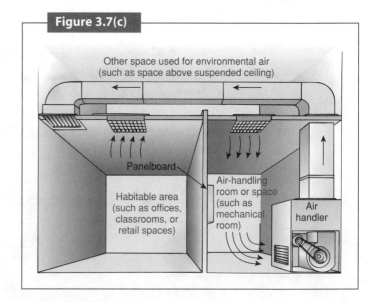

bles may only be used in ducts and plenums to connect to equipment that directly acts on or senses the contained air, but may be used freely in other space for environmental air, that is, the so-called plenum ceilings. Article 760 works essentially the same way. Chapter 8 is similar, but for communications cables, 300.22 is referenced directly in 800.154(A). On one hand, this is a significant restriction for communications, because it is highly unlikely that a communications circuit will connect to equipment that acts directly on the contained air in a duct. On the other hand, communications circuits are often installed in plenum ceilings, where they only have to meet the standards for low smoke and low flame spread (plenum-type cables, plenum raceways, or metallic wiring methods) and are not restricted because of how they are used.

Section 300.22(C)(2) covers electrical equipment other than wiring methods, such as transformers, boxes, controllers, and disconnects. Two types of equipment (and associated wiring) are permitted: (1) equipment with metallic enclosures, and (2) equipment with a "nonmetallic enclosure listed for the use and having adequate fire-resistant and low-smoke-producing characteristics." There are no special or additional requirements imposed by 300.22(C)(2) on equipment with metallic enclosures. That is, equipment with a metal enclosure need not be listed for use in air-handling spaces. However, it is assumed that the metallic enclosure is complete; in some cases, additional requirements may be imposed on specific equipment by other articles.

In the case of dry-type transformers, Section 450.13(B) also applies. Ready access is the primary concern of 450.13. Dry-type transformers with ratings that do not exceed 50 kVA are permitted in "hollow spaces of buildings, not permanently closed in by structure," such as suspended grid ceilings. Transformers in such spaces are not required to be readily accessible. However, the requirements for ventilation and spacing from combustibles must still be met. Small, totally enclosed or encapsulated transformers without ventilating openings can certainly meet all the requirements for installation in a ceiling space that is used for environmental air, but another factor must be considered for larger transformers. Many transformers rated up to 50 kVA are not totally enclosed or encapsulated. Most transformers in the range of 30 or 45 kVA will be ventilated. These transformers usually have ventilating openings near the top of the enclosure and a form of screen or mesh in the bottom to allow free air circulation for cooling. When the interior of the transformer and connecting wiring is open to the other space for environmental air, the interior parts and wiring exposed to the air must also be suitable for the location. Most such wiring that would be exposed to free air circulation is not "plenum-rated" and is not permitted in such spaces.

The point of restricting wiring in ducts, plenums, and other space for environmental air is to isolate sources of ignition and combustible materials from those spaces so that electrical equipment does not increase the spread of fire or products of combustion when it is installed in those spaces. This general requirement is given in 300.21, which also requires firestopping of electrical penetrations. Section 300.22 provides more specific requirements with regard to air-handling spaces.

Question 3.8

When are solid conductors allowed in raceways?

Related Question

- Is it a violation of Section 310.3 to install a 4 AWG solid copper conductor in a length of PVC conduit to make a connection to a grounding electrode?

Keywords

Raceway
Solid conductor

Keywords

Grounding electrode conductor
Sleeve
Physical protection

NEC References

250.64, 310.3, 326.104, 326.116, 680.23, 680.26

Answer

Section 310.3 prohibits solid conductors of size 8 AWG or larger from being installed in a raceway unless these larger solid conductors are permitted or required to be installed in raceways by other *NEC* rules. For the installation described in the question, the rigid nonmetallic conduit is being used as a sleeve to support and protect the grounding electrode conductor from an enclosure to the grounding electrode, probably to comply with 250.64(B). However, 250.64 does not require the grounding electrode to be solid, nor does it require a raceway. (The obvious alternative to a raceway for protection is choosing a routing where the conductor is not exposed to damage.) The fact that the grounding electrode conductor is permitted to be a solid conductor does not allow the requirement of Section 310.3 to be ignored where it is installed in a raceway system. In this case, however, the conduit is not a complete system from point to point—it is only a short length for physical protection, as shown in Figure 3.8—so the *NEC* does not prohibit this application.

If a raceway *system* or *wiring method* (as opposed to a sleeve or other short length for physical protection) is used for the grounding electrode conductor, it is necessary to comply with all applicable rules for raceway installations, including Section 310.3. The grounding electrode conductor is not treated differently from circuit conductors for the purposes of applying Section 310.3. As noted, grounding electrode conductors are not required to be solid conductors and often are not required to be installed in a raceway. For an example of where solid conductors are required to be installed in a raceway, see Sections 326.104 and 326.116, which cover the construction of integrated gas spacer cable. For a more commonly encountered example of such a requirement, Section 680.23(B)(2) requires conduit to be run to a wet-niche forming shell, and where nonmetallic conduit is used, "an 8 AWG insulated solid or stranded copper equipment grounding conductor shall be installed in this conduit." One other example is found in 680.26(C), where swimming pool bonding conductors are *required* to be solid (as opposed to grounding electrode conductors that are *permitted* to be solid). Therefore, if swimming pool bonding conductors are installed in a raceway, the exception to Section 310.3 is applicable. (These bonding conductors are re-

Conduit as a raceway system and conduit as a sleeve where larger solid conductors are permitted.

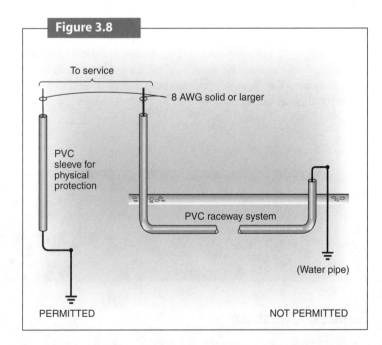

Figure 3.8

quired to be solid but are not required to be installed in a raceway. Nevertheless, in some cases, a raceway may be the only reasonable or practicable way to get the bonding conductor from the pool to the pump area.)

Question 3.9

Is a listing number adequate for identifying a manufacturer?

Related Questions

- Can a cable manufacturer be identified from a control number marked on the cable?
- If a manufacturer can only be identified by looking up a number or trademark in a book or on the Internet, is that considered "readily identified"?

Answer

Section 310.11(A)(3) does not say the manufacturer name must be printed on the cable. It does require that a "manufacturer's name, trademark, or other distinctive marking by which the organization responsible for the product can be readily identified." Let's assume the cable has an Underwriter Laboratories (UL) listing, so we can look at the UL marking requirements. Other listing agencies will have similar requirements.

The UL *Electrical Construction Equipment* directory ("The Green Book") provides the UL marking requirements. Under the general category Wires (ZGZX) and the specific categories Thermoplastic-Insulated (ZLGR) and Thermoset-Insulated (ZKST), the wire or cable is required to be marked with a listing mark (the UL symbol, in this case), a control number, and a product name. These markings are in addition to the required markings of size and specific type of insulation, and any optional markings, such as "Cable Tray Use" or "Sunlight Resistant."

The "listing number" can be used to identify a manufacturer, but it also identifies a product category. UL refers to this number as a "file number" or "control number." Figure 3.9 shows some of the markings on a cable. The control or file number is the seven-digit alphanumeric marking. A known control number can easily be matched to one of the manufacturers listed under the same category in the UL *Electrical Construction Equipment* directory. This information can also be obtained on a web page or by a phone call. Since the control number provides a ready means of identifying the manufacturer of a particular product, this marking meets the requirements of 310.11(A)(3). (Many or most products will also include the manufacturer's name on the product, but if the only marking visible is the marking shown in Figure 3.9, that marking is sufficient.)

Incidentally, the UL book also contains a listing of trademarks and trade names for products. This information can also be used to identify a manufacturer. However, a trademark or trade name is not always part of the marking, but the control number is required. Actually, it may be easier to identify the manufacturer with a control number than if only a trademark or trade name is used as permitted by listing standards, simply because the control numbers are a bit easier to look up than trademarks and logos.

Keywords

Marking
Identification
Manufacturer
Listing
Cable

NEC References

310.11

116   Chapter 3 Wiring Methods and Materials

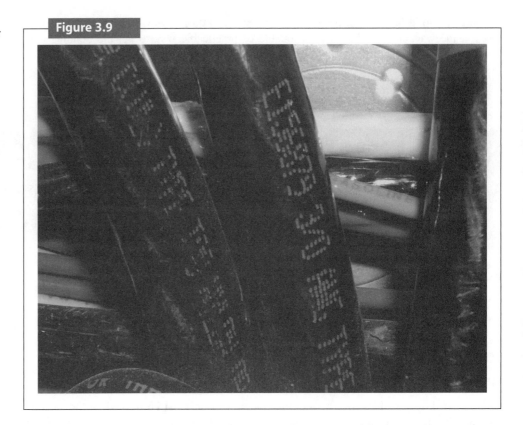

Figure 3.9 File number or control number marking on an insulated copper conductor.

Note that this discussion only covers the information required to be marked on a product, not the method of marking. Typically, the required markings for conductors are found on the cable insulation or multiconductor cable or cord jacket. However, for certain types of wire and multiconductor cables, the marking may be on a marker tape inside the cable or on a tag on the coil, reel, or carton. This information is found in 310.11(B).

Question 3.10 What color coding does the *NEC* require for conductors?

Keywords

Color code
Conductors
Identification
White
Gray
Green

Related Questions

- Does the *NEC* provide alternatives to color coding of conductors?
- How do *NEC* color codes apply to foreign-made equipment?

Answer

The most general or basic rules for conductor identification are found in Section 310.12, where we see that the required identification of conductors is not entirely based on color coding. Section 310.12(A) refers to 200.6 for identification of grounded conductors, and Section 200.6 requires that grounded conductors (usually neutrals) be identified with white, gray, or three white stripes along the length of the conductor. Section 310.12(B) refers to 250.119 for identification of equipment grounding conductors. This section requires equipment grounding conductors to be bare, green, or green with one or more yellow stripes. Section 250.119 and Article 250 do not specify any color for identifying grounding electrode conductors or most bonding jumpers. Outside of the requirements for grounded conductors and equipment grounding conductors, Section 310.12 refers to Sections 210.5(C) and 215.12 for identification of branch circuit and feeder conductors, but these sections do not specify any color coding for other conductors. The ungrounded conductors are required to be "finished to be clearly distinguishable from grounded and grounding conductors," according to 310.12(C). This means simply that ungrounded conductors may not be white, gray, green, green with yellow stripes, or have three white stripes. Most other color coding is optional under the *NEC*.

To understand color coding requirements, it may be helpful to look at two kinds of color coding requirements in the *NEC*. In some cases, colors are "assigned" to a specific type of conductor; in other cases, the colors are "reserved" for that type of conductor. For example, Section 200.6 assigns white, gray, and three white stripes to grounded conductors by requiring that where a grounded conductor is used, it must be identified by white, gray, or three white stripes. Section 200.7 then reserves white, gray, and three white stripes by stating that those means of identification may only be used for grounded conductors. Section 250.119 assigns the color code of green and green with yellow stripes, and then reserves those colors by prohibiting their use on grounded or ungrounded conductors. Section 310.12 also effectively reserves green and green with yellow stripes by requiring ungrounded conductors used for general wiring to be identified in other ways. (General wiring does not include special-purpose wiring such as remote control, signaling, fire alarm, or communications.)

White and gray are generally assigned and reserved for grounded conductors, but in certain cases, white or gray conductors may be used for other conductors, or grounded conductors may be identified in other ways. The specific cases where white and gray are not reserved are listed in 200.7 and include ungrounded circuits of less than 50 volts, multiconductor cables, and flexible cords. White or gray conductors in multiconductor cables may be used as ungrounded conductors, such as in switch loops or in 240-volt branch circuits to things like water heaters. In these cases, the white or gray conductors must be re-identified as ungrounded conductors where the conductors are visible and accessible. Colored tape, tags, or paint is usually used for re-identification. Section 200.6 applies only to insulated grounded conductors, so bare conductors such as those used in overhead service drops are not required to be marked gray or white. Also, 200.6 includes permission to identify grounded conductors by other means, such as a ridge on a conductor, in certain cables or flexible cords.

The *NEC* contains a number of other color coding rules that assign colors, but few that reserve any other colors. Consider the following requirements:

- Section 110.15 assigns the color orange to the "high leg" of a four-wire delta system. However, orange is not reserved for high-leg systems. Orange is commonly used in other systems or for other purposes, and is assigned to other systems as well. In fact, the orange color is only required where a grounded conductor is also present, so that the high leg will not be inadvertently or improperly used for loads that also connect to a grounded conductor.
- Section 424.35 assigns the colors yellow, blue, red, brown, and orange to the nonheating leads of electrical space heating cables, where each color represents a nominal voltage. These leads are also required to be marked with a voltage rating. Again, the colors are not reserved, just assigned, and this particular color identification does not apply anywhere else, even in Article 424.
- Section 504.80 permits intrinsically safe circuits to be identified by color coding. Some identification is required, and light blue is permitted to be used for the required

NEC References

200.6, 200.7, 210.5, 215.12, 250.119, 310.12, NFPA 79

identification. Light blue may also be used to identify raceways, boxes, and cable trays. This color coding is essentially a "permissive assignment"; other methods, usually labels, are permitted, but if color coding is used, light blue is reserved for that use and cannot be used for other conductors in that facility. Light blue is also reserved for enclosures and raceways if color coding rather than labeling is used to identify enclosures and raceways.

- Section 517.160(A)(5) requires circuit conductors of isolated systems in health care facilities to be identified with the colors orange, brown, and in some cases, yellow. Orange is used to identify the conductor that would have been grounded (white or gray or attached to the terminal of a receptacle that is identified for a grounded conductor) in a system that was grounded and not isolated. Again, these colors are assigned, but not reserved, and brown, orange, and yellow may be used for other purposes in the same building.

The code-making panels have repeatedly rejected attempts to mandate a specific color code for different voltage systems in a building, for good reasons. For one thing, a limited number of colors are commercially available for insulated wire. Also, a limited number of colors are clearly distinguishable from each other, especially under dimly lit or dirty conditions. If a facility has more than two 3-phase voltage systems, there are simply not enough colors to go around. Another reason that is often cited is that conductor identification is a design matter and the code should be flexible in this regard to deal with the many types of installations covered by the *NEC*. Yet another problem with mandating color codes is that they may create an idea that a system voltage can be identified by color, but such a practice is not safe. System voltage should be determined by testing or by examination of sources. Also, system voltages should not be limited to identification by an arbitrary color scheme, especially when there are millions of installations that have not adhered to that scheme in the past and where a newly imposed color scheme may actually increase hazards to personnel.

The *NEC* also contains requirements for identification of conductors of different systems. These rules apply when there are two or more voltage systems in a premises wiring system and are found in 210.5(C), which states that each ungrounded conductor of a branch circuit shall be identified by system. "The means of identification shall be permitted to be by separate color coding, marking tape, tagging, or other approved means and shall be posted at each branch-circuit panelboard or similar distribution equipment." A similar rule that applies to feeder conductors is found in 215.12. These rules have been modified from previous editions of the code (prior to 2005). Previously, such identification was only required for multiwire branch circuits.

A common color scheme is used in commercial buildings where two nominal utilization voltages are present. These two voltages are often derived from 480/277-volt, four-wire, grounded wye and 208/120-volt, four-wire, grounded wye systems. In many cases, the higher-voltage system is identified using brown, orange, yellow, and gray, and the lower-voltage system uses black, red, blue, and white. This method is a good way to identify two systems, but it is not the only way permitted by the *NEC*, nor the only safe way. Electricians and designers should not assume that the color code they are familiar with in one facility is applicable anywhere else. Figure 3.10 illustrates two of the many ways that conductors might be identified in a given facility. Notice that the method in Example 2 of Figure 3.10 provides for many more voltage systems in the same facility than Example 1. The method chosen must be posted at branch circuit distribution locations or, for feeders, at each feeder panelboard or similar feeder distribution equipment.

Where equipment such as industrial machinery is manufactured under other standards, such as IEC standards, the color coding of internal wiring may differ from that required by the *NEC*. NFPA 79, *Standard for Industrial Machinery,* also includes a color code. However, the NFPA 79 color code is an optional method of identifying conductors in which colors are used to distinguish between power, controls, low voltages and DC systems, and voltages from

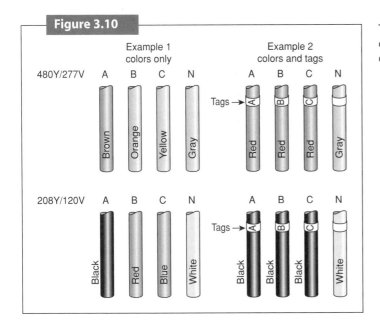

Figure 3.10 Two ways to identify the conductors of two different systems.

other sources. NFPA 79 does not conflict with the *NEC;* in fact, NFPA 79 is supplemental to the *NEC* and defers to the *NEC* on anything not specifically covered by NFPA 79. It also includes information about other color coding schemes that are not consistent with the *NEC* requirements. However, NFPA 79 does not permit these alternative color codes, nor does the *NEC*. Approval of such equipment is up to the authority having jurisdiction, as outlined in Sections 90.2, 90.3, and 90.4.

Question 3.11

Where can I find application information for conductors?

Related Questions

- How do I find voltage ratings for conductors?
- What does a dual rating such as THWN/THHN mean?

Answer

Section 310.8 lists the types of wire permitted for general wiring in dry, damp, and wet locations. All of the types listed in 310.8 have 600-volt ratings according to the Underwriters Laboratories (UL) directory. However, three types, RHH, RHW, and RHW-2, are also permitted for use as unshielded conductors in circuits up to 2000 volts with insulation meeting the requirements of Table 310.62. This is useful general information, but in order for an installer or inspector to be certain that a specific wire is suitable, certain markings are required.

The required markings are given in 310.11(A). Four items of information are required on all conductors for general wiring: rated voltage, type letter or letters, some means of

Keywords

Conductors
Wires
Voltage rating
Type
THHN
THWN
THW

NEC References

310.8, 310.11, 310.13, 310.16, 310.62

identifying the manufacturer, and the conductor size. As noted, the voltage rating of all wire types in 310.8 is at least 600 volts, but that information is required to be marked on the wire. Type letters are the types listed in 310.8 and Table 310.13, as well as in the ampacity tables, such as Table 310.16. Some common types are THHN, THWN, THW, XHHW, and XHHW-2, to name a few. The means of identifying a manufacturer may be the printed name, an abbreviation or brand name, a logo, or another identifier such as a listing file number. A manufacturer may be readily identified by a listing file number simply by calling the listing agency or by using its website. Conductor sizes are given in AWG (American wire gauge) for sizes 14 through 4/0 (0000) or in circular mils for larger conductors. Larger conductors are usually sized in thousands of circular mils, or kcmil. Although surface printing is the most common marking technique, other methods are permitted by 310.11(B).

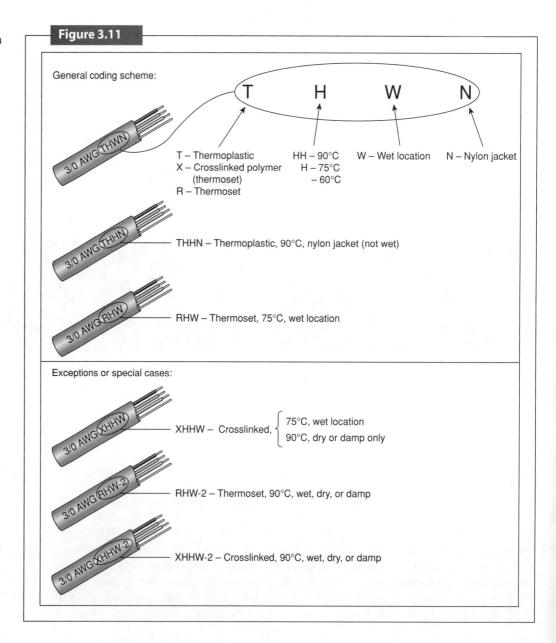

Example coding system for insulation types.

Figure 3.11

Table 310.13 lists applications and provides information about insulation types. Maximum operating temperatures and application provisions are the information most often sought by code users. For example, a Type THW insulation is generally a 75°C insulation type for use in dry, damp, and wet locations. A special 90°C type is available, but only for limited uses in electric-discharge lighting.

The letter system that describes and identifies various insulation types is based on a coding system that, although it has exceptions, can be helpful. For example, for the types mentioned in this answer, Figure 3.11 provides a key to the naming system. Note the effect of "-2" at the end of the type letters, which indicates an insulation type suitable for 90°C in wet locations. This application information can also be directly obtained from Table 310.13.

Some wires are marked with more than one type marking. A common example is a conductor marked THHN/THWN. This is both THHN and THWN, but when Table 310.13 is used, we see that THHN has a 90°C rating for use in dry and damp locations, whereas THWN is also usable in wet locations, but has only a 75°C rating. Thus, when this dual-rated type is used, it is a 75°C conductor in a wet location, such as in a raceway installed underground or outside on a rooftop. The same conductor may be treated as a 90°C conductor inside a building in a dry or damp location. A conductor that was only marked THHN could not be used at all in a wet location. A dual rating also may imply different ampacities, and may permit alternate uses. For example, some Type USE is also marked RHW. Type USE is for use outside only according to Article 338, but RHW may be used inside a building because it is recognized in 310.8.

All of the wire types in Table 310.13 are rated at least 600 volts, except for very small Type MI cables, which may be rated at 300 volts when used in certain control circuits. Some types may be over 600 volts, such as the special type of THW and the 2000-volt types mentioned previously. Also, like the special type of THW, some other types are intended for use only for specific applications. For example, Types TBS and SIS are only for switchboard wiring, and Type MTW is only intended for use as machine tool wiring, which is covered by Article 670 and NFPA 79.

Question 3.12

Do the special conductor sizes for dwellings apply to 208/120-volt, single-phase systems?

Related Questions

- When supplying dwelling units in an apartment building, may 4 AWG THWN copper be used as a single-phase, three-wire feeder derived from a three-phase, four-wire, 208Y/120-volt service?
- If Table 310.15(B)(6) does not apply, can the 85-ampere rating of a 4 AWG conductor from Table 310.16 be rounded up to 100 amperes?

Answer

Table 310.15(B)(6) is specifically permitted for *only* 120/240-volt service and feeder conductors. Figure 3.12 shows an example of the conductors to which these special ratings apply

Keywords

Ampacity
Dwelling
Service
Feeder
Single-phase
Three-phase
Current-carrying

NEC References

240.4, 240.6, 310.15

An application of the ampacity values in Table 310.15(B)(6) for 120/240-volt, single-phase systems.

Source: Based on *NEC Handbook*, 2005, Exhibit 310.8.

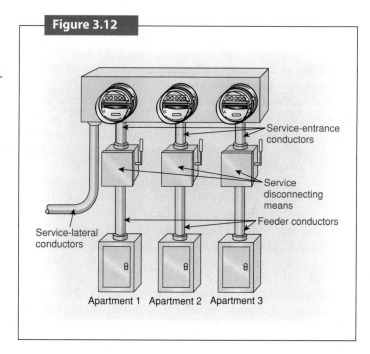

Figure 3.12

when the system is 120/240-volt, single phase, and the feeders and service-entrance conductors carry the entire load of each apartment.

The value that appears in the right-hand column of Table 310.15(B)(6) is the "rating" of the service equipment, not the ampacity of the conductor. So the ampacity is not changed by the rule, but the allowable rating of the equipment the conductors can supply is different. Section 240.4 requires that overcurrent protection be supplied in accordance with 310.15, which includes this rule.

As noted in one of the questions, 4 AWG THWN conductors, under the 75°C column, have an ampacity of 85 amperes. The current in the neutral of a three-wire, 120/240-volt, single-phase system carries only the unbalanced current of the ungrounded conductors and is zero when both legs are equally loaded, so the neutral is not considered to be current-carrying according to 310.15(B)(4)(a). (Dwelling units typically have loads that do not produce a high harmonic content, and the harmonics are not so much of an issue in single-phase systems as they are in three-phase systems.) Therefore, the 100-ampere service rating in Table 310.15(B)(6) is justified for the 85-ampere conductors partially because the neutral conductor acts as a heat sink rather than a heat producer with regard to the dissipation of heat from the two ungrounded conductors. This allows the service rating to be slightly greater than the normal allowable ampacity of the conductors. For example, if the ampacity of 4 AWG is determined using the formula provided in 310.15(C), the ampacity is approximately 102 amperes for the two current-carrying conductors in this application. (This method of determining ampacity is recognized under the *NEC* when the calculation is done under engineering supervision, using the full Neher-McGrath method, which is only summarized in Annex B.)

In an installation utilizing a single-phase feeder derived from a three-phase, four-wire wye system, where two phase conductors and the neutral are used, the situation is completely different. First, according to 310.15(B)(4)(b), the neutral is current-carrying. Second, based on Table 310.16, 4 AWG copper conductors have an ampacity of 85, Section 310.15(B)(6) does not apply, and the maximum rating of the overcurrent protection permitted by 240.4(B) is 90 amperes. The 90-ampere rating is based on Section 240.6, where the "standard sizes" are found.

As implied in the question, 310.15(B)(6) does not apply in this case. Also, 100 amperes is not the next standard size above 85 amperes. However, if the equipment has 75°C terminals, a 3 AWG conductor with a 75°C insulation would have adequate ampacity and could be protected by a 100-ampere overcurrent device.

Question 3.13

How are ampacity calculations coordinated with termination and continuous load requirements?

Related Questions

- How is ambient temperature determined for ampacity calculations?
- What is meant by "derating"?
- Can ampacity be increased for low ambient temperatures?
- How are electric heating loads calculated for feeders?

Keywords

Ampacity
Derating
Correction factor
Adjustment factor
Ambient temperature

Answer

Ampacity is defined in Article 100 as "the current, in amperes, that a conductor can carry continuously under the conditions of use without exceeding its temperature rating." The Fine Print Note (FPN No. 1) of Section 310.10 explains what the "principal determinants of operating temperature" are. These principal determinants include the conditions of use as well as the basic consideration of current flow in the conductor.

Ambient temperature is one of the "conditions of use" mentioned in the definition and is a "determinant" listed in the Fine Print Note to 310.10. The maximum ambient temperature that can reasonably be expected and to which the conductors would be exposed during normal operation is the temperature that should be used for temperature correction. This may be a relatively high temperature in some outdoor areas, a relatively constant standard temperature in many interior conditioned spaces, or a very low temperature if the load is seasonal (winter only) or located in cold areas, such as cold storage.

The other conditions of use or determinants are the number of adjacent load-carrying conductors and the rate at which heat dissipates into the ambient medium. Each of these conditions of use or determinants is covered in some way in the ampacity calculations and tables.

Each of the allowable ampacity tables referenced in 310.15 includes assumptions about the conditions of use and provides differing ampacities based on both those conditions and the insulation rating. The insulation temperature rating is the temperature rating mentioned in the definition of ampacity. The value from the table represents the load current determinant. The temperature correction factors account for the ambient temperature determinant. The adjustment factors found in 310.15(B)(2)(a) account for adjacent current-carrying conductors. The fourth determinant from 310.10 is rate of heat dissipation. Heat dissipation rates are also accounted for in the assumptions in each table: For Table 310.16, it is "in raceway or cable or directly buried," and for Table 310.17, "free air." Also, 310.15(B)(2)(b) requires spacing to be maintained between raceways, tubing, or conduits generally. (The *NEC* does not provide a minimum spacing except for medium-voltage applications as covered in Figure 310.60. Figure 310.60 is for use with Tables 310.77 through 310.86.)

NEC References

Article 100, 110.14, 210.19, 215.2, 220.51, 230.42, 310.10, 310.13, 310.15, Table 310.16, Figure 310.60, 424.3

Other factors that do not change conductor ampacities but do influence conductor selection are continuous loads and terminal temperature ratings. These considerations are separate from the ampacity calculation, but will dictate a minimum-size conductor for a given load based on the characteristics of the overcurrent device and terminations on each end of a conductor. These factors, especially terminal temperature provisions, may influence the final selection of an ampacity for a wire, but do not change the ampacity calculation based on 310.15. Figure 3.13(a) illustrates the three issues that must be considered in selecting a conductor and in determining its ampacity.

Example 1

A feeder supplies a heating load that has been calculated at 190 amperes using Article 220. The load is only on during cold weather, and is controlled by an ambient thermostat that allows the load to operate only at temperatures below 40°F. The conductors are located in unconditioned spaces, so they will carry load only when the ambient temperature is below 40°F. The conductors are THWN copper, so they are rated 75°C in wet, dry, or damp locations according to Table 310.13. No more than three current-carrying conductors will be installed in any one raceway.

First, let's determine the smallest conductor that could supply this load under the specified conditions of use. To determine the minimum ampacity of a conductor considering the conditions of use, we must account for any required derating. "Derating" is a rather broad term that is sometimes used in the *NEC* to refer to the application of correction factors or adjustment factors, or both. Ampacity is a continuous value by definition, so we only need to consider the calculated load from Article 220 to determine this rating. No adjustment factor will be applied to the three conductors. However, since the conductors will only carry current when the ambient temperature is 40°F or less, we can apply a correction factor. The lowest temperature range given in Table 310.16 is 70°F to 77°F, which gives a correction factor of 1.05 from the 75° column. (To use a higher factor, we would have to do the ampacity calculation under engineering supervision using the formula given in 310.15(C)). This correction factor gives us an ampacity of 175 A × 1.05 = 183.75 A for a 2/0 AWG conductor, 200 A × 1.05 = 210 A for a 3/0 AWG conductor, 230 A × 1.05 = 241.5 A for a 4/0 AWG conductor, and 255 × 1.05 = 267.75 A for a 250 kcmil conductor. From this we can see that we would need at least a 3/0 AWG conductor to be sure the conductor is big enough for the load under the specific conditions of use, and it would not matter whether the load was continuous or noncontinuous. However, this size does not consider the requirements for the overcurrent device or the terminations.

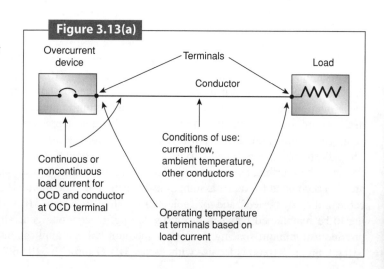

Considerations in the selection of a conductor and its ampacity.

According to Section 424.3(B), fixed electric space heating is to be considered a continuous load. Although this section is about sizing branch circuits, it also applies to sizing of feeder conductors supplying fixed electric space heating.

For a feeder, the minimum conductor size is based on the load calculated under Article 220, and is increased by 25 percent for continuous loads. This rule is found in 215.2(A)(1). Nearly identical rules are found in 210.19(A) for branch circuit conductors and in 230.42 for service conductors. In each rule, an exception allows for conductors to be sized at 100 percent of continuous loads if the overcurrent device is rated for continuous operation, so we can easily see that the rule about continuous loads is for compatibility with the overcurrent device and not about any problem with the conductors themselves. According to Section 220.51, fixed electric space heating loads on feeders or service conductors are calculated at 100 percent of the total connected load, but this is only for the load calculation, not for the sizing of overcurrent devices.

The minimum feeder conductor rating in 215.2 is determined "before the application of correction or adjustment factors." Therefore, we can find this size simply by looking in the allowable ampacity tables. Many heating loads are cycling loads that are controlled by thermostats and may never be continuous, depending on the design of the heating system. Others, such as self-regulating heat-tracing cables, may fluctuate continuously but at a level somewhat below the full current rating of the equipment. Still others, such as fixed-wattage heat-tracing cables, will be on at full load until the ambient temperature rises above the thermostat set point. Although we know that the space heating is considered to be continuous, we will consider both continuous and noncontinuous loads in this example.

For a continuous load, the minimum conductor ampacity is 190 A × 1.25 = 237.5 A. For a noncontinuous load, the minimum conductor ampacity is simply the connected load, or 190 A. Therefore, for a continuous load, we would need at least a 250 kcmil conductor (255 A) from the 75° column of Table 310.16. For a noncontinuous load, the same table tells us we would need at least a 3/0 AWG (200 A) conductor. Note that these are minimum sizes based on feeder rules, and, as we have seen, are intended to make sure the conductors are compatible with the overcurrent devices for a specific load. The operating temperature of the conductors at some point in the circuit other than at the overcurrent device does not affect the overcurrent device, so in this context we are only concerned with the temperature where the conductor connects to the overcurrent device. (The listing test of the overcurrent device is based on only a 6-foot length of conductor.)

The minimum conductor rating is also the minimum overcurrent device rating for a continuous load. From the standard overcurrent device ratings given in 240.6, the minimum overcurrent device ratings would be 250 A for a continuous load and 200 A for a noncontinuous load.

We must also consider terminal temperature ratings. For conductors larger than 1 AWG or overcurrent devices rated over 100 A, both of which apply in our case, the terminals are assumed to be rated at 75°C. This rule is found in 110.14(C)(1). The load is 190 A. This load need not be increased because it is continuous. The 190 A load is the current that will generate internal heat in the conductor and will directly affect the temperature of the terminations. Therefore, to comply with 110.14(C)(1), we must select a conductor size from the 75° column of Table 310.16 that has an ampacity at least equal to the 190 A load. In other words, we are looking for a conductor that can carry 190 A without exceeding 75°C, so connected terminals will not operate at over 75°C. From Table 310.16, we see that this minimum size is 3/0 AWG.

Now we have found three (or four) minimum sizes for conductors to supply the specified load under the specified conditions. Two of these sizes do not consider conditions of use (derating factors are not applied), and two do not consider continuous loading because continuous loads are a factor only for overcurrent devices. In this example, the largest of the three minimum sizes is 250 kcmil for a continuous load, or 3/0 AWG for a noncontinuous load. The conductor sizes and ampacities are summarized in Table 3.13(a). See Figure 3.13(b).

Table 3.13(a) Considerations in Ampacity and Conductor Selection, Example 1

	Minimum Conductor Size	
Consideration	190 A Continuous Load	190 A Noncontinuous Load
Conditions of use (310.15)	3/0 AWG (200 A)	3/0 AWG (200 A)
Overcurrent device (215.2)	250 kcmil (255 A)	3/0 AWG (200 A)
Terminations (75°C) (110.14)	3/0 AWG (200 A)	3/0 AWG (200 A)

Considerations in ampacity and conductor selection, Example 1.

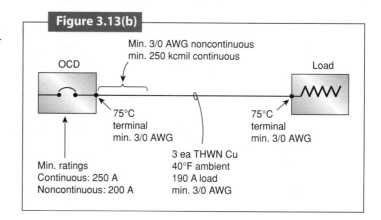

Figure 3.13(b)

In this example, a 3/0 AWG conductor satisfies all conditions if the load is noncontinuous, but 250 kcmil is required if the load is continuous; in this case, the *NEC* says the heating load is considered a continuous load. Note that in the discussion of this example we found that a 4/0 AWG conductor was adequate for the continuous load because it had a corrected ampacity of over 125 percent of the 190 A load. However, this value was determined through the application of correction factors—in this case, a correction factor greater than 1.00—and 215.2 required the minimum size to be determined before application of correction or adjustment factors. Note also that although we solved our problem by addressing the issues in a particular order, each minimum size can be determined separately, so these sizes can be determined in any order.

To further illustrate the three different types of requirements and how they are applied, two shorter examples follow.

Example 2

Given: Three THWN copper feeder conductors in a raceway, 95 A calculated load, 27°C ambient temperature, and standard terminal temperature ratings.

A. Conditions of use

No correction or adjustment factors apply. (Actually, correction factor and adjustment factors are both 1.00.) From Table 310.16, 75° column, 3 AWG is adequate for the load, whether continuous or noncontinuous.

B. Overcurrent device

For a continuous load: 95 A × 1.25 = 118.75 A minimum, or 1 AWG (130 A).

Overcurrent device minimum size: 125 A

For a noncontinuous load: 95A minimum or 3 AWG (100 A)

Overcurrent device minimum size: 100 A (from standard sizes in 240.6)

C. Terminations

For a continuous load, a 125 A overcurrent device would be required, and this would be assumed to have 75°C terminals. However, the terminations at the other end (the load end) of a 1 AWG conductor would be assumed to be rated 60°C according to 110.14(C)(1)(a). Therefore, the minimum conductor size for a 95 A load with 60°C terminals would be 2 AWG (95 A). If the load were noncontinuous, the minimum overcurrent device would be 95 A (100 A if we used only standard sizes), so terminations on both ends would be assumed to be rated 60°C, and the minimum size would be 2 AWG.

The required sizes in this example (largest of three minimums) would be 1 AWG for a continuous load or 2 AWG for a noncontinuous load. Conductor selection for Example 2 is summarized in Table 3.13(b). See Figure 3.13(c).

Table 3.13(b) Considerations in Ampacity and Conductor Selection, Example 2

	Minimum Conductor Size	
Consideration	95 A Continuous Load	95 A Noncontinuous Load
Conditions of use (310.15)	3 AWG (100 A)	3 AWG (100 A)
Overcurrent device (215.2)	1 AWG (130 A)	3 AWG (100 A)
Terminations (60°C) (110.14)	2 AWG (95 A)	2 AWG (95 A)

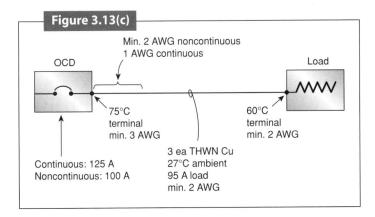

Considerations in ampacity and conductor selection, Example 2.

Example 3

Given: Four XHHW-2 copper feeder conductors in a raceway, all considered to be current-carrying. The calculated load is 75 amperes; the ambient temperature is 110°F, but all terminations are rated 75°C.

A. Conditions of use

We need a conductor that is adequate for the 75 A load after the application of correction and adjustment factors. The conductor in this example is a 90°

conductor in a wet, damp, or dry location (Table 310.13). We can see that the correction factor for a 110°F ambient temperature is 0.87 in the 90° column, and the adjustment factor for four conductors is 80 percent, or 0.80, from Table 310.15(B)(2)(a). By trial and error, we can see that a 6 AWG conductor won't be big enough because its ampacity before derating is 75 A. Then we find that 4 AWG is not big enough because its ampacity of 95 A becomes 66.12 A after derating. Then we see that a 3 AWG conductor has an ampacity of 110 A. The derated ampacity is 110 A × 0.87 × 0.80 = 76.56 A. Therefore, a 3 AWG XHHW-2 conductor is adequate for the 75 A load.

For an approach that does not rely on trial and error, divide the needed ampacity by the product of the derating factors. The result will be the minimum ampacity that will work from the table. For example, in this case, 75 A / (0.87 × 0.80) = 75 A / 0.696 = 107.76 A, or 108 A. The smallest conductor from the 90° column that exceeds 108 A is 3 AWG.

B. Overcurrent device

For a continuous load: 75 × 1.25 = 93.75 A minimum or 3 AWG from the 75° column. For practical purposes, the smallest overcurrent device would be 100 A. We cannot use the value from the 90° column for this purpose because 110.14(C) says that unless the terminals are rated for the higher temperature, we can only use the higher temperature ratings (ratings over 75°C) for ampacity correction, adjustment, or both. We used the 90° value for ampacity correction in this case. The assumed rating for a 100 A breaker is 60°C, but it is given in this example that all terminations are rated 75°C.

For a noncontinuous load: 75 A minimum or 4 AWG conductors with a minimum standard overcurrent device of 80 A.

C. Terminations

Since all terminations are 75°C, we simply choose a conductor whose ampacity in the 75° column is at least equal to the load current. We see that 4 AWG is the smallest size that could carry 75 amperes while connected to 75°C terminals.

The required sizes in this example would be 3 AWG if the load were continuous and 3 AWG if the load were noncontinuous. Table 3.13(c) provides a summary of this example. See Figure 3.13(d).

The three examples given here did not include one in which a larger conductor was required for the terminations than for the overcurrent device or the conditions of use, but such cases are also possible.

Table 3.13(c) Considerations in Ampacity and Conductor Selection, Example 3

Consideration	Minimum Conductor Size	
	75 A Continuous Load	75 A Noncontinuous Load
Conditions of use (310.15)	3 AWG (110 A derated to 77 A)	3 AWG (110 A derated to 77 A)
Overcurrent device (215.2)	3 AWG (100 A)	4 AWG (85 A)
Terminations (75°C) (110.14)	4 AWG (85 A)	4 AWG (85 A)

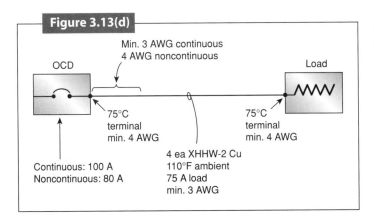

Considerations in ampacity and conductor selection, Example 3.

Question 3.14

Do the ampacity tables of Article 310 apply to both AC and DC conductors?

Related Questions

- Do conductors carrying only DC currents have higher ampacities?
- When can I use the formula for calculating ampacities instead of the ampacity tables?

Answer

The requirements of Article 310, including Sections 310.15 and 310.60, apply to conductors used for general wiring in both AC and DC circuits. Since the majority of conductor installations under the *NEC* are for AC circuits, people often think that the ampacities in Article 310 apply only to AC circuits. This is not the case. The ampacity for conductors used in DC services, feeders, and branch circuits is selected from the applicable ampacity table in Article 310. For systems up to 2000 volts, Section 310.15 and Tables 310.16 through 310.20 apply. Table 310.16 is most often used. For systems of 2001 through 35,000 volts, Section 310.60 applies along with Tables 310.67 through 310.86.

You may argue that the ampacity of a conductor should be higher for a direct current than for an alternating current because of the presence of reactance and skin effect in AC circuits. This is taken into account if you wish to use the formula from 310.15(C) or 310.60(D). Note that this formula includes a variable YC for the "component ac resistance resulting from skin effect and proximity effect." You can also see from this formula that where $YC = 0$, the ampacity will be slightly higher. However, this formula must be used along with the entire Neher-McGrath method as outlined in Annex B. (Annex B is generally not adequate for these calculations, but it does provide examples and the necessary references.) Also, the formula method may only be used under engineering supervision and may require information that is not always readily available.

The ampacity tables, especially those for up to 2000 volts, are not intended to be exact values of current-carrying capacity. The most-used table, Table 310.16, is titled "Allowable

Keywords

Ampacity
AC
DC
Allowable ampacity
Neher-McGrath method
Engineering supervision

NEC References

310.15, Table 310.16, 310.60

Ampacities . . ." and provides just that—not the exact values, but allowable safe values for a variety of conditions within the parameters stated in the table. The tables for over 2000 volts reference specific conductor arrangement details and are very specific to the arrangements found in 310.60.

Question 3.15

What are the differences in requirements for boxes in combustible and noncombustible materials?

Keywords

Boxes
Combustible
Noncombustible
Plaster ring
Drywall
Fire resistance

NEC References

300.21, 314.16, 314.20, 314.21, 314.25

Related Questions

- Why do boxes in combustible materials have to be flush or extend from the surface?
- Must plaster rings be flush and without gaps in drywall?
- How do I know whether drywall or other materials are combustible or noncombustible?

Answer

Boxes are intended to protect wiring and contain any arcing and sparking that may be produced where wiring is spliced or terminated or where devices are installed. Where a box is installed in a wall, and the box is not flush with the wall surface (or extending from the wall surface), the surface material essentially becomes part of the enclosure or becomes an extension of the enclosure. For a box or its extensions to contain arcing and sparking without igniting the surrounding material, the enclosure must be largely noncombustible. For this reason, boxes and associated fittings such as plaster rings that use a flush-type cover or faceplate may be recessed up to 1/4 inch from the finished surface only where the surrounding surface is noncombustible, as shown in Figure 3.15(a). The 1/4-inch allowance is simply a reasonable dimension that recognizes that making boxes perfectly flush in drywall, plaster, masonry, or concrete is usually not likely or possible. However, where the surface finish is combustible, the box must be flush or extend from the finished surface so that no part of the combustible surface becomes an extension of the box. The box can be made flush through the use of plaster rings, extension rings, or listed box extenders.

The same reasoning applies to the finish on surfaces where canopies are used, because the canopy is also an extension of the box. In fact, according to Section 314.16(A), the volume of a "domed cover" such as a luminaire canopy may be counted for the purposes of determining the volume of a box. Therefore, the surface surrounding a box or between a box and the edge of a canopy must be noncombustible. Section 314.20 covers the location of the box in a wall or ceiling, and 314.25(B) covers the requirements at luminaire canopies. In the case of luminaire canopies, a combustible surface may be covered with the glass fiber pads that are commonly supplied with luminaire canopies, but the requirement of 314.20 still applies. Section 314.25(B) applies only where the surrounding material is combustible and is therefore required to be covered within the canopy. Section 314.20 applies to all boxes in walls and ceilings and provides different rules for combustible and noncombustible materials.

The *NEC* does not define precisely what is combustible and what is noncombustible. However, Section 314.20 provides a short list of common ceiling and wall materials that are considered noncombustible. These include concrete, tile, gypsum, and plaster. Other wall ma-

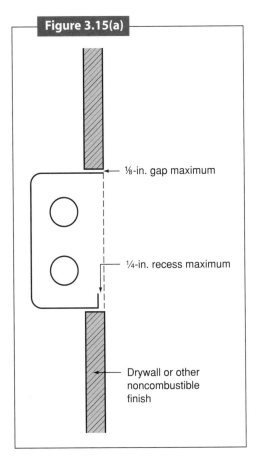

Figure 3.15(a)

Box installed with maximum 1/4-inch clearance in noncombustible finish.

Source: *Electrical Inspection Manual*, 2005, Figure 3.24.

terials or wall finishes that would be noncombustible include glass, masonry, stone, and metal. Walls finished with ordinary wood paneling or wood siding would certainly be considered combustible. Note that the rule in 314.20 refers to the surface of the wall or ceiling, so a wall or ceiling constructed with combustible wood framing but covered with noncombustible gypsum board would be considered noncombustible for the purposes of this rule. A more precise definition of a noncombustible material is any material that passes the ASTME 136-test procedure for noncombustible materials. The manufacturer of a construction material should be able to provide information on the combustibility of the material.

Some walls or ceilings are constructed with noncombustible materials so that the assembly will have a fire rating. Section 314.20 does not specifically address this issue. Fire-rated assemblies are permitted by building codes to have a limited number of penetrations (such as boxes) of a limited area. In some cases, boxes may not be permitted on opposite sides of a wall in the same framing space or "stud space." Local building codes should be consulted on this issue, since the requirements for fire-rated construction are not covered in the *NEC*. However, the *NEC* does have a specific rule for boxes and a more general rule for all types of penetrations. The specific rule for boxes is found in 314.21, and the more general rule that requires maintenance or restoration of fire ratings is found in 300.21. Section 300.21 says: "Openings around electrical penetrations through fire-resistant-rated walls, partitions, floors, or ceilings shall be firestopped using approved methods to maintain the fire resistance rating." Typically this requires selection of an approved system for firestopping penetrations of various types. Section 314.21 requires that repairs be made where necessary so that there are no gaps or open spaces greater than 1/8 inch at the edge of a box or fitting (such as a plaster ring) that is installed in a plaster, drywall, or plasterboard surface and that is intended to be used

132 Chapter 3 Wiring Methods and Materials

A box installed in a combustible surface must be flush.

Source: *Electrical Inspection Manual*, 2005, Figure 3.25.

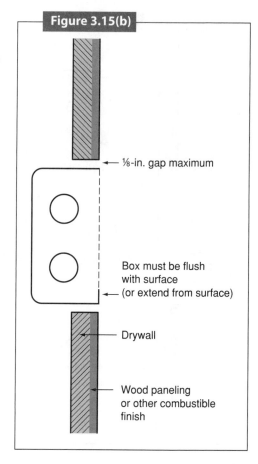

Figure 3.15(b)

with a flush-type cover or faceplate. Section 314.21 applies even when an assembly is not fire-rated. This requirement is illustrated in Figure 3.15(a) as well as Figure 3.15(b), which also illustrates the requirement for a box to be flush with (or extend from) a surface finished with combustible materials. Repairs in this case may usually be made with plaster or drywall compound. Note that the rule is about the edge of the box or fitting, so where repairs are needed, an oversized plate may be used to conceal the repairs, but not to make the repairs.

Question 3.16

Are box extensions the only way to shield combustible material at boxes?

Keywords

Boxes
Extensions
Combustible finish
Noncombustible finish

Related Questions

- How may a box be made flush in a wall if a combustible finish is added?
- How can I fix a box that is recessed more than 1/4 inch in a noncombustible wall finish?

Answer

Section 314.20 requires the front edge of most boxes in noncombustible surfaces to be no more than 1/4 inch from the finished surface, as shown in Figure 3.15(a). Most boxes installed in combustible surface material must be flush or must project from the finished surface, as shown in Figure 3.15(b). In effect, any wall surface that extends beyond the front edge of a box is an extension of the box. Since boxes are intended to contain any arcing or sparking that originates in the box, a "box extension" made of combustible material is prohibited. If a box is properly installed in a noncombustible wall, such as gypsum board, and another layer of finish is added, the 1/4-inch allowance will likely be exceeded or, if the new finish is combustible, the box will probably not be flush. This problem may be fixed in three basic ways: extend the front edge of the box, relocate the box, or replace the box.

Plaster rings or extension rings (extension boxes) may be used to make the box deeper so it becomes flush or projects from the surface, but these may not be available in the exact depth needed and usually require some damage (and repair) of the finish. Figure 3.16(a) is a photograph of a plaster ring, and Figure 3.16(b) is a photograph of an extension box. Relocating the box may also involve damage and repair. In some cases, a box may be replaced with an "old-work box," like the one shown in Figure 3.16(c), that has external ears or clips for attachment as permitted by 314.23(C), but this is often difficult and may also require some surface repair. Some old-work boxes have integral devices that hold the box in place in the wall, some may be nonmetallic, and some use separate holding straps like those shown in Figure 3.16(c). Another option is a box extension that only requires removal and replacement of the device. These extensions provide a new interior surface that extends from the box to the finish of the wall, shields the combustible surface from the contents of the box, and can be adjusted to differing depths. These are now available as listed products like the one shown in Figure 3.16(d).

Adjustable box extensions are available in metallic and nonmetallic construction and are probably the simplest method to extend the box so that it will be flush with the combustible

NEC References

314.20, 314.23

Figure 3.16(a)

A plaster ring for a 4-inch square box.

An extension ring or extension box for a 4-inch square box.

Figure 3.16(b)

Old-work box with separate holding straps for installation in an existing wall.

Figure 3.16(c)

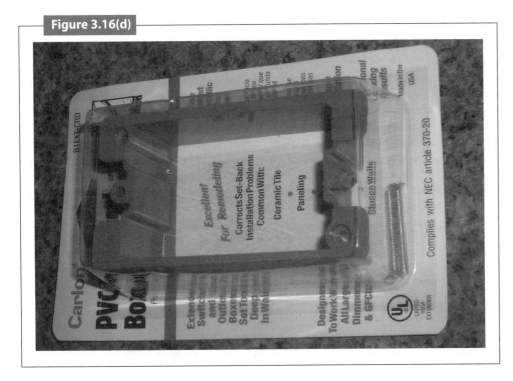

Figure 3.16(d)

Nonmetallic box extension for device boxes that are recessed more than allowed by Section 314.20 (or 370-20 in earlier editions of the *NEC*).

wall finish. They may also be used where the setback in a noncombustible surface is more than 1/4 inch. Such extensions have been available for many years but have not been available as listed products until fairly recently. (Article 314 provides construction specifications for boxes and does not require that all boxes be listed, but many authorities have questioned the use of an unlisted extension on what are usually listed device boxes.) Moving the box is typically a less desirable (and more expensive) alternative because it will invariably involve wall repair. Of course, it is sometimes possible to remove the wiring from the old box, break that box out of place, and install a new ("cut-in") box using the devices that are designed for installation of a box in a finished wall. This or the use of an extension box is an alternative that works if there is a need to enlarge the box because of additions to the wiring system.

Question 3.17

Where is Type NM cable permitted?

Related Questions

- Is NM cable permitted in buildings over three floors above grade?
- Is NM cable permitted in a wet location if it is installed in raceways?
- Is NM cable permitted in a building that also includes a place of assembly?
- May NM cable be used in medical and dental offices and clinics?

Keywords

Type NM cable
Construction type
Combustible construction
Noncombustible construction
Type III

Keywords

Type IV
Type V
Assembly
Health care
Patient care area
Raceway
Wet location

NEC References

334.10, 334.12, 511.3, 517.2, 517.13, 518.2, 518.4, Annex E

Answer

The uses that are and are not permitted for Type NM (nonmetallic-sheathed) cable are listed in 334.10 and 334.12, respectively. These sections also cover the uses of NMC (nonmetallic sheathed corrosion-resistant) and NMS (hybrid nonmetallic sheathed with signal or data conductors) cable, but type NM is the most widely used, by far. These permitted uses are modified by some jurisdictions, but the *NEC* no longer places a limit on the use of NM cable based on building height. Essentially, NM cable is permitted in one- and two-family dwellings, multifamily dwellings, and other buildings, but additional restrictions are placed on its use in multifamily dwellings and other buildings. In other than one- and two-family dwellings, the use of NM cable is restricted to combustible types of construction. Also, in one-, two-, and multifamily dwellings, NM cable may be installed exposed or concealed, but in nondwelling occupancies it must be concealed behind a thermal barrier with a 15-minute finish rating.

Many of the uses that are not permitted for NM cable are based on occupancy type. For example, NM cable may be run exposed in a dropped or suspended ceiling in dwelling units, but not in other occupancies. This rule in 334.12 is really just a restatement and clarification of the permitted uses in 334.10. "Exposed" (as applied to wiring methods) is defined in Article 100 as "on or attached to the surface or behind panels designed to allow access." Thus, an installation in a suspended "hard" drywall ceiling is concealed and, because of the finish rating of gypsum board, would generally be permitted in nondwelling occupancies. However, an installation in a suspended grid ceiling is exposed, by definition, and is permitted only in dwelling occupancies.

Section 334.10(A)(1) also restricts NM cable to normally dry locations, so it may not be used where exposed or subject to excessive moisture or dampness. Type NM cable may be permitted in a raceway, but if the raceway is in a wet location, the NM cable is also in a wet location, so the NM cable would not be permitted. The inside of a conduit is in the same location as the outside, so contained wiring must be suitable for the same dry, damp, or wet location.

Some other areas where NM cable may not be used include motion picture studios, storage battery rooms, hoistways, hazardous locations except for wiring of nonincendive or intrinsically safe circuits, and theaters. Most of these could be parts of other buildings that also may contain occupancies where NM cable is permitted. For example, 334.12(A)(4) prohibits NM cable "in theaters and similar occupancies, except where permitted in 518.4(B)." Most assembly occupancies are required to be of fire-rated construction, but according to 518.4(B), NM cable may be used in nonrated portions of an assembly occupancy or in nonrated portions of a building that contains an assembly occupancy. The requirements for fire ratings in these areas are determined from building codes. An assembly area for fewer than 100 persons that is an incidental part of another occupancy (that is, where the assembly area is not the primary purpose of the occupancy) is subject only to the requirements for the primary occupancy. This rule is found in 518.2(B).

The rules mentioned in Article 334 are not the only restrictions on the use of NM cable. A prime example is patient care areas in health care facilities. Most hospitals are of noncombustible or fire-resistive construction, where NM cable is not a permitted use according to 334.10. In addition, patient care areas must be wired with a method that includes an insulated equipment grounding conductor in the cable or raceway, and another equipment grounding conductor in the form of a metallic raceway or suitable metallic cable armor. Most NM cable meets neither of these requirements, and even if it had an insulated equipment grounding conductor, the nonmetallic sheath fails the second requirement. Although many smaller medical and dental clinics are of combustible construction, the patient care areas in those clinics must have the "redundant grounding" required by 517.13 for all patient care areas. Patient care areas are defined in 517.2 as "any portion of a health care facility wherein patients are intended to be examined or treated," and the definition of "health care facility" specifically mentions "clinics" and "medical and dental offices." Areas of such buildings that are not for patient care may be wired with NM cable if the building is of combustible construction as described and the NM cable is concealed in accordance with 334.10. Many of the uses that are and are not permitted for NM cable are illustrated and summarized in Figure 3.17.

Although the *NEC* has no specific height restriction for buildings wired with NM cable, the buildings that are allowed to be of combustible construction are often defined partially by height. Thus, height may be an indirect restriction. For example, high-rise condominium buildings are multifamily dwellings or mixed occupancies, but are not likely to be permitted in the Type III, IV, or V construction where NM cable is permitted. Annex E provides additional information on building construction types, but the applicable building code will define what kinds of buildings are required or permitted to be of each type of construction.

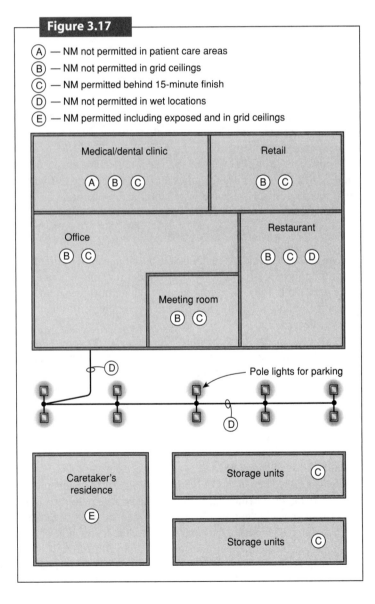

Hypothetical mixed-occupancy buildings (combustible construction), showing areas where NM cable is and is not permitted.

Question 3.18 — When and where is protection required for cables in attics and roof spaces?

Keywords

Attics
Type NM
Type AC
Type MC
Protection
Residential
Dwelling units
Stairs
Access
Guard strips
Running boards

NEC References

320.15, 320.23, 320.30,
330.10, 330.23, 330.30,
334.15, 334.23, 334.30

Related Questions

- What type of protection is required for cables in attics?
- In what area of an attic is protection required for cables?
- What types of cables require protection in attics?
- When is protection for cables required throughout the attic?
- What type of support or protection is required for cables?
- Must cable runs in attics be fastened in place?
- Are the protection requirements for attics and unfinished basements different?

Answer

Type NM (nonmetallic-sheathed cable), Type AC (armored cable), and Type MC (metal-clad cable) are three types of cables that are frequently used in dwelling units. The three cable types have significant differences in their construction and, of the three, Type NM is most commonly associated with "residential wiring." However, the ways the cables are installed and used in dwellings are quite similar. Also, the uses of attics and roof spaces are fairly consistent in dwellings. Therefore, the requirements for the installation and protection of Types NM, AC, and MC cables in attics are the same. In fact, the rules spelled out for Type AC in Section 320.23 are used for the other types, as required by Section 330.23 for Type MC and Section 334.23 for Type NM.

According to a dictionary, an attic is the space directly below a roof and above other habitable areas. Some attics have little or no access and are not convenient for storage, but attics that have convenient access are very likely to be used for storage or other purposes, and protection of cables in those spaces is important. For this reason, only accessible attics and roof spaces are covered by the rules in Section 320.23.

Section 320.23 contains two types of rules. First, Section 320.23(A) covers cables run either across the top of floor joists or across the face of rafters or studding and within 7 feet of the floor. In either case, the cables are subject to physical damage, so guard strips are required to protect the cables in these locations. Physical damage may result directly from user contact or from use of the space for storage. Second, Section 320.23(B) covers cables run parallel to the sides of rafters, studs, and floor joists, but not across their faces. These cables are less exposed and susceptible to damage, so only the general rules of 300.4(D) apply. Section 300.4(D) is primarily concerned with spacing of cables back from the faces of framing members to reduce the chance of damage by nails and screws.

As noted, the protection rules discussed here apply only in accessible attics. In general, according to the initial language of 320.23(A), the rules apply throughout an attic if the attic is accessible. However, if the attic or roof space is "not accessible by permanent stairs or ladders, protection shall only be required within 1.8 m (6 ft) of the nearest edge of the scuttle hole or attic entrance." This rule does not specify that the permanent stairs or ladder must lead directly to the attic. The issue is really ease of access. For example, an attic space may be part of a level that also includes finished spaces. The attic access may well be a door or similar entry from a level that is served by a permanent stairway. In such cases, no portable ladder is required to gain access. Where the form and location of the attic access facilitate the use of that space for storage or similar purposes, cable protection is required in all accessible portions of the attic, not just in the 6-foot space around the entry. These applications are illustrated in Figure 3.18(a).

An attic may well include portions that are accessible, such as the unobstructed central area with a high roof, and portions that are inaccessible, such as the areas near the eaves of a roof with a shallow slope. The specified protection methods do not apply to those areas that are inaccessible. However, the judgment of which areas may be accessible and which are not is the responsibility of the authority having jurisdiction.

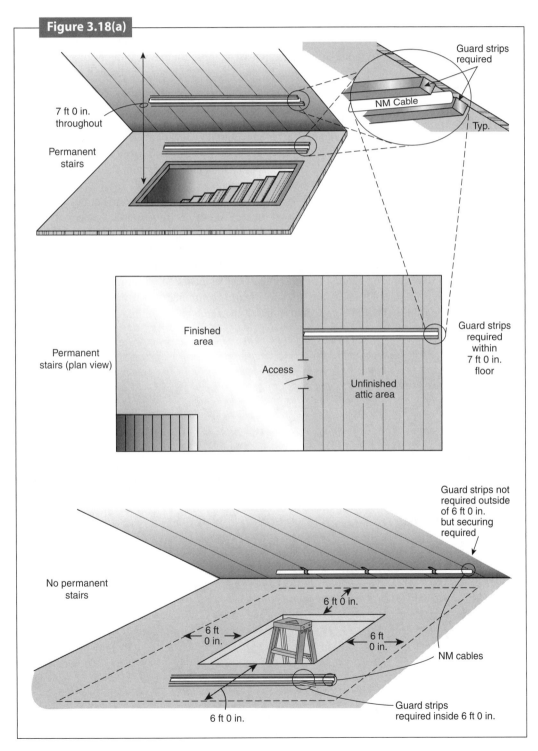

Figure 3.18(a)

Three attic spaces accessible by permanent stairs or by ladders, showing areas where cable protection is required.

Another point should be made in this context. The requirements for securing and supporting are not relaxed in attics in general unless the only method of installation is fishing in concealed spaces where supporting is impracticable. (Impracticable means "cannot be done in practice.") The precise language of the rules permitting unsupported cables varies slightly in Sections 320.30(B), 330.30(B), and 334.30(B), but the effect is essentially the same in new construction, because most areas are open during new construction and fishing is less likely to be necessary. The general rules of 320.30, 330.30, and 334.30 apply in attics in new construction. Cables run through attics in new construction should be supported and secured in place without regard to the eventual accessibility of the attic space if installing such secure support is practicable at the time of installation, as it usually is.

Although the rules for protection of NM, MC, and AC cables in attics are the same, these rules differ somewhat for other unfinished areas. The most common examples of such an area are basements and crawl spaces. Essentially, these are areas where cables will be "exposed" (as applied to wiring methods) as defined in Article 100. There are no special rules for installations of Type MC cable in basements. The rule for MC cable is found in 330.10(A) and 330.10(A)(4), where it says that Type MC cable may be used where exposed or concealed if it is not subject to physical damage. Where subject to physical damage, Type MC cable must be protected, according to 330.12, but specific methods of protection are not mentioned. For Type AC or NM cables, 320.15 and 334.15 apply. Both sections require the cables to closely follow the surface of the building finish or of running boards. Section 334.15(B) lists wiring methods that may be used to protect Type NM cable from physical damage, but "other approved means" may also be used. Exposed runs of AC cable are permitted to be installed on the underside of joists if supported at each joist and not subject to physical damage, but similar methods are permitted for NM cable only if the cable contains at least two 6 AWG or three 8 AWG conductors. Smaller sizes of NM cable that are run across the bottoms of joists must be supported on running boards. An example of cables installed on running boards is shown in Figure 3.18(b). Crawl spaces are not specifically mentioned, but if the crawl space is accessible, the cables should be protected in much the same way as in basements or attics.

Type NM cable run across bottom of joists on running boards.

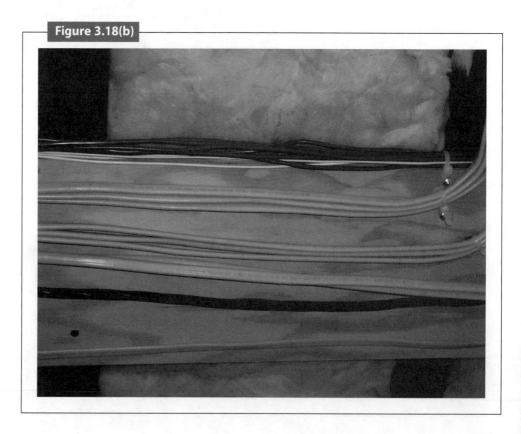

Figure 3.18(b)

Question 3.19

Do the extended support lengths for RMC apply where three-piece couplings or unions are used?

Related Question

- Can a 20-foot extended support spacing be used for size 4 RMC with three-piece malleable iron unions ("Ericksons") instead of standard steel conduit couplings?

Answer

Unions of the type described are called three-piece couplings in most catalogs, and are available in aluminum, steel, malleable iron, and as listed explosion-proof fittings. "Erickson" is a brand name. Three-piece couplings that are specifically intended for use on RMC (rigid metal conduit) or IMC (intermediate metal conduit) can be used in place of a standard (one-piece) threaded coupling for joining together lengths of rigid metal conduit. Figure 3.19 shows an example of a three-piece coupling. These fittings are commonly available in malleable iron or steel. They are usually used where the conduit cannot be turned into the fitting. Threadless couplings could also be used where the conduit cannot be turned to make the threads wrenchtight. However, when the conduit is supported in accordance with the extended support distances in Table 344.30(B)(2), the accompanying language in Section 344.30(B)(2) requires that the coupling device be threaded.

A listed three-piece union or coupling is a threaded device that has been evaluated in the same manner as a standard rigid metal conduit coupling for mechanical strength and grounding continuity. Keep in mind that although you may judge a malleable iron coupling to be less strong than the rigid steel conduit you are using, the rules apply to all types of rigid metal conduit, so these special support rules may also be used with aluminum, silicon-bronze, or stainless steel RMC, as well as to listed couplings made of various materials. An additional requirement is that the supports be designed and located so as to "prevent transmission of stresses to termination where conduit is deflected between supports." It should be recognized

Keywords

RMC
Union
Coupling
Support
Threadless
Threaded
Termination

NEC References

344.30

Figure 3.19

A three-piece coupling for use with threaded conduit where the conduit cannot be turned into the fitting.

Source: *NEC Handbook*, 2005, Exhibit 344.4. (Courtesy of Appleton Electric Company, EGS Electrical Group.)

that couplings are not conduit terminations. In fact, 344.30(A) requires support within 3 feet (5 feet in special cases) of terminations regardless of conduit trade size. The idea is to provide supports that are adequate to prevent the conduit from being supported primarily by the termination fittings.

Question 3.20 Are aluminum conduits and fittings permitted to connect to steel enclosures?

Keywords

Dissimilar metals
Aluminum
Steel
Conduit
Fitting
Enclosure

NEC References

110.14, 342.14, 344.14, 358.12

Related Questions

- Are steel conduits permitted to be used with aluminum fittings?
- Do requirements for wiring terminations with dissimilar metals apply also to different metals used for raceways or enclosures?

Answer

Section 344.14 addresses dissimilar metals in rigid metal conduit (RMC) installations. This section specifically permits the use of aluminum fittings and enclosures with steel RMC and permits steel fittings and enclosures to be used with aluminum RMC. The question does not specify the type of conduit being used, but permission can be found in Section 342.14 for the use of aluminum fittings with intermediate metal conduit (IMC), which is a steel product. Also, in 358.12(6), Exception, aluminum fittings and enclosures are permitted for use with electrical metallic tubing (EMT) "where not subject to severe corrosive influences." Based on the *NEC*'s permissions to use aluminum fittings with steel conduit and aluminum conduit with steel fittings, it is reasonable to conclude that an aluminum fitting is also acceptable for connecting conduit to a steel wireway. The Underwriters Laboratories (UL) guide card information on conduit fittings does not contain any restriction on connecting aluminum conduit fittings to steel enclosures. The key to the use of such fittings is to identify any limitations on their use that may be marked on the product carton. Many commonly used conduit bodies and weatherproof cast boxes are made of aluminum alloys, as are many types of enclosures and fittings for use in hazardous (classified) areas. Figure 3.20 shows examples of aluminum fittings that are often used with steel conduit.

The question mentions the difference between the rules for conductors and the rules for wiring methods with regard to dissimilar metals. The reference to conductors and electrical connections is found in 110.14, where the devices are expected to be carrying current normally and where galvanic action will be more likely, more severe, more destructive, and more dangerous than for contacts between dissimilar metals where current is not normally expected. Also, the likelihood of corrosive effects due to galvanic action depends on the relative positions of dissimilar metals in the galvanic series. The galvanic series arranges metals and alloys in a list that ranges from most likely to corrode to least likely to corrode, or vice versa. The farther apart two metals are in the series, the more likely it is that galvanic action will take place. Steel and aluminum are nearby neighbors in this series, and therefore galvanic action is much less likely in this case than it is for copper and aluminum, because copper and aluminum are widely separated in the series. (More information on this subject is widely available in engineering handbooks and textbooks on properties of materials or corrosion, as well as in ASTM standards and on the Internet.)

Figure 3.20 Examples of listed aluminum fittings that may be used with aluminum, steel, or other types of conduit.

Question 3.21

In what occupancies are cable trays permitted?

Related Questions

- Are cable trays permitted in other than industrial occupancies?
- Are there any restrictions on how a cable tray may be used in a specific occupancy?
- Are cable trays permitted in air-handling spaces?
- Is cable installed in cable tray required to be listed and labeled for cable tray use?
- What is an industrial occupancy?

Keywords

Cable tray
Industrial
Occupancy
Single conductors
Cables
Type TC
Type PLTC

NEC References

300.22, 334.10, 392.3, 392.4, 725.61, *Life Safety Code*

Answer

According to Section 392.3, "cable tray installations shall not be limited to industrial establishments." Wiring methods listed in Table 392.3(A) are permitted in any occupancy, subject to any specific limitations placed on the wiring method by the referenced articles for the wiring method. However, single conductors and Type MV (medium-voltage) cables are limited to industrial occupancies under 392.3(B).

The authority having jurisdiction is responsible for classifying the occupancy type based on the applicable building code or NFPA 101, *Life Safety Code*. The *Life Safety Code* defines an industrial occupancy as "an occupancy in which products are manufactured or in which processing, assembling, mixing, packaging, finishing, decorating, or repair operations are conducted." The *Life Safety Code* appendix includes the following as examples of industrial occupancies: dry cleaning plants, factories of all kinds, food processing plants, gas plants, hangars (for servicing/maintenance), laundries, power plants, pumping stations, refineries, sawmills, and telephone exchanges. The following commentary is also included in the appendix: "In evaluating the appropriate classification of laboratories, the authority having jurisdiction should treat each case individually based on the extent and nature of the associated hazards. Some laboratories are classified as occupancies other than industrial; for example, a physical therapy laboratory or a computer laboratory." Buildings such as convention centers are typically classified as assembly-type occupancies, not industrial occupancies, so size is not the primary factor.

Type TC (tray cable) cable is not the only cable wiring system permitted to be installed in cable trays. As noted previously, Section 392.3(A) lists the wiring methods that can be installed in a cable tray. One must also review the "uses permitted" in the article applicable to the cable type or other wiring method to determine if any are specifically permitted or restricted from being installed in a cable tray. Type NM (nonmetallic) cable, for example, is permitted in a cable tray provided that the cable is identified for that use (see Section 334.10(4)). For another example, remote control and signaling cables used in cable trays must be Type PLTC (power-limited tray cable) if installed outdoors, but ordinary cable types that are not specifically identified for cable trays may be used indoors, according to 725.61(C).

Section 392.4 prohibits the use of cable tray in environmental air spaces unless the wiring method that the tray supports is suitable for use in the environmental air space. To comply with 300.22, the cable tray would also have to be metallic unless there is a nonmetallic tray that is identified for that specific application (a fiberglass tray may be an example). Certain types of cable trays have been specifically designed for data and communications wiring installed in suspended ceiling spaces, and those uses are permitted as long as the cables are also suitable for the location. Figure 3.21 illustrates some of the uses and restrictions on the use of cable trays according to the type of occupancy and the contents of the tray.

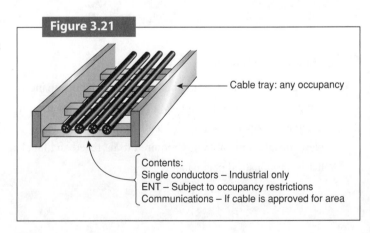

Use of cable trays depends on the contents of the cable tray.

Figure 3.21

Question 3.22

What type of single conductors or multiconductor cables may be used in cable tray?

Related Questions

- When are single conductors permitted in cable tray?
- Is it permitted to put single conductors of size 12 and 16 AWG in a cable tray with power cables?
- Are all conductors or cables used in cable trays required to be listed or marked for installation in a cable tray?
- How are cables identified for use in cable trays?
- What is the minimum size of multiconductor cable that can be used in a cable tray?

Keywords

Cable tray
Tray cable
Type TC
Type PLTC
Listed
Marked
Identified
Sunlight resistant
Multiconductor cable

Answer

Section 392.3 says that cable tray installations are not limited to industrial installations. However, the wiring methods that may be used are variously restricted, either in 392.3 or in the specific articles listed in Table 392.3(A). According to 392.3(A), the wiring methods in the table may only be used "under the conditions described in their respective articles and sections." Table 392.3(A) does not include any single conductors. Single conductors, even those used as equipment grounding conductors, are only permitted in 392.3(B) for industrial occupancies. (Short bonding jumpers are permitted to take the form of single conductors to comply with 392.6.)

Section 392.3(B)(1) reads as follows:

Single conductor cables shall be permitted to be installed in accordance with (B)(1)(a) through (B)(1)(c).

(a) Single conductor cable shall be 1/0 AWG or larger and shall be of a type listed and marked on the surface for use in cable trays. Where 1/0 AWG through 4/0 AWG single conductor cables are installed in ladder cable tray, the maximum allowable rung spacing for the ladder cable tray shall be 230 mm (9 in.).
(b) Welding cables shall comply with the provisions of Article 630, Part IV.
(c) Single conductors used as equipment grounding conductors shall be insulated, covered, or bare and they shall be 4 AWG or larger.

Thus, single conductors must generally be 1/0 AWG or larger, and single conductors smaller than 4 AWG are not permitted in cable trays for any purpose or any occupancy. Where small conductors are to be installed in cable trays, they must be in a multiconductor cable or one of the other wiring methods permitted in Table 392.3(A). The table includes many types of multiconductor cables. Sections 392.6(E) and (F) restrict the types of cables that may be mixed together in the same cable tray, and 392.9 provides fill limits for various types and mixtures of cables. The types of cables that may be mixed, the separations required, and the methods of providing separations are covered by certain other articles, such as Article 760 for fire alarm cables, Article 725 for remote control and signaling cables, and Article 800 for communications cables. (See Sections 725.26, 725.55, 760.55, and 800.133.)

NEC References

310.8, 392.3, 392.6, 392.9, 330.10, 334.10, 336.10, 725.61, 800.133

Single conductors in cable trays are limited to industrial occupancies so that the cables will only be used under the higher degree of maintenance and supervision that is usually associated with these occupancies. A cable tray is a support system for other cable and raceway methods. Where single conductors are used in a cable tray, the tray becomes more like a wiring method than a support system. This exposes the conductors to a higher potential for physical damage than if the conductors were installed in a raceway or as part of a cable assembly. The idea is that in an industrial facility more control will be exercised over who works on, services, or exposes the conductors to damage than would be likely in commercial or institutional occupancies. (Classification of the occupancy is a function of the building code and the authority having jurisdiction.)

The articles and sections listed in Table 392.3(A) may also require that cables used in cable trays be of certain types or have certain markings. As noted earlier, where single conductors are permitted they must be marked for use in cable trays. Article 334, Section 334.10 allows Type NM cable to be installed in cable trays, but only where the cables are identified for that use. Section 330.10 permits Type MC (metal-clad) cable in cable trays without any additional markings unless additional markings or construction characteristics are required for the location, such as for a wet location, a hazardous location, or outdoors. Section 336.10 obviously permits Type TC (tray cable) to be installed in cable trays, but also permits its installation in raceways, or outside raceways or cable trays in certain circumstances. Sections 725.61(C) and 800.133(B) simply list the types of cables that may be installed in cable trays; these cables are not required to be additionally marked as suitable for cable tray use. Section 725.61(C) requires Type PLTC (power-limited tray cable) in outdoor applications but permits other types indoors.

The smallest types of conductors and cables that are commonly used are those for remote control, signaling, and communications applications. The *NEC* does not specify minimum sizes for these cable types. Minimum sizes are mandated by the product standards used for testing and listing the cables. The minimum size is not necessarily specified by size, but is often determined by other tests, such as a test of tensile strength. Typically, the smallest conductor size in cables for Class 2, Class 3, or communications circuits is 24 AWG, but, based on the tensile strength of the multiconductor cable, the size may be as small as 30 AWG.

Where an additional marking is required for cable tray use, the cable must be marked "for cable tray use," "for CT use," or "for use in cable trays." These markings are not required on Type TC cable or Type PLTC cable, but other additional markings may be required. For example, regardless of occupancy, Section 392.3 requires that insulated conductors and jacketed cables exposed to direct rays of the sun be *identified* as sunlight resistant. This is a slightly more specific requirement than the more general rule found in 310.8(D) for outdoor installations, which requires that "insulated conductors and cables used where exposed to direct rays of the sun shall be of a type listed for sunlight resistance or listed and marked 'sunlight resistant.'" The requirement for listing (and identification) is concerned with the fact that many plastics will deteriorate if exposed to ultraviolet radiation, and additives or special plastics are required to resist such damage. According to 725.82(E), sunlight and moisture resistance is a basic requirement for Type PLTC cable, but 336.12(3) says Type TC may not be used where exposed to sunlight unless identified for the use. This is consistent with Underwriters Laboratories (UL) marking guides that say Type PLTC is sunlight resistant, so additional markings to that effect are optional; however, Type TC must have additional markings that say "sunlight resistant" or "sun res" if it is to be exposed to the sun.

Figure 3.22 summarizes some of the uses, restrictions, and marking requirements for cables in cable trays.

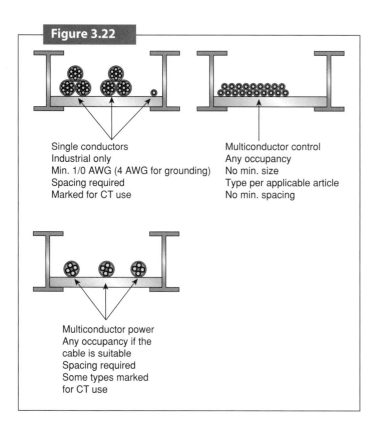

Uses of single conductors and multiconductor cables in cable trays.

Chapter 4

Equipment for General Use

Question 4.1

Are green or white conductors in flexible cord permitted to be re-identified?

Keywords

Flexible cord
Grounded conductor
Equipment grounding conductor
Identified
Re-identified
Green
White

NEC References

200.6, 200.7, 400.2, 400.7, 400.8, 400.22, 400.23

Related Questions

- May a white insulated conductor in flexible cord be used for an ungrounded phase conductor if the conductor is marked at all terminals with colored tape on the surface?
- May a green insulated conductor in a flexible cord be used for a phase conductor if the conductor is re-identified at all terminals?

Answer

Sections 400.22 and 400.23 of the *National Electrical Code* provide requirements for identification of grounded conductors and equipment grounding conductors in flexible cords. These special rules for grounded conductors in flexible cords are also recognized in 200.6(C). Section 400.22(C) states that white insulation is a means of identifying the grounded conductor, and 400.23 states that a green insulated conductor shall not be used for other than equipment grounding purposes. Neither of these sections specifically permits remarking of the grounded conductor or the equipment grounded conductor for purposes other than that stated within these sections. In fact, in the case of 400.23, the rule states that a green insulated conductor *may not* be used for any purpose except equipment grounding. However, Section 400.22, which covers grounded conductor identification, is not so definite. It only says how a grounded conductor is to be identified, and does not say such a conductor may only be used as a grounded conductor. This is because other rules of the *NEC* apply as well.

Section 400.2 says other articles of the code apply to flexible cords, so we may also apply the rules found in Article 200, titled "Use and Identification of Grounded Conductors." In general, Section 200.6 says grounded conductors are to be identified by white, gray, or three white stripes, and Section 200.7 says conductors identified in these ways shall be used only for grounded conductors. However, Section 200.7(C)(3) provides another special rule for flexible cords. The special rule says that the white or gray conductor, the conductor with three white stripes, or the conductor that is otherwise identified in accordance with 400.22 as a grounded conductor in a flexible cord may be used as an ungrounded conductor ("other than a grounded conductor") for connecting equipment or appliances as permitted in 400.7. This means that if the flexible cord is used in a permitted manner in accordance with 400.7, "Uses Permitted" (and, of course, not in violation of 400.8, "Uses Not Permitted"), the white conductor may be used as an ungrounded "phase" conductor. No re-identification is required in this case. The white conductor will be connected on each end to ungrounded conductors or terminals intended for ungrounded conductors, as shown in Figure 4.1, and those terminations provide adequate identification.

For example, let's say we want to connect a three-phase motor with flexible cord in order to facilitate frequent interchange or to prevent the transmission of noise or vibration, and we choose to use a four-conductor flexible cord. The colors of the conductors in the cord are likely to be black, red, white, and green, as shown in Figure 4.1. The green conductor may only be used for equipment grounding, but the white conductor, along with the red and black conductors, may be used for the three ungrounded conductors. You may choose to re-identify the white conductor with, say, blue tape, but re-identification is not required.

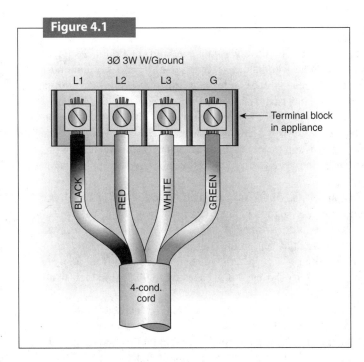

White conductor in flexible cord is permitted by 200.7(C)(3) to be used as an ungrounded conductor without re-identification.

Where can I find information on the proper and improper uses of flexible cords and extension cords?

Question 4.2

Related Questions

- May I run flexible cords through holes in cabinets or desktops?
- Are flexible cords permitted to run through holes in a raised floor?
- Are flexible cords permitted for permanent wiring to vibrating pumps?
- May I use flexible cords for permanent wiring to movable machines?
- Is a cord permitted to extend power to a garage door opener?
- What rules in the *NEC* cover replacement of cords on listed equipment?
- Where can I find rules relating to paralleling of flexible cords?

Keywords

Flexible cord
Extension cord
Cord set
Listed
Portable
Permanent wiring
Parallel conductors

NEC References

110.3, 240.5, 400.7, 400.8, 501.140, Article 590

Answer

The primary rules for the use of flexible cords are found in Sections 400.7 and 400.8. Section 400.7 covers the permitted uses, and Section 400.8 covers the uses that are not permitted. Obviously, an application of flexible cord must be permitted by 400.7 and not be prohibited by 400.8. Many misapplications of cords are based only on finding a permitted use, without consulting the prohibited uses. Some of the most common violations involve fastening a cord in place or otherwise attaching it to a building or running cords through walls, ceilings, floors, or through windows or doors.

Although flexible cord may not be run through walls, ceilings, or floors, Section 400.8 does not say anything about holes in desks or cabinets. Section 400.14 does require bushings or similar fittings where the cord passes "through holes in covers, outlet boxes, or similar enclosures." This section also does not mention desks or cabinets. Certainly a very general rule like that found in 300.4 applies. This rule requires all conductors to be adequately protected where subject to physical damage. This is the primary concern where cord is run through a desk or countertop, as is often the case in offices and banks. Many commercial furniture systems are available with cord grommets specifically for this purpose. An example is shown in Figure 4.2.

Many of the provisions of 400.7 and 400.8 are modified for specific types of equipment or occupancies. For example, in an information technology equipment room that meets all of the requirements of 645.2 (Article 645 does not apply otherwise), flexible cords with attachment plugs may be up to 15 feet long, may run on the floor surface if properly protected, and may run through openings in the floor if the openings provide protection from abrasion according to 645.5. This article can modify Article 400, according to 90.3, so permission to do this is not needed in 400.7.

Section 400.7 allows flexible cord for portable equipment, to facilitate frequent interchange, to prevent transmission of noise or vibration, and for connection of moving parts. These permissions apply to such things as portable and movable pumps and motors. In one case, Section 501.140(A)(3) permits an electric submersible pump to be connected with flexible cord. Although these pumps are essentially stationary and vibration is usually not the

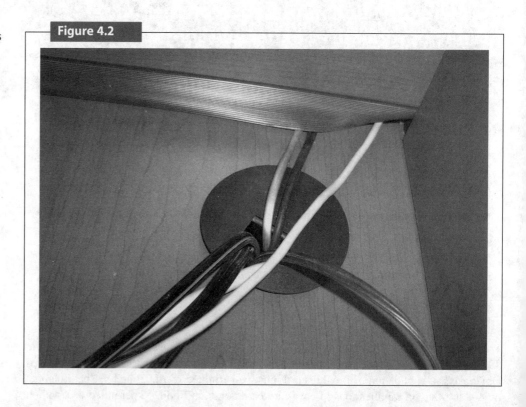

Figure 4.2

Flexible cord and other cables run through a grommet in a desktop in commercial office furniture.

issue, they are considered to be portable for the purposes of 501.140, and the cord is even permitted to be installed in a raceway between the wet pit and the power source, an application that would normally be a violation of 400.8(6). Again, Chapter 5 has the effect of modifying Chapter 4, but in this case, this is an example of where it is "otherwise permitted in this *Code*," as mentioned in 400.8(6).

Extension cords, or "cord sets," as they are usually called in the *NEC* and listing standards, are intended primarily for temporary use to extend the cord of an appliance or tool to a branch circuit receptacle outlet. They are not permitted to take the place of permanent wiring of a structure and are not permitted as an extension of the branch circuit. Permanent uses are more restricted by Article 400 than temporary uses. Temporary wiring is covered by Article 590. Under that article, flexible cords and cord sets are permitted for use in temporary branch circuits and feeders, but only for a limited time—typically 90 days or for the duration of a construction project. Extension cords that are left in place for long periods are usually in violation of the *NEC*. For example, cords used for permanent decorative holiday lighting should be replaced with permanent wiring methods. Similarly, a cord may not be used to extend power to a new garage door opener where there is no receptacle in an appropriate location. Extensions of permanent wiring methods to an appropriate outlet location should be used in such cases.

The issues here are not just related to the use of cords, where the cord is not permitted as a substitute for fixed wiring and is not permitted to be attached to the building. In most occupancies, 210.50 also applies, which requires that a receptacle outlet be installed wherever flexible cords with attachment plugs are used, and, in dwelling units, requires that the receptacle for a specific appliance be located within 6 feet of the intended location of that appliance. Where all surfaces are finished and an appliance is added, there may be a good application for exposed wiring of some type, including surface raceways, but it is not a permitted application of flexible cord.

Cord sets must also be protected from overcurrent in accordance with Section 240.5. A cord may not be used to supply a load in excess of its rating, but overcurrent protection is not always required to be selected to limit current to the rating of the cord. For example, a field-assembled extension cord set as small as 16 AWG is permitted to be protected by a 20-ampere overcurrent device.

Of course, as with all listed equipment, an extension cord or cord set must also be used in accordance with its listing and any instructions or markings that come with the product. This rule is found in Section 110.3(B) and is reiterated in 240.5(B)(3) for listed extension cord sets. For example, cord sets used or stored outdoors must be labeled for that use, and cords should have all wrappings removed, be extended to their length, and not left coiled or hanked as they came from the manufacturer. Replacements or repairs of cords must be in accordance with the manufacturer's listing and instructions in order to avoid violations that would nullify the listing. Repairs of some cord types are permitted by 400.9, providing the cords are of the hard-service or junior hard-service type, the conductors are 14 AWG or larger, and the characteristics of the cord are restored by the splicing method.

Where Article 400 does not provide specific information on an application of flexible cord, 400.2 says the "applicable provisions of other articles of this *Code*" apply. If, for example, an application of portable, movable, or vibrating equipment required that flexible cord be connected in parallel, the provisions of 310.4 would apply. This section would require that all paralleled conductors be the same length, the same size, the same material, have the same insulation type, and be terminated in the same manner, and the conductors could not be smaller than 1/0 AWG. There are exceptions to the size requirement, but the exceptions apply primarily to control equipment and elevators.

Question 4.3: Is marking required for GFCI-protected receptacles in new installations?

Keywords

GFCI
Replacement
Receptacle
Marking
Grounded
Grounding type
Nongrounding type

NEC References

406.3

Related Questions

- Where should the "No Equipment Ground" marking be used on receptacles?
- When are receptacles protected by upstream GFCIs required to be marked "GFCI Protected"?

Answer

The labeling requirement referenced in the question is found in Section 406.3(D)(3). It applies to replacement receptacles only. According to this section, when a nongrounded outlet is replaced with a grounded outlet protected by a ground-fault circuit interrupter (GFCI), the new outlets and all new downstream outlets shall be marked "No Equipment Ground" and "GFCI Protected." The replacements it applies to are only those where a grounding-type GFCI receptacle or a grounding-type receptacle protected by a GFCI is used because no grounding means exists at the outlet, as shown in Figure 4.3(a). The marking "No Equipment Ground" applies to all such replacements unless another nongrounding-type receptacle is used. The "GFCI Protected" marking only applies to replacement receptacles supplied through an upstream GFCI. Other receptacles protected by GFCIs are permitted but not required to have the "GFCI Protected" label. Markings on new grounded GFCI-protected receptacles as shown in Figure 4.3(b) are very helpful to users and inspectors, but the markings are not required.

Since all new receptacles are required by 406.3(A) and (B) to be grounding type and to be grounded, the "No Equipment Ground" label should have no other use except for ungrounded replacements. Existing ungrounded receptacles may also be replaced with new ungrounded receptacles in some cases, but the lack of an equipment ground is obvious in those cases. Where a grounded receptacle is used without a grounding means, the required GFCI will provide shock protection for persons using that receptacle. However, the user must be

Requirements for marking of grounding-type receptacles used for replacements.

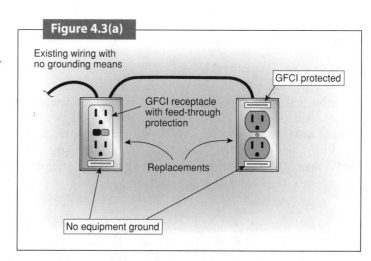

Figure 4.3(a)

Figure 4.3(b) New grounding-type receptacle marked as GFCI protected but not required to be so marked.

warned when there is no equipment ground because the ungrounded GFCI will not provide the function of an equipment ground that may be necessary for proper operation and protection of equipment. For example, equipment grounding conductors are often used for signal reference in certain types of equipment and may be used to protect the equipment from damage due to static electricity.

Question 4.4

When is a service panel required to have a main overcurrent device?

Related Questions

- May a panel with two to six main disconnects be used for service equipment?
- May the sum of the main overcurrent devices exceed the rating of the service equipment and service conductors?
- When is overcurrent protection required ahead of a panelboard?
- What is the difference between a lighting and appliance panelboard and a power panelboard?

Keywords

Panelboard
Power panelboard
Lighting and appliance branch-circuit panelboard
Overcurrent protection
Service disconnect

NEC References

215.2, 230.71, 230.90, 250.24, 408.34, 408.36

Answer

Section 230.71 permits a service to have up to six disconnecting means grouped together or in a single enclosure, usually a panelboard. This rule is most useful for larger services and is very useful in reducing the required rating of any one service disconnect. However, the rule may also apply to some smaller services, and the rule does not take into account all the requirements for protection of panelboards.

To provide an illustration, we will assume that the panelboard in question is listed as suitable for use as service equipment. The panel is rated 100 amperes, single phase, three wire, and is provided with six or more spaces for overcurrent devices. However, there are only two double-pole overcurrent devices installed in the panelboard: one 20 amperes and one 30 amperes. Each feeds three-wire branch circuits (two ungrounded conductors and one grounded conductor).

The first question that must be answered is: Is the panelboard a "power panelboard" or a "lighting and appliance branch-circuit panelboard" or something in between? Section 408.34 covers the classification of panelboards and answers this question.

The panelboard in our example falls within the parameters of a "lighting and appliance branch-circuit panelboard" because both circuits (the rule is 10 percent or more) supply lighting and appliance branch circuits. According to 408.34, lighting and appliance branch circuits are those branch circuits that include a neutral and are rated 30 amperes or less. This definition is illustrated in Figure 4.4(a), which also shows that other panelboards are called "power panelboards."

Section 408.36(A) requires a lighting and appliance branch-circuit panelboard to be protected by not more than two sets of fuses or circuit breakers whose combined rating (the sum of the ratings) is not more than the rating of the panelboard. These overcurrent devices are for the purpose of protecting the bus in the panelboard. The required overcurrent protection can be located within the panelboard or in the supply conductors to the panelboard.

In our example, the two circuit breakers cannot be used to provide the required protection. The two circuit breakers are connected to the panelboard bus, but they do not provide overcurrent protection for the bus. The purpose of the two circuit breakers is to protect the branch circuits. Although the two circuit breakers could provide some overload protection for the panelboard, they cannot also provide short-circuit and ground-fault protection for the

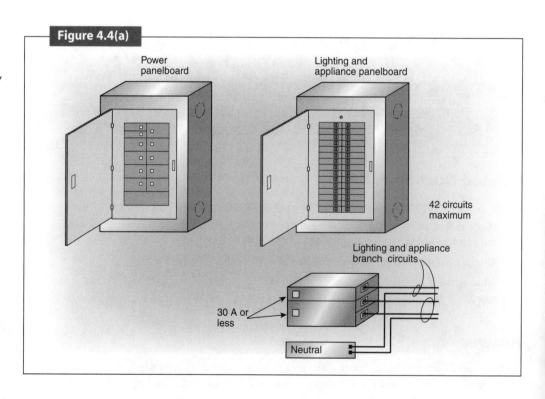

Illustration of panelboard definitions.

Source: 1999 *NEC Changes*, p. 221.

panelboard as required. Note that in this case, the two circuit breakers could be used to provide overcurrent protection for the service conductors because only overload protection is required by 230.90(A). However, the requirements for protection of lighting and appliance panelboards are separate and different, and this example does not comply with 408.36(A) because the overcurrent devices are not on the supply side of the panelboard.

If we modify our assumptions a bit, our answer may change. Let's say that the two breakers are both 50 amperes supplying three-wire feeders, but the additional spaces still remain. Now we have a power panelboard as shown in Figure 4.4(a), and overcurrent protection matching the rating of the panelboard is not required. The problem remains that additional overcurrent devices could be added, but as long as no more than six devices are installed, and all are over 30 amperes or all supply circuits without neutrals, no overcurrent device for the panelboard is required. Installations like this are permitted by Article 408 and Article 230.

Let us look at one other possibility with this same panel. In this case, assume that the panelboard includes two 30-ampere overcurrent devices from a system with a grounded neutral, but neither supplies a circuit that includes a neutral. However, there is still a neutral in the panel, as required by 250.24(B). Now we have a power panelboard because although 10 percent or more of the devices are rated 30 amperes or less, they are not lighting and appliance branch circuits because the branch circuits don't include neutrals. Generally, Section 408.16(B) would require a main overcurrent device for this panel, as shown in Figure 4.4(b), because of the small overcurrent devices and the presence of a neutral in the supply. But the exception to this rule exempts power panelboards with multiple disconnects that are used as service equipment.

A possible problem with this third scenario is that if a 30-ampere or smaller branch circuit is added and that circuit includes a neutral, then the panel becomes a lighting and appliance branch circuit panelboard, and protection for the panel is required. The addition of such a circuit would create the code violation in this case, so elimination of the circuit or addition of a feeder in its place would probably make more sense than requiring overcurrent protection for the panel. Again, 408.36(A), Exception No. 2, exempts panelboards used as service equipment, but only where they are existing in residential occupancies.

Typically, power panelboards are used in larger installations where more capacity is needed (more than the 100 amperes in the previous examples). In many such panels, there are no provisions for more than six overcurrent devices and there may be no provisions for small

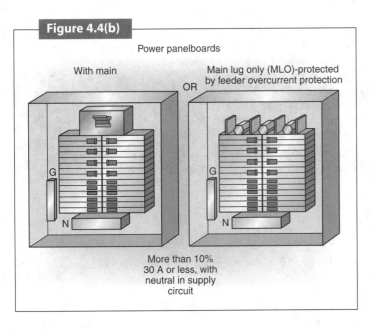

Figure 4.4(b)

Overcurrent protection is required for some power panelboards with neutrals and small overcurrent devices.

Source: 1999 *NEC Changes*, p. 223.

An example of a power panelboard used for service equipment where the sum of the ratings of the overcurrent devices is permitted to exceed the ampacity of the service conductors, and over-current protection is not required for the panelboard.

Source: Based on *NEC Handbook*, 2005, Exhibit 230.27.

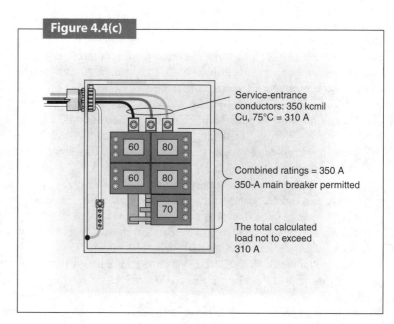

Figure 4.4(c)

overcurrent devices, so the panels cannot easily become lighting and appliance branch circuit panelboards. As shown in Figure 4.4(c), the sum of the service overcurrent devices is permitted to exceed the rating of both the panel and the service conductors as long as the calculated load does not exceed the rating of the panel or the service conductors. This information is found in Section 230.90, Exception No. 3.

In summary, in addition to any requirements for overcurrent protection of conductors or other equipment, some panelboards are required to be protected against overcurrent. Specifically, a panelboard will be required to have overcurrent protection if it is supplied by a feeder that includes a neutral, and if 10 percent or more of its overcurrent devices supply branch circuits rated 30 amperes or less. Some of these panelboards may be the lighting and appliance type, and some may be power panelboards. Different rules apply to panelboards used as service equipment. The rules are also different for panels that supply only feeders and for panels that supply branch circuits.

Question 4.5

Are luminaires permitted to be connected by a cord and plug?

Keywords

Luminaire
Cord
Cord and plug
Receptacle
Electric-discharge lighting

Related Questions

- Are luminaires permitted to be connected with cord but without plugs or receptacles?
- Is a plug with a NEMA-type configuration required for connecting electric-discharge lights?

Answer

Flexible cord is permitted for wiring of luminaires by 400.7(A)(2). This section does not always require an attachment plug for wiring luminaires. Section 410.30 also permits cord to be used for connecting luminaires, but is more specific about the requirements. According to 410.30(B), a hard-usage or extra-hard-usage cord may be used without an attachment plug if the luminaire requires adjustment or aiming after installation. Section 410.30(C)(1) also permits a cord to be used with a canopy and strain relief rather than an attachment plug cap. Most other applications of cord for luminaire connections, other than listed cord-pendant luminaires, require the use of an attachment cap or connector body as shown in Figure 4.5. Such receptacle-and-cord assemblies may not be used where concealed or where located above dropped or suspended ceiling in violation of 400.8.

A cord-connected luminaire with a connector body (female attachment cap) or an attachment plug (male plug cap) does not necessarily create a receptacle outlet for general use. The outlet is still a lighting outlet. In many cases, a cord, plug, and receptacle assembly is part of a complete listed fixture assembly rather than being partially composed of separate components as shown in Figure 4.5. Ordinarily, a receptacle is required to be of the same rating as the branch circuit in accordance with 210.21. However, when electric-discharge lighting is used with mogul-base screw-shell lampholders or flanged surface inlets, 410.30(C)(2), 410.30(C)(3), and Exception No. 2 of 210.21(B)(3) permit a receptacle with a rating lower than the rating of the branch circuit. For example, a group of high-bay HID (high-intensity discharge) luminaires may be connected to a 40-ampere branch circuit in accordance with 210.23(C), but the receptacles may have ampere ratings of only 15 or 20 amperes. Each receptacle (and plug) rating must be at least 125 percent of the full-load current rating of the luminaire. The cord or internal fixture wiring may require supplementary overcurrent protection to comply with 240.5.

When cord-and-plug connections are provided by the luminaire manufacturer, the plug and receptacle configurations are often the same for various luminaire voltages, and the connected voltage may even be determined by the installer at the time of installation. These receptacle configurations are not the same as the National Electrical Manufacturers Association (NEMA) standard configurations for specific voltages and current ratings and are not interchangeable with those standard configurations. NEMA standard configurations may be used

NEC References

210.21, 210.23, 400.7, 400.8, 410.30

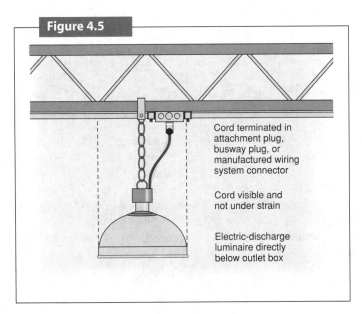

Figure 4.5

Illustration of a cord-and-plug-connected electric-discharge luminaire.

Source: Based on *NEC Handbook*, 2005, Exhibit 410.5.

Cord terminated in attachment plug, busway plug, or manufactured wiring system connector

Cord visible and not under strain

Electric-discharge luminaire directly below outlet box

within their ratings, but NEMA configurations are not required for the luminaire connections permitted in 410.30. For that matter, 410.30(C)(1)(2)(c) permits the use of a listed assembly of cord with a manufactured wiring system connector in lieu of the usual plug assembly. As noted, the use of an attachment plug and receptacle does not create a receptacle outlet in the usual sense, and receptacles suitable for other cord-and-plug-connected loads are not required and often are not desirable.

Question 4.6 Must a low-voltage lighting system come entirely from one manufacturer?

Keywords

Low voltage
Lighting
Wiring methods
Listing

NEC References

110.3, Article 411, 411.3, Chapter 3, Chapter 4, Chapter 7

Related Questions

- What is a "low-voltage lighting system"?
- What wiring methods are intended to be used with low-voltage lighting?
- Must the wire, transformer, luminaires, and all other components come from the same manufacturer?

Answer

As used in the *NEC* and Article 411, a "low-voltage lighting system" is one that operates at 30 volts or less. The *NEC* does not require all components of a low-voltage lighting system to come from the same manufacturer. In fact, many manufacturers of luminaires, track, or similar lighting system fittings and components do not also make transformers, switches, and wire or cable. Section 411.3 does require that a low-voltage lighting system be listed, and 110.3(B) requires that all instructions included with the listing or labeling be followed. The manufacturer's instructions should specify what components must be used to comply with the listing requirements. For example, many installation instructions for low-voltage lighting say that dimmers may not be used unless specifically listed for use with low-voltage lighting transformers, but such dimmers are often available only from other manufacturers. Figure 4.6 shows fixtures and fittings that come from one manufacturer used with lamps that are made by another manufacturer; the other parts (not shown), such as the supply wiring, the power supply, and switching devices, may also come from other manufacturers.

If we take Article 411 in the context of Chapter 4, we should expect that the article covers "Equipment for General Use," which is the title of Chapter 4. Article 411 does not say anything about the wiring methods used to connect a low-voltage lighting transformer to its source or to the lighting loads it supplies. Since no special rules are provided for wiring methods in Article 411, the regular rules of Chapter 3, "Wiring Methods and Materials," must be followed. Thus, in many residential applications, the wiring method may be Type NM (nonmetallic) cable; in other applications, perhaps conduit and wire, Type MC (metal-clad) or AC (armored) cable, or the method used for ordinary wiring in the area would be appropriate. In some cases, where the lighting load is very limited and can be (and is) supplied by a Class 2 power source, the wiring may be installed in accordance with Article 725.

Figure 4.6

Low-voltage (nominal 12 volt) under-cabinet lighting.

Source: *Limited Energy Systems*, 2002, Figure 2.1.

According to the Underwriters Laboratories (UL) *Electrical Construction Equipment* directory, power supplies are listed under a different standard than the luminaires or "fixtures" themselves. Also, the directory states that the listing for "Low Voltage Incandescent Fixtures and Fittings" covers "low voltage fixtures, and fixture fittings that are parts and/or subassemblies intended for final assembly into low voltage fixtures in the field." The directory goes on to say that the fixtures and fittings are rated for 30 V or less, "for connection to an isolating type power supply listed for the purpose and installed using fixed wiring methods." As noted previously, both the power supply and the lighting fixtures/fittings must be listed, but the power supplies are usually not listed under the same standard as the fixtures and fittings and might not be from the same manufacturer. Some manufacturers make only fixtures, some make only fittings, some make "systems," and some make all components, including power supplies. Obviously, a separate listed power supply must be provided for a manufacturer that supplies only under-cabinet fixture fittings and under-cabinet fixtures and does not make power supplies.

The fact that the system and wiring is "low voltage" does not mean that the cable types intended for remote control and signaling are appropriate. Class 2 cable, for example, is listed for use only on Class 2 circuits or on some audio circuits or intrinsically safe circuits for which Class 2 cable is specifically permitted. Low-voltage lighting may not use the special-purpose methods of Chapter 7, but must be wired using a wiring method from Chapter 3 unless there is a recommended product that is specifically listed for low-voltage lighting systems, or the lighting system power supply is listed as Class 2. (Most Class 2 power supplies could supply no more than one 50-watt lamp, but may be adequate for a few 5-watt lamps.) A special cable type suitable for direct burial and for outside use with low-voltage lighting—usually landscape lighting—is listed and available for those uses, but it is not intended for indoor use unless indoor use is included in the listing and instructions.

Question 4.7: Must all disconnects be readily accessible?

Keywords

Disconnect
Disconnecting means
Cord and plug
Attachment plug
Receptacle
Appliance
Accessible
Readily accessible
Unit switches

NEC References

422.16, 422.31, 422.33, 422.34, 430.102, 430.107, 430.109, 440.3

Related Questions

- Can a cord and plug be used for a disconnect if the cord and plug is not readily accessible?
- How do disconnecting requirements apply to appliances?

Answer

In general, at least one disconnect for any electrical equipment must be readily accessible. Other disconnects, such as a disconnect that is nearby or "in sight" as defined in Article 100, may not be required to be readily accessible. The specific rules and the *NEC* references that apply depend on the type of equipment, and in some cases, the occupancy type.

Consider a refrigerated display case in a supermarket or grocery store. Often, the refrigeration compressors are located in a machine room in the rear or on a mezzanine of the store. The display cases are located in the sales area of the store and contain only a lighting circuit and one circulation (evaporator) fan circuit per system, along with some control wiring. A switch to control the power to the lighting and a plug with a receptacle to disconnect each circulating fan motor is usually provided by the manufacturer. The cord-and-plug disconnects are accessible, but not readily accessible because they require the removal of panels for access. Store owners don't want these disconnects to be accessible to customers because inadvertent or malicious disconnection could result in considerable losses.

In this case, the compressors are covered by Article 440, but according to 440.3(B), Article 430 applies to evaporator-type refrigeration equipment or fan coil–type cases (refrigeration equipment using remote compressors). A disconnect for this type of equipment is required to meet the requirements of Article 430, Part IX. According to Section 430.109(F), a horsepower-rated attachment plug and receptacle is permitted to serve as the disconnect. Section 430.102 covers the location of the disconnect and requires a disconnect in sight of a motor in most cases, but does not say this "local" disconnect must be readily accessible. However, according to 430.107, "one of the disconnecting means shall be readily accessible." So, if there is another disconnecting means, such as the branch-circuit circuit breaker, that is readily accessible, the cord-and-plug connection used as a disconnect at a motor is not required to be readily accessible.

Similar rules apply to household appliances. Section 422.33 allows the use of a "separable connector or an attachment plug and receptacle" as a disconnecting means. However, where this disconnect is not accessible, another disconnect must be supplied that meets the requirements of 422.31. According to 422.31, if the appliance is not over 300 VA or 1/8 horsepower, the branch circuit overcurrent device is all that is required. However, for larger appliances, the branch circuit device must be in sight or lockable in the off position if it is to serve as the disconnecting means, according to 422.31(B). The locking means must remain in place even when a lock is not installed.

Note that the requirement for attachment plug and receptacle disconnects for appliances is being "accessible," not "readily accessible." Accessible (as applied to equipment) does not preclude moving an appliance, such as a refrigerator, to make the disconnect accessible. In the case of an electric range that is built in and not easily moved for service, 422.33 considers a receptacle and plug at the base of the range to be accessible from the front by removal of a drawer. Figure 4.7 shows an example of a range that is not built in and is easily moved

Figure 4.7 Accessible, but not readily accessible, cord-and-plug connection used as a disconnect for an electric range.

for service, with an accessible cord-and-plug connection at the back of the range. Dishwashers, trash compactors, and waste disposers are specifically permitted to be built in and connected by a cord and plug, but the receptacles are required to be accessible according to 422.16(B)(1)(4) and 422.16(B)(2)(5). These accessible receptacles, often made accessible by removing a toe-kick plate at the front of the appliance, meet the requirements of 422.33, so the branch circuit switch or circuit breaker is not required to be in sight or lockable.

Appliances with unit switches that have marked off positions are required to have additional disconnecting means, typically the branch circuit overcurrent device, according to 422.34. The form and location of the other required disconnecting means depends on the type of occupancy.

Question 4.8

Is a pool pump motor treated as an appliance or as a motor?

Related Question

- Are motor-operated appliances treated differently from motors?

Keywords

Appliance
Motor
Nameplate
Horsepower

Keywords

Table values
Load calculations

NEC References

220.14, Article 422, 430.6, 430.22, 430.24, Table 430.248

Answer

An appliance is defined in Article 100 as "utilization equipment, generally other than industrial, that is normally built in standardized sizes or types and is installed or connected as a unit to perform one or more functions such as clothes washing, air conditioning, food mixing, deep frying, and so forth."

Equipment such as a pool pump motor may be part of an appliance or may be just a separate motor. The answer to whether it is a motor or an appliance depends on whether it comes as an assembly and is listed as an appliance. An appliance has a nameplate with an ampere rating or wattage that is used for load calculations and conductor sizing. In some cases, an appliance may be marked with a minimum wire size or circuit rating. Motors that are separate from the pump assembly and not covered by an overall listing for the assembly also have nameplates with the information required by 430.7, including horsepower ratings, but they are treated as separate motors.

Appliances are subject to the requirements of Article 422, where determinations of ratings for circuit elements and load calculations are based on the nameplate. Nameplate information is used for sizing conductors for motor-operated appliances according to Section 430.6(A)(1), Exception No. 3. According to 430.6(A), conductor sizing and ratings of other circuit elements for separate motors are typically based on horsepower ratings and the full-load current values in Tables 430.247 through 430.250. "Table values" are also used for load calculations according to 220.14(C), which refers to 430.22, 430.24, and 440.6, whereas nameplate values are used for load calculations for appliances in accordance with 220.14(A), which refers to Article 422.

Some types of equipment may seem to fall into more than one category, such as a window air conditioner or a waste disposer. However, if the equipment is listed as an assembly, as is a waste disposer or an air conditioner, it is an appliance by definition, and load calcula-

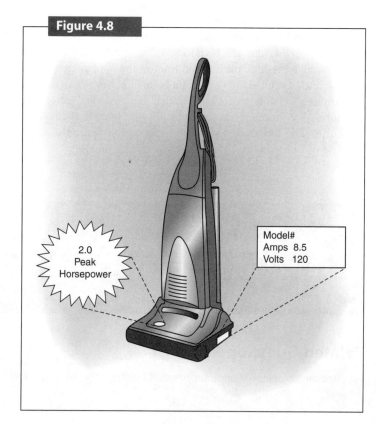

Use of appliance horsepower ratings and nameplate values.

Source: 1999 *NEC Changes*, p. 267.

tions or other circuit elements will be based on the nameplate information. Loads or circuit elements based on Article 430 or 440 calculations will usually not be significantly different from loads based on appliance nameplates, because manufacturers use similar or identical methods to determine what should be on the nameplate of an appliance. Nevertheless, appliances are intended to operate in a specific way under specific conditions, so where an appliance is used, the nameplate information is given precedence over a field calculation. Also, in many cases, the horsepower rating stated on an appliance may be a peak rating rather than a continuous rating and may be more useful for marketing purposes than for sizing circuits. In these cases, the nameplate ampere rating or power rating provides more accurate information about the load of an appliance.

For instance, Figure 4.8 shows an example of an appliance with an advertised "peak" horsepower rating and the nameplate information from the same appliance. According to Table 430.248, a 2-horsepower motor on a 120-volt system should be expected to draw up to 24 amperes, and circuit elements and load calculations would be required to be based on that value if it were a separate motor with a 2-horsepower rating. However, the nameplate shows a full-load current rating of 8.5 amperes, which means the vacuum motor is probably really closer to 1/2 horsepower than 2 horsepower. This portable appliance is intended to be connected to an ordinary 15- or 20-ampere branch circuit, and its load is accounted for either in the general lighting load or in the allowance assigned to receptacles in 220.14.

Question 4.9

What is the purpose of code letters on motors?

Related Questions

- Where can I find information on locked-rotor code letters for motors?
- How is a locked-rotor code letter used in the *NEC*?
- How are design letters used in the *NEC*?

Answer

Motors draw significantly more current when they are starting than they do while running. The maximum starting current can be determined by applying rated voltage to a motor while the rotor is prevented from turning, thus the name "locked rotor." This starting current is also called the "inrush current" or the "stalled rotor current" in other standards or contexts. A locked-rotor indicating code letter provides a way to estimate the starting current in any motor that is started across the line, that is, without reducing the voltage during starting. These code letters are separate designations from the letter designation of the design type, such as Designs A through D. Design B energy-efficient motors are currently the most common type for general-purpose applications. Figure 4.9 shows the approximate relationship between speed and current in typical motors with locked-rotor code letters C, G, and M.

Design letters refer to torque and starting current characteristics of motors as defined by the National Electrical Manufacturers Association (NEMA). Code letters are more direct representations of the amount of current required to start a motor by directly connecting it across

Keywords

Code letter
Locked rotor
Stalled rotor
Starting current
Inrush current
Energy-efficient motors
Disconnect rating
Controller rating

NEC References

430.6, 430.7, 430.52, 430.110, Tables 430.251(A) and (B), 440.12, 440.41, 455.8

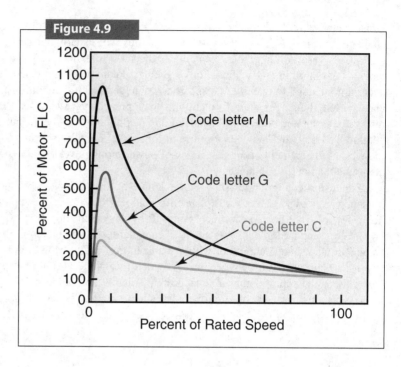

Figure 4.9 Approximate starting currents of AC motors. Source: *NEC* Seminar.

the line to a power source. Since the code letter provides a way to estimate the starting current of a motor, it also provides a way to estimate the voltage drop or other effects of starting a motor. Excessive voltage drop may cause problems with other connected equipment, perhaps causing lights to flicker, dim, or extinguish themselves. For acceptable performance, high starting currents may have to be lowered through the use of various types of reduced-voltage starters. In some cases, utilities may require large motors to be started using reduced-voltage methods. Generally, these are design considerations, and the *NEC* does not specify what voltage drop is permitted for general motor applications. However, the code letter becomes critical to the selection of overcurrent devices in fire pump circuits because 695.7 does specify a maximum voltage drop for a fire pump circuit during both starting and running conditions.

Up until 1996, the *NEC* used code letters in the selection of motor branch circuit, short-circuit, and ground-fault protective devices. Table 430.52 no longer uses code letters, although it does consider design letters. In this case, energy-efficient Design B motors are different from other motors only in the maximum rating permitted for an instantaneous-trip circuit breaker used in a combination controller-disconnect device.

Starting currents are also considered in the selection of disconnects and controllers for motor circuits. Special tables, Tables 430.251(A) and (B), are provided for the purpose of estimating the starting current, but the values in these tables are only for use with Sections 430.110, 440.12, and 455.8(C) for determining disconnect ratings when more than one motor is connected to the disconnect or with Section 440.41 for controller ratings for hermetic refrigerant motor compressors that do not have horsepower ratings. Design letters are used to estimate starting currents in Table 430.251(B), but these values are generalizations that are useful only for a specific purpose. Locked-rotor code letters provide a more accurate way to estimate starting currents.

Following is an example of the use of Table 430.7(B): Assume that a motor is rated 10 horsepower (HP), 460 volts, three phase, Design B, code letter G, with a nameplate full-load amperes (FLA) rating of 12 amperes. The full-load current (FLC) from Table 430.250 for sizing all circuit elements except overload devices is 14 amperes for this motor (see 430.6).

From Table 430.7(B), a motor with code letter G ranges from 5.6 to 6.29 kVA per horsepower with locked rotor. If we use the higher value in this range, the starting kVA for a 10-horsepower motor is 10 HP × 6.29 kVA/HP = 62.9 kVA or 62,900 VA.

With a system voltage of 480 volts, three phase, the approximate starting current is 62,900 VA divided by 480 V × 1.732, or 62,900 VA/831 V, which equals 75.7 A.

Thus, the 10-horsepower motor in this example takes 12 amperes to run at full load, but requires up to 76 amperes to start.

Question 4.10

How is overcurrent protection provided for motors and motor circuits?

Related Questions

- Where are the overcurrent protection requirements for motors found?
- How is a motor provided with overcurrent protection if the motor has a long starting time or an exceptionally high starting current?
- When can overload protection for motors be omitted?
- How are overcurrent requirements for motors different from other equipment?

Keywords

Overcurrent
Ground fault
Short circuit
Overload
Motor
Branch circuit

Answer

Overcurrent protection normally provides for sensing and interrupting of short circuits, ground faults, and overloads. The circuit breakers and fuses that are most commonly used for overcurrent protection of most feeders and branch circuits provide all three types of protection in one device. The same method may also be used for motors, if it will work. However, motors usually draw high currents when they are started. This high current is often called the "starting current," "inrush current," or "stalled-rotor current," but is called "locked-rotor current" in the *NEC*. Article 430 permits methods of providing overcurrent protection that will not operate on locked-rotor currents while still providing short-circuit, ground-fault, and overload protection for the motor and motor circuit conductors. Often, this is accomplished by using one device for overload protection and another device for short-circuit and ground-fault protection, as illustrated by Figure 4.10(a).

Overload protection opens a circuit at relatively low currents that are a little higher than the rating of the motor or conductors. Overload currents are likely to cause damage only if they persist for a few minutes or more, so the overload device may be a relatively slow-acting device. Overload devices are intended to prevent damage to the motor or motor circuit due to overloads and may be located anywhere in a circuit as long as they are in series with the entire load. A common location for the overload protection is as shown in Figure 4.10(a)—the last element before the conductors that connect directly to the motor terminals. The requirements for motor and motor circuit overload protection are found in Part III of Article 430, mostly in Section 430.32.

The device that provides short-circuit and ground-fault protection must operate very quickly on high-current faults, but must allow the locked-rotor current to continue long enough to allow the motor to start. Short-circuit and ground-fault protective devices cannot prevent a short circuit or ground fault from occurring in a circuit or motor. Instead, they react to the high current associated with a short circuit or ground fault that has occurred and then deenergize the damaged circuit. Thus, short-circuit and ground-fault protection can only be provided for circuits and equipment that are on the load side of the protective device. The re-

NEC References

240.4, 430.6, 430.31, 430.32, 430.51, 430.52, 430.53, 430.62, 430.63

Overcurrent protection for motor circuits often consists of more than one element: one for short-circuit and ground-fault protection, and another for overload protection.

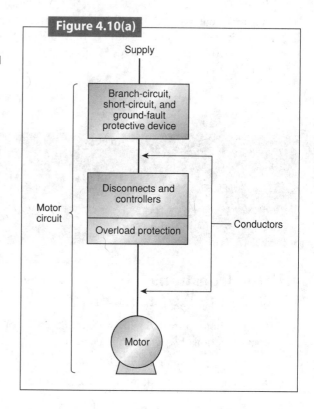

Figure 4.10(a)

quirements for short-circuit and ground-fault protection of motor circuits are found in Parts IV and V of Article 430. Part IV and Section 430.52 provide the main rules for motor branch circuits, and Part V and Sections 430.62 and 430.63 provide the rules for motor feeders. In order to protect all elements of a motor circuit, the short-circuit and ground-fault protective device must be at the beginning of the circuit, as shown in Figure 4.10(a).

According to Section 430.6, motor overload protection is the only element of a motor circuit that is typically sized based on nameplate current values. All other elements, such as short-circuit and ground-fault protection, disconnecting means, and conductors, are sized based on Tables 430.247 through 430.250 as applicable. The applications of FLA (nameplate) and FLC (table) values are summarized in Figure 4.10(b).

The objective of overload protection is "to protect motors, motor-control apparatus, and motor branch-circuit conductors against excessive heating due to motor overloads or failure to start," according to 430.31. This basic requirement and many special cases are covered in 430.32. In some cases, special methods of providing overload protection will be required in order to meet *NEC* requirements or to meet operational requirements. In other special cases, overload protection is permitted or required to be omitted. Large motors (over 1500 horsepower) are permitted to use embedded temperature protectors to detect and cause interruption of motor overloads. (These are often used in conjunction with other devices for better protection.) Motors that take a long time to accelerate, perhaps due to high-inertia loads, may trip overload devices on startup. In such cases, slower-acting devices may be used, or sometimes the overload device may be removed from the circuit during starting and reconnected after the motor is running, as permitted by 430.35. According to 430.32(D), some small motors (1 horsepower or less) may use the short-circuit and ground-fault protective device as overload protection. In reality, these applications use the operator as the overload protector because this permission only applies where the motor is not permanently installed, is manually started, and is in sight of the controller location, that is, in sight of the person manually starting the motor. In the case of fire pumps or other loads where the operation of a motor over-

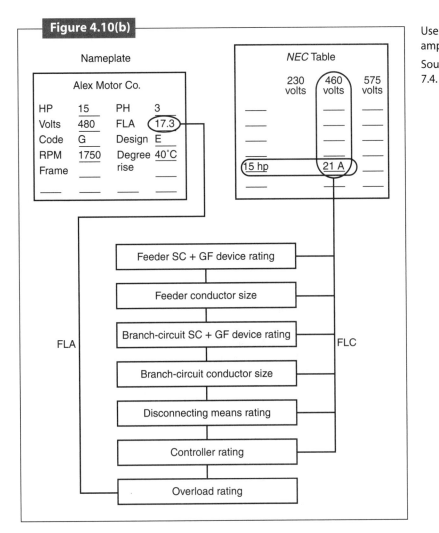

Figure 4.10(b)

Uses of full-load current (FLC) and full-load amperes (FLA) current values.

Source: *Electrical Inspection Manual*, 2005, Figure 7.4.

load device may actually increase hazards, overload protection is not required or permitted at all, and only short-circuit protection is permitted. (See 240.4(A), 430.31, and 695.6(D).)

Overload protection is commonly provided by an overload relay unless a motor is inherently protected. Overload relays may be solid-state devices with adjustable trip settings or "heaters" that are sized for a specific range of current values and that react to current by melting or by interacting with related temperature sensors. Section 430.32 provides maximum ratings for the overload devices. The maximum ratings are based on the service factor and temperature rise characteristics of the motor. For most motors, the maximum rating of the overload device is 125 percent of the nameplate full-load ampere rating. Although Section 430.32 provides these calculation values and other values for thermal protectors, a calculation usually should not be done and is not needed where a "heater" is selected from a table for the specific overload relay using nameplate current directly. The 125 percent (or 115 percent) from 430.32(A)(1) is incorporated in the table ranges provided. Where a solid-state adjustable relay is used, the setting should also be made directly from the motor nameplate value, and the percentage of that value at which the device should trip may be separately selected. The calculated value is used only where a device such as a fuse is used for overload protection.

Often, more than one form of overload protection is used. This may be done to provide better or redundant protection, as mentioned earlier for very large or very high-value motors. Different forms of overload protection may also be provided for different purposes. For example, in hazardous (classified) areas, the surface temperature of a motor may be a critical consideration. Some motors intended for operation in classified areas are supplied with integral thermal devices to limit surface temperatures, while overload relays are used for the more ordinary overload function.

As noted previously, short-circuit and ground-fault protection is installed at the beginning of a branch circuit. According to 430.51, this is "to protect the motor branch-circuit conductors, the motor control apparatus, and the motors against overcurrent due to short circuits or grounds." Section 430.52 has two basic requirements. First, according to 430.52(B), each short-circuit and ground-fault protective device is required to be "capable of carrying the starting current of the motor." This is simply a performance requirement—the motor must be able to start.

The second requirement is found in Section 430.52(C), which establishes maximum ratings for different types of devices based on the horsepower ratings and full-load current values from Tables 430.247 through 430.250. Maximum ratings may also be established by overload relay tables—essentially manufacturer's instructions for the use of the overload relays and controllers. Exceptions to 430.52(C) also allow rounding up to the next standard size or increasing ratings for special cases where motors are difficult to start without tripping the short-circuit and ground-fault device. Section 430.52(D) provides requirements for torque motors (motors such as some damper motors that are intended to operate in a locked-rotor condition, producing torque but not turning), and 430.53 provides for situations where more than one motor may be connected to the same branch circuit.

Part V of Article 430 provides similar requirements for feeders. In the case of feeders, the rules of 430.62 and 430.63 permit the feeder overcurrent device to exceed the rating of the feeder conductors where necessary to allow supplied motors to start without operating the feeder overcurrent device.

Consider the illustration in Figure 4.10(c). This example shows the ratings of some elements of a motor circuit.

1. The motor is rated at 25 horsepower, 460 volts, three phase, service factor 1.15. This motor has a nameplate full-load ampere (FLA) rating of 30 amperes. According to Table 430.250, the "table value" of full-load current (FLC) for this motor is 34 amperes.

2. The minimum conductor rating for this motor is $1.25 \times 34 = 42.5$ amperes, according to 430.22. This could be an 8 AWG or 6 AWG conductor, depending on the terminal temperature ratings, so the conductor ampacity would be 50 or 55 amperes from Table 310.16 (assuming that no other derating is required for the conditions of use).

3. Where overload protection is provided by a conventional thermal overload relay, the sensing device or heater would be selected from an overload relay table based on the nameplate current, and in operation could not exceed 125 percent of the nameplate FLA current. In this case, the maximum rating of the overload device is $1.25 \times 30 = 37.5$ amperes. Note that this device will also protect the conductors both upstream and downstream from the overload relay by limiting the current in the conductors to 37.5 amperes because the conductors must be rated for at least 42.5 amperes, and, if an overload occurs, the load will be removed from the conductors.

4. The short-circuit and ground-fault device is shown as an inverse-time circuit breaker in this example. For such a device, Table 430.52 limits the rating of the inverse-time circuit breaker to 250 percent of the FLC ampere rating of the motor. Thus, the maximum rating for this device is $2.50 \times 34 = 85$ amperes. However, according to 240.6, 85 amperes is not a standard size, so this maximum value may be rounded up to the next standard size of 90 amperes. Note that this maximum value is over twice as large as the required ampacity of the conductors. However, this overcurrent device is only for short-circuit and ground-fault protection of everything in the circuit; overload protection is provided separately. The short-circuit and ground-fault protection may be much smaller and still operate effectively. In fact, the lowest-rated device that will allow the motor to start and run

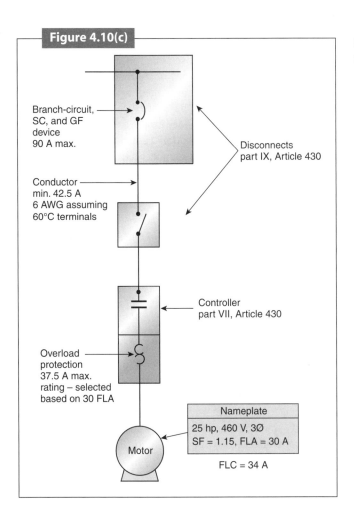

Example of sizing some elements of a motor circuit.

and still meet or exceed the minimum ratings for disconnects will provide better protection in most cases, and in some cases, the maximum rating marked on the controller or for use with a specific overload device may also be smaller than this calculated value. Manufacturers' instructions may provide a more restrictive requirement than the *NEC*, and must be followed in accordance with 110.3(B).

Question 4.11

Do small motors such as damper motors require local disconnects?

Related Questions

- Is a local disconnect required for each small motor (less than 1/8 horsepower)?
- What if the motors are installed in suspended ceiling spaces?
- If a mechanical code requires a disconnect to be provided in sight of equipment when the supply voltage exceeds 50 volts, does this rule apply to the electrical installation?

Keywords

Motor
Circuit breaker
Disconnect
Snap switch
Mechanical code

NEC References

400.8, 430.102, 430.109,

Answer

Section 430.109 covers the requirements for the type of disconnect in a motor circuit, and both 430.109(B) and 430.109(C) provide for alternative types of disconnects for small stationary motors. For any motor, 430.109(A)(2) permits the disconnect to be a listed molded-case circuit breaker. The disconnect for a small motor may be the branch circuit overcurrent device, a general-use switch, a snap switch, or certain manual motor controllers. However, none of the rules in 430.109 amend the disconnect location requirements of Sections 430.102(A) and (B). Section 430.109 only lists the type of disconnect required for these small motors. The branch circuit overcurrent device is permitted as the disconnect, provided that it also is in sight of the motor, as required in 430.102(B).

Prior to the 2002 *NEC*, Section 430.102 included very narrow exceptions to the requirement for having a disconnect within sight of a controller and broadly applicable exceptions for having a disconnect in sight of a motor. Now, disconnects must be in sight of motors with very narrow exceptions that apply primarily where it is not practicable (that is, cannot be done in practice), where a disconnect in sight of the motor would increase hazards, or in industrial facilities with written lock-out/tag-out policies. Simply put, in most cases, a disconnect is required in sight of a motor, as shown in Figure 4.11.

Required disconnects are not necessarily large safety switches. Often, small motors such as damper motors are wired through junction boxes. With small motors, the disconnecting means is permitted to be a general-use snap switch that could be installed within sight of the damper motor, perhaps even in the same box that would be used for a junction box. In some cases, a cord and plug may be used as the disconnect; however, flexible cord is prohibited by Section 400.8(5) from being installed in suspended ceiling spaces. In addition, 300.22(C) limits the use of flexible cords in other spaces used for environmental air.

The *NEC* does not amend or modify the mechanical code. Any additional requirements in the mechanical code or any other code adopted by the jurisdiction must also be met.

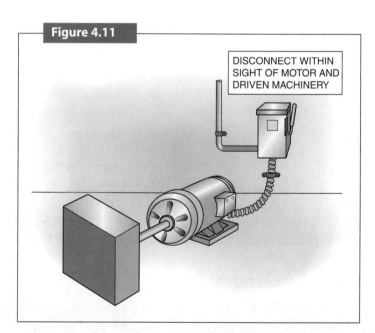

Disconnects are required to be in sight of most motors.

Source: 2002 *NEC* Changes Seminar.

Question 4.12

When are motors permitted to be connected by a cord and plug?

Related Questions

- Are cord-and-plug connections permitted as motor disconnects?
- Are stationary motors permitted to be connected by flexible cord?
- May I use pin-and-sleeve connectors as motor disconnects?
- Is flexible cord permitted for connections of movable motors and machines?

Keywords

Motor disconnect
Cord and plug
Flexible cord
Attachment plug
Receptacle
Pin and sleeve

NEC References

210.50, 368.56, 400.7, 400.8, 400.14, 406.6, 406.7, 406.8, 430.108, 430.109, 430.110, 501.11

Answer

These questions address primarily two issues: the use of flexible cords, and the use of cord-and-plug connections as disconnecting means. Where there is no plug-in connection, the answer is based only on the permissible uses of flexible cords. Where plug-in connections are used, the use of the cord as a connection method and the use of the plug as a disconnect must both be addressed.

Article 400 covers flexible cords. For flexible cord to be used in compliance with the *NEC*, the use must be permitted by 400.7 and not prohibited by 400.8. Some typical reasons for using cord to connect motors are permitted by 400.7(6), (7), and (9). These rules allow a motor (or some other equipment) to be connected with cord in order to facilitate frequent interchange, prevent the transmission of noise or vibration, or connect to moving parts. Other permissions may apply to motor-operated appliances according to 400.7(8). Note that the permitted uses include portable or easily exchanged motors and stationary or fixed motors. Some motors may be essentially fixed but move somewhat for adjustment or during operation, such as some motors associated with belt drives.

As noted, the use of flexible cords must be both permitted and not prohibited. With regard to motors, this means, for one thing, that the cord may not be used as a substitute for fixed wiring; however, where the cord is used for the purposes permitted in 400.7, it is not likely to be "fixed" wiring. Section 400.8 also specifies that flexible cords may not be run through holes in or concealed behind walls, floors, or any kind of ceiling. They may not be run through doors or windows. Also, in general, cords may not be attached to building surfaces or installed in raceways, either of which would tend to make the cord fixed wiring. (Section 368.56(B) permits one connection to a building for drops from a busway to a motor or other equipment. Section 501.140 permits a cord to a submersible pump in a Class I area to be installed in a raceway. The submersible pump motor is treated as if it were portable, and this justifies the use of the cord, both under 400.7 and under 501.140. Section 400.14 includes other limited permission for cords to be installed in raceways in industrial establishments.)

Plug connections may be used on cords if the use of the cord is permitted, as illustrated by Figure 4.12. Article 430 does not restrict the use of plug connections as disconnects to portable or movable equipment. Where used to facilitate frequent interchange, or where connected to a pendant (also permitted by 400.7), a plug and receptacle or plug-and-cord connector will aid such connections. However, where a plug is used, a few requirements for receptacles should also be noted.

Attachment plugs and receptacles may serve as disconnecting means for motors where the use of the cord is permitted by Article 400.

Source: Based on *NEC Handbook*, 2005, Exhibit 430.26.

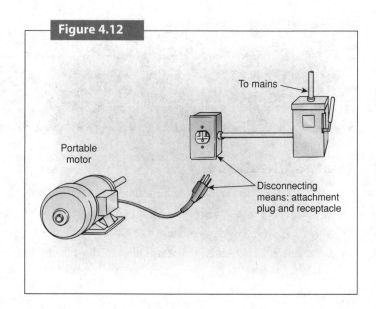

Figure 4.12

Section 210.50 has some general requirements for locations of receptacles. Some are common sense, such as the requirement of 210.50(B) that where a cord-and-plug connection is used, a receptacle outlet must be installed. However, where flexible cords are permitted to be permanently installed, such as for prevention of transmission of noise or vibration, the receptacles are not required, because an attachment plug is not used. Section 210.50(A) also says that a cord pendant is to be considered a receptacle outlet. Section 210.50(C) requires a receptacle to be located within 6 feet of an appliance location, but this rule applies only to appliances (not to individual motors) and only to dwelling units (not to commercial or industrial applications).

Other requirements for receptacles also apply, such as those found in 406.6, 406.7, and 406.8. These rules require that listings and markings be on the receptacles and attachment plugs, that the attachment plugs not be interchangeable with receptacles of different ratings, and that the receptacles and connections be suitable for damp or wet locations when used in such locations.

If a cord-and-plug connection is used with a motor, the attachment plug and receptacle will provide a means of disconnecting the motor. Section 430.108 requires "every disconnecting means" from the feeder to the motor to be of a suitable type and suitable rating to comply with 430.109 and 430.110. Thus, every receptacle and attachment plug connection in a motor branch circuit must meet the requirements of 430.109 and 430.110.

Section 430.109(F) permits a cord-and-plug connection to be used as a motor disconnect. Except for small portable motors (less than 1/3 horsepower) and certain appliances and room air conditioners, the attachment plug and receptacle must be rated in horsepower for the motor type and horsepower used. The *NEC* does not distinguish in these rules between straight-blade, twist-lock, and pin-and-sleeve devices. It simply requires an adequate horsepower rating so that the motor disconnecting means can safely interrupt the motor current.

In the case of cord-and-plug connections, Section 430.110 does not really add anything to the requirements of 430.109 except to provide the way of determining equivalent horsepower ratings for combination or multimotor loads. The equivalent horsepower could be relevant to the selection of an attachment plug and receptacle if, for example, a cord-and-plug connection were used for a multimotor machine. In most cases, each motor is likely to be separately connected where cord-and-plug connections are used, although multimotor appliances such as window air conditioners connected by a cord and plug are also common.

The *NEC* allows cord-and-plug connections to motors in many applications, as long as the rules for the flexible cords are not violated and the ratings of the plugs and receptacles are adequate. "Frequent interchange" is a common reason for using such connections, but these applications are still limited by the permitted uses of cords. For example, as noted previously, a cord-and-plug connection is not permitted if the cord is located above a dropped ceiling, even if the motor is portable and the receptacles and attachment plugs are properly rated.

Question 4.13

How are conductors sized for motors over 600 volts?

Related Questions

- Are conductors for medium-voltage motors required to be sized at 125 percent of motor current?
- Where do I find current ratings for motors over 600 volts?

Answer

A key *NEC* reference for answering this question is Section 430.221. For motor circuits rated over 600 volts, nominal, Part XI of Article 430 amends or adds to the other requirements of Article 430. In the case of sizing the circuit conductors to a motor or motors rated over 600 volts, Section 430.224 does modify the requirements of Sections 430.22 and 430.24. According to Section 430.224, the minimum conductor size in circuits supplying motors over 600 volts is based on the current rating at which the overload device will function. Section 430.226 sets the maximum overload device rating at 115 percent of the continuous current rating of the controller, which will typically be sized for the motor nameplate load. The 125 percent provision of Section 430.22 is not applicable in a motor circuit rated over 600 volts. Conductors may be larger if desired, of course. Maximum conductor sizes are restricted by practical considerations such as termination size and space.

Section 430.6 requires that the values from Tables 430.247 through 430.250 be used for sizing all elements of a motor control circuit except overload devices. Overload devices are sized based on nameplate information. Table 430.250 contains some values for 2300-volt motors, but no values for higher-voltage motors or motors over 500 horsepower. However, since it is the overload device rating that provides the basis for sizing conductors for motors over 600 volts, nameplate ratings may be used and table values are not needed. Unlike many smaller motors, higher-voltage motors are not so common or generic or as likely to be off-the-shelf, so the specific ratings of individual motors are more appropriate for sizing circuit elements than table values, and actual values must be known for these motors.

Although Article 210 covers branch circuits, and Section 210.19(B) contains provisions for branch circuits over 600 volts, Article 210 defers to Article 430 for branch circuits supplying motors. This information can be found in 210.1. Similarly, for motor load calculations, Section 220.14(C) refers to 430.22 and 430.24 for motor load calculations, but 430.22 and 430.24 are modified as discussed for motors over 600 volts.

Keywords

Motors
Over 600 volts
Conductors
Branch circuits
Current ratings
Medium voltage

NEC References

210.1, 210.19, 220.14, 328.10, 430.6, 430.22, 430.24, 430.221, 430.224, 430.226, Tables 430.247 through 430.250

The term "medium voltage" is often applied to motors over 600 volts. Most such motors actually operate in the range of 2000 volts to 15,000 volts. Medium-voltage cables (Type MV, Article 328) are permitted for power systems from 2001 volts to 35,000 volts according to Section 328.10, although the term "medium voltage" may be applied to systems up to 45,000 volts or higher in contexts other than the *NEC*.

Question 4.14

May the circuit breaker for an air-conditioning unit also serve as the disconnect?

Keywords

Air conditioner
Disconnect
In sight
Circuit breaker

NEC References

Article 100, 430.109, 430.110, 440.3, 440.14

Related Questions

- If a circuit breaker is nearby and in sight, is a safety switch disconnect at the air conditioner also required?
- Is the circuit breaker sufficient as both overcurrent protection and disconnect?

Answer

Section 440.14 provides the requirements for the location of air-conditioning equipment disconnecting means. It reads as follows:

> Disconnecting means shall be located within sight from and readily accessible from the air-conditioning or refrigerating equipment. The disconnecting means shall be permitted to be installed on or within the air-conditioning or refrigerating equipment.

There are two exceptions to this rule, but the exceptions apply only to industrial occupancies and cord connections. The Fine Print Note mentions that Parts VII and IX of Article 430 also apply. According to Section 440.3, Article 430 applies except as modified by Article 440. In Article 430, Part VII applies to controllers and Part IX applies to disconnecting means. Section 430.109 is in Part IX, and Section 430.109 permits a listed circuit breaker to be used as a disconnect. The minimum rating of the disconnect is given in 430.110, but most air conditioner nameplates specify a maximum rating for a circuit breaker or fuse, and some also specify a minimum rating, as shown in Figure 4.14(a).

Nameplate information on an air conditioner—in this case, a heat pump.

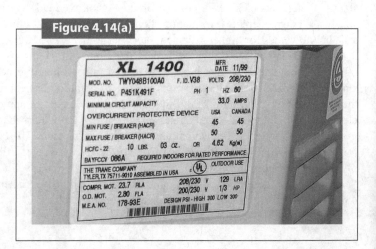

Figure 4.14(a)

The disconnect is also required to be "in sight from" the air-conditioning equipment, according to 440.14. The phrase "in sight" is defined in Article 100, which says that the disconnect is required to be in sight and within 50 feet of the air-conditioning equipment.

If the circuit breaker is in the line of sight, readily accessible, within 50 feet, and complies with the nameplate on the air conditioner, no additional disconnect or overcurrent device is required. Figure 4.14(b) shows a code-compliant application as described.

Figure 4.14(b)

Air-conditioning unit with a circuit breaker that also serves as the required disconnect.

Question 4.15

Can a circuit for HVAC equipment be tapped to supply separate integral circuit breakers?

Related Questions

- How does the *NEC* classify types of power conductors?
- Can an internal overcurrent device be used as a disconnect on HVAC equipment?
- What is the difference between a branch circuit and a feeder?
- What is a tap conductor?
- Where are taps permitted?

Keywords

HVAC equipment
Tap
Tap conductors
Branch circuit
Feeder
Service conductors

Keywords

Overcurrent protection
Overload
Disconnect

NEC References

Article 100, 210.19, 240.21, 430.28, 440.3, 440.4, 440.14

- Are taps permitted from branch circuits?
- Are taps permitted to be made inside utilization equipment in the field?

Answer

The conductors used for distribution of power are classified in the *NEC* as service conductors, feeder conductors, and branch circuit conductors. These terms are defined in Article 100. Service conductors are the conductors between the service point (the utility connection point) and the service disconnecting means. Feeders are the conductors between the service equipment or other power source and the final branch circuit overcurrent device. Branch circuit conductors are those between the final overcurrent device and the outlet or outlets. The term "outlet" is also defined in Article 100 as a point on a wiring system at which current is taken to supply utilization equipment.

These definitions are important to the discussion of circuits supplying heating, ventilating, and air-conditioning (HVAC) equipment because they establish that an outlet is more a place than a thing, that an outlet is not necessarily located in a box, and that an overcurrent device added to a branch circuit may change part of the circuit into a feeder. Also, there may be additional conductors between an outlet and the utilization equipment itself. Many types of utilization equipment include additional conductors and supplementary overcurrent devices as part of the equipment. Typically, the *NEC* covers the conductors up to the utilization equipment, and may cover other conductors, such as flexible cords and fixture wires, if they are not an integral part of the utilization equipment. The distinctions among service conductors, feeders, and branch circuits are critical because different rules apply to each type of conductor. Branch circuits in general are covered by Article 210, feeders by Article 215, and services by Article 230. Special rules may apply to specific types of branch circuits, such as motor circuits that fall within the scope of Article 430, and many items of HVAC equipment include motors or are primarily motors.

Overcurrent protection for feeders and branch circuits must usually include short-circuit, ground-fault, and overload protection. For an overcurrent device to provide short-circuit and ground-fault protection, it must be located at the "beginning" of a conductor, or the point where it receives its supply, in accordance with 240.21. However, 240.21 permits overcurrent devices to be located at other than the point of supply, and 240.4 permits overcurrent protection to be provided at other than the conductor ratings in some cases. These rules are usually referred to as the "tap rules." (Tap rules are covered in more detail in Chapter 2 of this book.)

Branch circuit taps are permitted by Section 240.21(A) in accordance with 210.19. Section 210.19 provides four "branch circuit tap rules" as exceptions, two of which (210.19(A)(3), Exceptions No.1 and 2) are applicable only to household ranges and cooking appliances. Two other exceptions to 240.21(A)(4) permit conductors that are smaller than the branch circuit conductors for specific purposes. Each of these is related primarily to conductors feeding known loads where overloading is very unlikely, so the branch circuit overcurrent device is essentially providing only short-circuit and ground-fault protection for the branch circuit taps. These branch circuit tap rules are usually not applicable to HVAC equipment.

The rules of 240.21(B) apply only to feeders, and most of these rules require the conductors to terminate in overcurrent devices that will provide overload protection for the conductor. Overload protection can be provided by a device anywhere in a conductor as long as the device is in series with the entire load. The overcurrent device ahead of the feeder tap is expected to provide short-circuit and ground-fault protection in most cases. (Three similar tap rules are provided for motor feeders in 430.28 that, according to 440.3(A), also apply to hermetic refrigerant motor compressors.) The overcurrent devices mentioned here—the ones that are used to define feeders and branch circuits—are those that are part of the premises wiring, not those that may be part of utilization equipment.

Internal overcurrent devices are sometimes intended to also serve as disconnecting means. Many manufacturers of HVAC equipment offer integral internal disconnects as an option that is permitted by 440.14, especially in rooftop packaged units. According to 440.14, a local external disconnecting means is not required in this case. The only way to be sure of the intended connection method, number of supplies, and requirements for disconnecting means and overcurrent protection is to read the nameplate information that is required by 440.4 for multimotor and combination-load equipment. Sometimes additional information and wiring requirements are provided in other installation and labeling instructions and installation details. Listing and labeling instructions are part of the requirements of the *NEC,* according to 110.3(B). For example, an integral disconnect may or may not satisfy the requirement for overcurrent protection of the equipment, depending on the requirements provided on the nameplate. Usually a built-in disconnect comes in the form of a circuit breaker, but the nameplate may require fuses—even fuses of a specific type and rating. The fuses in such cases would not have to be installed within sight of the air-conditioning unit, but would have to be installed in the circuit that supplies the air conditioner.

Consider the following examples and the related diagrams. Each of these illustrates the supply to an HVAC unit that contains a hermetic refrigerant motor compressor covered by Article 440, and each has subdivided circuits within the unit for different loads. For simplicity, not all of the details are shown. One load may be a compressor and fan motor(s), and the other may be supplementary heat or another compressor. In fact, this arrangement is quite common in packaged heat pump applications with supplemental heat, but it could also apply to two separate pieces of utilization equipment that had integral disconnecting means. All of these examples assume that the integral overcurrent devices are properly rated and intended as disconnects.

In the first example, illustrated by Figure 4.15(a), a conductor is brought to the equipment and tapped to two separate disconnects that include overcurrent protection. This shows

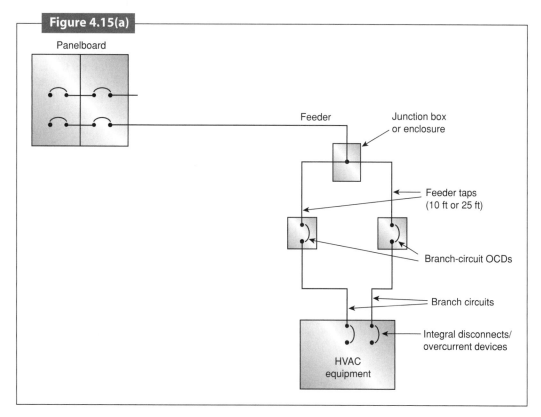

Two separate branch circuits supplied by taps to a common feeder through separate overcurrent devices.

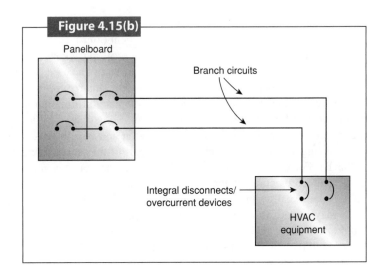

Two separate branch circuits supplied from separate branch circuit overcurrent devices.

a feeder being tapped to create two branch circuits, because the branch circuit overcurrent devices are the final overcurrent devices in each circuit and are also essentially the outlets. This arrangement is permitted by either the 10-foot tap rule (240.21(B)(1) or 430.28(1)) or the 25-foot tap rule (240.21(B)(2) or 430.28(2)), but all other aspects of those rules must also be followed.

In the second example, Figure 4.15(b) shows two sets of conductors brought to the HVAC equipment and terminated on the two overcurrent devices. Since the two overcurrent devices at the equipment are part of the utilization equipment rather than the premises wiring, the two sets of conductors are branch circuits, and the two overcurrent devices are supplementary overcurrent devices from the standpoint of the *NEC*. From the standpoint of the HVAC equipment, this is the same as Figure 4.15(a) because both schemes supply the HVAC equipment with two separate branch circuits.

The third example is illustrated by Figure 4.15(c), in which one set of conductors is brought to the unit, and a tap is made in the field and within the HVAC equipment to the two overcurrent devices. This is generally *prohibited,* because the supply conductor is a branch circuit, the two overcurrent devices are supplementary overcurrent devices, and the utilization equipment is probably not intended as a tap enclosure. This arrangement is not a permitted use of a branch circuit tap according to 210.19(A)(4), but it would be permitted if the conductors were not reduced in size beyond the tap point (not tap conductors) *and* the branch circuit rating

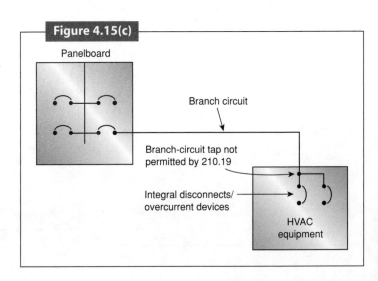

Violation: Two circuits supplied by branch circuit taps from a single branch circuit.

(the rating of the branch circuit overcurrent device) was adequate for both loads and did not exceed the maximum rating marked on the equipment for one load, which is unlikely.

Internal wiring of utilization equipment is usually not intended for modification in the field unless specific instructions are provided with the equipment. Usually, when utilization equipment is supplied with two separate sets of terminations, the intent is to supply two separate branch circuits. However, HVAC and similar equipment is often provided with a single termination point, and the circuits are internally subdivided or tapped within the equipment by the manufacturer. (This is very common with electric heating equipment where the total load exceeds 48 amperes. See 424.22.) These internal overcurrent devices may or may not be intended for use as disconnects. This question can be answered by consulting the installation instructions and labeling.

If a single termination point is provided or taps are otherwise supplied internal to listed equipment, that termination represents the end of the branch circuit. Taps and terminations that are integral to the equipment must meet a product listing standard and are not subject to *NEC* requirements. Figure 4.15(d) shows an example in which a single branch circuit is provided with an external disconnect because the internal overcurrent devices are not intended to be disconnecting means. This drawing represents the most common arrangement for multimotor and combination-load HVAC equipment.

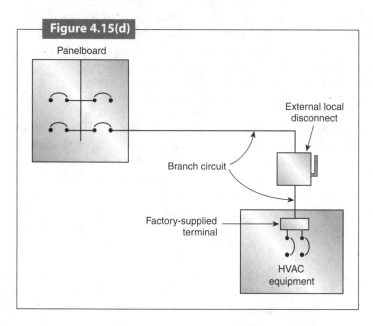

A single branch circuit supplying a single piece of utilization equipment with internal taps to multiple overcurrent devices.

Question 4.16

What restrictions does the *NEC* impose on the locations of transformers?

Related Questions

- What is the required spacing between transformers and walls?
- Are transformers permitted on combustible floors?

Keywords

Transformer
Clearance

Keywords

Combustible
Ventilation
Panelboard
Work space
Dedicated space
Insulation system
Insulation class

NEC References

110.3, 450.1, 450.9, 450.11, 450.21, 450.22, 450.27

- May transformers be hung above panelboards?
- What clearances are required between transformers and building openings?

Answer

Although Section 450.1 says Article 450 applies to the installation of all transformers, there are eight exceptions for specific transformer applications. The exceptions are for transformers that are used for special purposes or are part of other apparatus, that is, transformers not used for power distribution in premises wiring. For example, the small transformers used to create Class 2 control systems or power-limited fire alarm circuits are not covered by Article 450, and neither are the transformers used in ballasts or integral to low-voltage lighting luminaires. Clearances for transformers used for signs and lighting are covered in Articles 600 and 410, respectively. All types of transformers are also subject to whatever installation requirements are included in the manufacturer's listing and labeling instructions.

The *NEC* requires two types of clearances for transformers in Article 450: (1) clearances for ventilation necessary for proper operation of the transformer, and (2) clearances to reduce fire hazards that may be created by normal or abnormal operation of the transformer.

Transformers are heat-producing equipment, so they should be expected to get relatively hot, that is, "hot" relative to what is comfortable for people to touch. The typical specification for the surface temperature of enclosed transformers that are operating normally is about 40°C above the ambient temperature. At normal room temperatures, this translates to a surface temperature of about 70°C, or over 150°F. These temperatures are not fire hazards in most cases, as long as the transformer is not used as a shelf to store combustibles and is not installed too close to combustible building finishes. Adequate ventilation is necessary to keep the normal operating temperatures from being exceeded. (Unless specially marked, dry-type transformers are not intended for operation in ambient temperatures exceeding 40°C or 104°F.)

Internal temperatures in operating dry-type transformers may well be up to 220°C in some types. Obviously, combustible materials must be separated from equipment at these high temperatures. Most open-type transformers or exposed-core transformers are industrial control transformers and are required to be installed in other enclosures and separated from combustible materials. General-purpose power transformers like the one shown in Figure 4.16(a) are typically furnished with enclosures. These enclosed transformers are air cooled, but may or may not have ventilating openings. The enclosures provide a heat-resistant barrier between the high-temperature internal parts and building components or personnel.

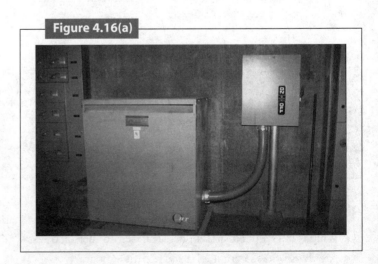

Dry-type ventilated transformer for general-purpose power distribution use.
Source: *NEC* Seminar.

Section 450.9 requires ventilation of the space where a transformer is located so that the heat produced by the transformer can be dissipated and the transformer will not be overheated. Transformers are very efficient when operating normally, so the full-load losses are typically less (often much less) than 5 percent of the transformer rating. Transformers rated around 30 to 45 kVA are not likely to produce more heat than a portable electric heater, an amount of heat that is often readily dissipated by normal infiltration and convection. Larger transformers may require mechanical ventilation, depending on the location and size of the area. Ventilation is a requirement of the *NEC*, but as in other heating and cooling loads, it is a problem for mechanical (HVAC) design rather than electrical design, in most cases.

Section 450.9 also requires that transformers with ventilating openings be spaced from obstructions in accordance with the clearances that are required to be marked on the transformer. The requirement for marking of the needed clearance was added to the *NEC* in the 1990 edition, so older transformers may not include this marking. A good rule of thumb for transformers that are not marked with a clearance dimension is that the distance from the wall or other obstruction should not be less than the height of the ventilating opening. A ventilating opening can be seen at the top of the transformer in Figure 4.16(a).

Clearances from combustibles for dry-type transformers are covered in 450.21. Transformers rated 112.5 kVA and smaller require clearance from combustibles of at least 12 inches or must be separated by "a fire-resistant, heat-insulated barrier." Transformers over 112.5 kVA must be installed in a room of fire-resistant construction. The "transformer room" must be built to have a fire rating of at least one hour. Both of these rules exempt transformers that are completely enclosed except for ventilating openings. To be exempted, the smaller transformers must also be rated 600 volts or less and may be "with or without ventilating openings," and the larger transformers must have a Class 155 or higher-rated insulation system. Some small transformers that are totally enclosed with encapsulated windings have no ventilating openings and are designed to be mounted directly to walls. Ventilating clearances are required only for transformers with ventilating openings, according to 450.9 and 450.11. In short, ventilation clearances apply to transformers with ventilating openings, and clearances from combustibles apply to transformers that are not completely enclosed. The insulation class must be considered, but most modern general-purpose power transformers have Class 155 (Class F) or Class 220 (Class H) insulation systems. Some manufacturers use Class 220 insulation on all of their power transformers. Less common types, special-purpose types, and types with insulation rated less than Class 155 must meet all installation requirements discussed here as well as any specific instructions included with the transformer.

Do not confuse insulation class temperatures with temperature-rise classifications. Many manufacturers make 80° rise, 115° rise, and 150° rise ratings, all with Class 220 insulation. The temperature-rise rating is an indirect measure of the heat losses in the transformer at full load. The insulation temperature class is an indication of the amount of heat the insulation can withstand. Insulation systems with higher temperature classes also tend to be more fire resistant.

The transformers most commonly installed in buildings for power distribution are dry-type transformers, rated 600 volts or less, that are totally enclosed or totally enclosed except for ventilating openings, like the one shown in Figure 4.16(a). For these transformers, the enclosure provides the heat-resistant barrier. The required clearances from combustibles also apply to transformers mounted on combustible floors. However, for transformers that are completely enclosed, the clearance restrictions usually do not apply, unless the insulation class conditions are not met, as explained previously. However, many transformers are marked with required clearances from combustible materials. These instructions must be followed, in accordance with 110.3(B), even (or especially) where the marking requirement is more restrictive than the code requirement.

The bottom of a ventilated transformer enclosure is often a mesh or screen. This may not seem to create a complete enclosure, but it is really a screened "fingerproof" ventilating opening. Some installers, designers, or inspectors may prefer to have a noncombustible surface below these openings, but such a barrier is not specifically required by the *NEC*, unless required by the manufacturer.

Subject to structural considerations, all dry-type transformers may be mounted on floors or walls or may be hung from structures. According to Section 450.13 (taken in its entirety), transformers must be accessible, but not necessarily readily accessible. Most transformers will not be serviced (have live parts exposed) while energized, but those that are must have work space as defined in 110.26(A). A transformer may be hung in the open above a panelboard, as shown in Figure 4.16(b), as long as it does not interfere with the work space of the panelboard. The transformer is part of the electrical installation, so it is not "foreign equipment" and may be in the dedicated space for the panelboard. The required dimensions and clearances are illustrated and summarized in Figure 4.16(b).

Dry-type transformers installed outdoors must have weatherproof enclosures in accordance with 450.22. The other requirements for clearances from combustibles are also summarized in this section. Note that only transformers over 112.5 kVA are mentioned because the enclosure requirements for smaller transformers in the exception to 450.21(A) have been met by the required weatherproof enclosure. The language of this rule is pretty much the same as 450.21(B) with Exception No. 2.

A transformer may be installed above a panelboard if all required clearances for work space and ventilation are provided.

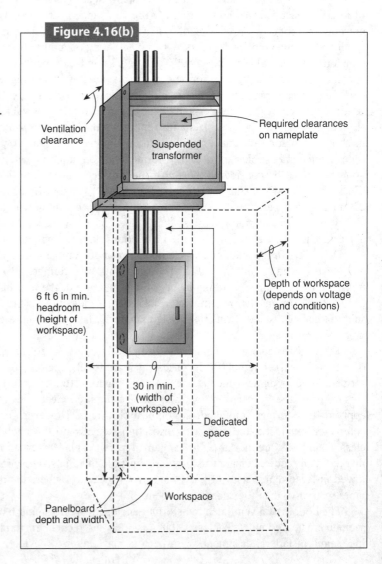

Figure 4.16(b)

The *NEC* contains no specific rules with regard to the clearances of outdoor oil-filled transformers from buildings or building openings. Section 450.27 does require that "combustible material, combustible buildings, and parts of buildings, fire escapes, and door and window openings be safeguarded from fires" from oil-filled transformers installed on or near buildings. This section requires the use of one or more types of safeguards to protect the building. The safeguards include space separations, fire-resistant barriers, automatic fire suppression systems, and oil-containment enclosures. Figure 4.16(c) illustrates two of these methods (containment and space separations), where actual dimensions depend on an evaluation of the hazard. The *National Electrical Safety Code* (*NESC*), which applies primarily to utilities, lists similar methods to be used to reduce fire hazards due to oil-filled transformers, but also does not recommend specific dimensions. According to Section 152(A)(2) of the *NESC*, one or more of the following methods should be used: "less flammable liquids, space separation, fire-resistant barriers, automatic extinguishing systems, absorption beds, and enclosures." Factory Mutual is a source for information on loss prevention with regard to transformers, and local utilities usually have rules, sometimes published, about how to place oil-filled transformers with respect to buildings and building openings.

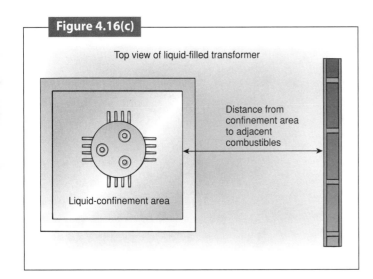

Oil-filled transformer in a containment area. Required spacing from combustibles is based on the capacity of the tank in this example.

Source: Based on *NEC Handbook*, 2005, Exhibit 450.15.

Chapter 5

Special Occupancies

Question 5.1 Where do I find the classification of a hazardous location?

Keywords

Hazardous
Classified
Location
Area classification
Class I
Class II
Class III
Division
Zone
Woodworking
Munitions
Explosives
Diesel

NEC References

500.1, 500.2, 500.5, 500.6, 500.7, 505.5, 506.5, NFPA 497, NFPA 664

Related Questions

- What is a "hazardous location"?
- How are classified areas defined?
- Where can I find descriptions of classes, divisions, and groups?
- Can an area be classified at some times and not other times?
- Where are terms relating to area classification found?
- How are areas classified if they are used for woodworking?
- How are munitions and explosive storage areas classified?

Answer

A complete answer to these questions is beyond the scope of this book, but they will be answered here in a general way. For a more complete discussion of these topics, see the other standards mentioned in the Fine Print Note (FPN) of Section 500.1. The National Fire Protection Association (NFPA) book *Electrical Installations in Hazardous Locations* is also recommended. In addition, many NFPA standards are available for specific industries and types of processes and products.

The broad definition of a hazardous or classified area is found in the scope statement of Section 500.1. This section, taken in context, may be summarized to say that a hazardous (classified) area is one where "fire or explosion hazards may exist due to flammable gases or vapors, flammable liquids, combustible dust, or ignitible fibers or flyings." Since this must also be taken in the context of the *National Electrical Code*, classification of areas is meaningful only where there is also some electrical equipment that could provide an ignition source for the flammable, combustible, or ignitible materials. In other words, the objective of area classification under the *NEC* is to identify areas where an electrical system or electrical equipment could ignite hazardous (flammable, combustible, or ignitible) materials, and to keep that from happening. The problem may be solved in various ways, such as by using special electrical equipment, by separating or isolating the electrical equipment from the hazardous materials, by removing the hazardous materials themselves, or by reducing the likelihood that the hazardous materials will be released or be present in ignitible quantities or mixtures.

The *NEC* sets up a classification scheme based on the type of material (the class) and the likelihood that the materials will be present (the division). Groups are used to describe the characteristics of liquids, vapors, gases, and dusts. Section 500.2 includes definitions of a number of protection techniques, and these techniques and their application are further described in Section 500.7. An alternative classification scheme that uses some different protection techniques is based on zones rather than divisions. This scheme and the corresponding techniques are described in Articles 505 and 506.

Locations classified by class and division are defined and described in Section 500.5 of the *NEC*. Section 500.6 covers material groups for division-classified areas. Definitions and descriptions of locations classified by zone are found in 505.5 for Class I areas and in 506.5 for areas involving dusts, fibers, and flyings. The zone classification uses a similar method of describing material groups in Class I areas, but uses different names for those groups.

Class I areas are classified because of the presence of flammable liquids, gases, or vapors, and Class II areas are those where combustible dust is present. Ignitible fibers and flyings are covered by Class III. Divisions 1 and 2 and Zones 0, 1, and 2 describe in a general way the probability that an ignitible gas or vapor will be present. (Area classification under Article 506 uses Zones 20, 21, and 22 for areas classified because of dust, fibers, or flyings, but the basic idea is the same.) In either scheme, the lower the zone or division number, the more likely that the material will be present at any given time. If the hazardous atmosphere or conditions exist only infrequently or are unlikely under normal conditions, the area may be Division 2 or Zone 2, but that does not mean it is classified at some times and not at others. The area will be classified as long as those conditions exist—even occasionally. A classified area may become unclassified if the use of the area changes completely or some other condition, such as a change in ventilation, is made, but an area will not fluctuate between being classified at some times and not at others.

The simple presence of a "hazardous" material is not sufficient to cause an area to be classified. The specific properties of the materials, the quantities present or available, and the frequency or likelihood that they will be present are among the factors that are used to classify areas. For example, in spite of the name, sawdust is usually classified as a flying or fiber rather than a dust, and woodworking facilities may employ dust collection and housekeeping techniques that prevent the accumulation of hazardous quantities. NFPA 664 covers wood processing and woodworking facilities. For another example, many solvents, diesel fuels, and fuel oils are used in areas where they are below the temperature that would cause them to produce an ignitible mixture in air (the flash point) and therefore do not create classified areas. These materials are distinguished from "flammable" liquids and are often "combustible" liquids by definition, but they can cause an area to be classified if they are handled above their flash points. Classification of liquids, definitions of terms such as "flammable" and "combustible," and general recommendations for Class I areas can be found in NFPA 497, *Recommended Practice for the Classification of Flammable Liquids, Gases, or Vapors and of Hazardous (Classified) Locations in Chemical Process Areas*. Figure 5.1 is an example of a generic area classification drawing of the sort that may be found in NFPA 497.

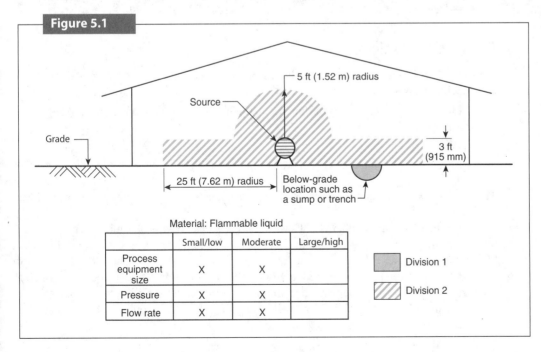

Figure 5.1

Leakage source located indoors, at floor level. Adequate ventilation is provided. The material being handled is a flammable liquid.

Source: NFPA 497, 2004, Figure 5.10.1(c).

The *NEC* is most concerned with preventing flammable, combustible, and ignitible materials from being ignited by the electrical system. In some situations, materials may be subject to ignition by other sources, such as static electricity, flame, or even simple contact with air or moisture. In such cases, the areas need not be classified under the *NEC*. For example, according to 500.5(A), "Where pyrophoric materials are the only materials used or handled, these locations shall not be classified." Pyrophoric materials are materials that may ignite on contact with air. For similar reasons, areas where explosives are manufactured or munitions are stored are generally not classified. Other materials, such as ammonia, may be flammable, but are typically well contained, well ventilated, and relatively difficult to ignite, so ammonia refrigeration systems might be located in unclassified areas. Other unclassified areas include those where flammable materials are often found but experience has shown have a low incidence of fire from flammable materials ignited by electrical sources, such as residential garages and the areas around common gas appliances.

The function of the *NEC* with regard to area classification is primarily to define the terms and provide for safe installation methods and protection techniques. The *NEC* also includes extracted material from other standards that provide specific methods and area classification information for certain very common types of installations. For example, Article 514, "Motor Fuel Dispensing Facilities," covers a very common occupancy, and Article 511 covers commercial garages for repair and storage (but not parking garages). These articles include the classifications and dimensions of classified areas for these common occupancies and may also clearly describe the areas that are not classified. In those occupancies not specifically covered in Articles 511 through 517, other standards and recommendations must be used to determine the extent and classification of hazardous areas.

Question 5.2

Must areas where solvents, fuel oil, or diesel is used be classified?

Keywords

Classified area
Class I area
Flammable
Combustible
Flash point
Solvent
Fuel oil
Diesel

NEC References

500.5, NFPA 30, NFPA 497

Related Questions

- Where can I find information on classification of areas based on the characteristics of a specific material?
- If I have a solvent that has a flash point of 143°F, what area classification would apply to this material?

Answer

The *NEC* defines area classifications, but is not adequate for assigning classifications to areas in most cases.

According to Section 500.5(B), "Class I areas are those in which flammable gases or vapors are or may be present in the air in quantities sufficient to produce explosive or flammable mixtures." "Flammable gases or vapors" are, by definition, those that have flash points not over 100°F. The flash point is defined in NFPA 497 as "the minimum temperature at which a liquid gives off vapor in sufficient concentration to form an ignitible mixture with air near the surface of the liquid, as specified by test." Therefore, a liquid solvent or a fuel that is kept below its flash point will not give off significant quantities of vapor or form an ignitible mixture in air.

Class IA, IB, and IC liquids are flammable liquids. Class II and Class IIIA and IIIB liquids are called combustible liquids. These classes of liquids (do not confuse classes of liquids with classes of areas) are described in NFPA 30, *Flammable and Combustible Liquids Code*, as well as in NFPA 497, *Recommended Practice for the Classification of Flammable Liquids, Gases or Vapors and of Hazardous (Classified) Locations for Electrical Installations in Chemical Process Areas*. Both documents are excellent resources for definitions and guidelines for determining what areas should be classified. Other National Fire Protection Association (NFPA) standards are available for specific industries or processes, such as liquified petroleum gas (LPG) storage or dry cleaning establishments.

In accordance with the definitions of the terms "flammable liquid" and "combustible liquid" found in NFPA 30, diesel fuel is a Class II combustible liquid because its flash point (125°F for No. 2 diesel) falls in the range between 100°F and 140°F. NFPA 30 also stipulates that Class I area classification is applicable to Class II combustible liquids only where these liquids are stored or handled above their flash point. In most areas, diesel fuel is not stored or handled at temperatures exceeding its flash point, especially where a diesel generator is installed in an indoor, conditioned space, so such areas are not classified. Figure 5.2 illustrates an indoor generator with an attached fuel tank that does not create a classified area.

Based on the flash point temperature, the solvent in the question is a Class IIIA combustible liquid. Class IIIA liquids are those with a flash point at or above 140°F (60°C) but below 200°F (93°C). Therefore, this solvent is even less likely to create a classified area than diesel fuel, and an area where the solvent could create a hazardous atmosphere would be too hot for extended human occupancy. Any area that would be classified because of the presence of this solvent is likely to be inside a process or containment vessel, where there is no electrical equipment.

This general information should answer the specific questions, and addresses the common questions about diesel fuel. In other cases, one should review the material safety data sheet (MSDS) for the product or contact the manufacturer for information on fire and other hazards that may be associated with the specific product.

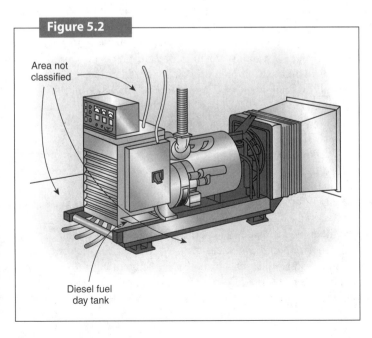

Figure 5.2 Indoor installation of a diesel-fueled emergency or standby generator with a base-mounted "day tank" where the area is not classified.

Question 5.3

How is equipment identified as suitable for hazardous locations?

Keywords

Hazardous
Classified
Class I
Class II
Class III
Equipment
Motor
Luminaire
Temperature
Explosionproof
Dust ignitionproof
Division
Zone
Identified
Approved
Flammable
Combustible
Ignitible

NEC References

110.2, 110.3, 500.5, 500.7, 500.8, 501.125, 502.125, 503.5, 503.125, 503.130, 505.5, 505.8, 505.9, 505.20

Related Questions

- Is "higher rated" Class I equipment also suitable for Class II areas?
- What is the difference between motors or other equipment for Class I and Class II areas?
- Is a motor suitable for Class I or Class II also suitable for Class III?
- Are temperature ratings required for luminaires in areas where combustible materials are used?
- Is there a way to use ordinary equipment in a classified area?
- Are ordinary motors ever permitted in Class I areas?
- What protection techniques are permitted in classified areas?
- What constitutes an "other protection technique"?
- Is equipment for division-classified areas suitable for zone-classified areas?

Answer

All equipment installed under the rules of the *NEC* must be "approved" according to 110.2. "Approved" is defined as "acceptable to the authority having jurisdiction." Some general guidelines for approval are found in 110.3(A), but in the case of equipment for use in hazardous (classified) locations, additional and more specific information is provided in 500.8. Most electrical equipment for use in hazardous locations is required to be identified for that use, but 500.8 expands on this requirement as well.

According to Section 500.8(A)(1), "Equipment shall be identified not only for the class of location, but also for the explosive, combustible, or ignitible properties of the specific gas, vapor, dust, fiber, or flyings that will be present." Section 500.8 goes on to say that surface temperatures must be limited. For Class I and II areas, the surface temperature is limited to below the ignition temperature of the gas or vapor that will be present, and in the case of dusts (Class II areas) the temperature is additionally limited to below 165°C or 329°F. These specific temperature limits also apply to equipment in Class III areas, but according to 503.5, Class III equipment is further limited to below 120°C or 248°F if the equipment is subject to overloading. Thus, surface temperatures may be higher on certain equipment for Class I areas than for similar equipment in Class II or Class III areas. Keep in mind that the purpose of classifying areas in the *NEC* is to identify those areas where normal electrical equipment could be a source of ignition and then to require special types of equipment where necessary to eliminate the electrical system as a source of ignition in those areas.

The protection techniques used for classified areas depend on the nature of the flammable, combustible, or ignitible materials. Class I equipment is often "explosionproof," which assumes that the gas or vapor will enter enclosures and perhaps be ignited by internal sources. The enclosure must contain the resulting explosion and control the release of hot or flaming gas to the exterior so that the electrical equipment is not a source of ignition for the flammable gas or vapor outside the enclosure. Class II equipment, or "dust ignitionproof" equipment, is very different. Because particles of dust are much larger than the molecules or atoms of gases and vapors, the dust may be more effectively excluded from the enclosures, cables, and raceways, so the function of the enclosures is to operate in such a way that dust, including dust blankets, on or around the equipment will not be ignited. As noted, temperatures may be more restricted, but the enclosure is not required to withstand an internal explosion because the dust will be excluded from the enclosures. Class III equipment is more easily built to exclude the larger fibers and flyings but is more restricted in surface temperatures because the fibers and flyings may be even more likely to build up on equipment and be more readily ignited after being exposed to the heat of the equipment.

In other words, equipment for Class I, Class II, and Class III does not form a hierarchy of "higher rated" to "lower rated," but instead represents different requirements, so that equipment suitable for, say, Class I application is not necessarily suitable for Class II or Class III. Equipment suitable for Class I or Class II is likely to be suitable for Class III, but only if it meets the lower temperature limitations for Class III. Some types of equipment, such as the junction box shown in Figure 5.3(a), may be suitable for all three types of areas because it is not heat-producing equipment and is built to exclude dust, fibers, and flyings in addition to being explosionproof.

Motors provide a good example of the differences in requirements for Class I, Class II, and Class III areas. Generally, in Class I, Division 1 locations, motors must be identified for Class I, Division 1 areas. In addition, they must be ventilated with a source of clean air, be totally enclosed and filled with an inert gas, or be submerged in liquid in accordance with 501.125(A). Section 501.125(B) requires motors with arcing parts that are used in Class I, Division 2 locations to be identified for Class I, Division 1 or have the arcing parts suitably enclosed. However, motors that are not identified for Class I locations may be used in Class I, Division 2 locations if the motors have no arc-producing devices. In short, *some* "ordinary" motors may be used in *some* Class I areas. Similar rules may be found in 502.125 for Class II areas where surface temperatures are a primary consideration. In Class II, Division 1 areas, motors must be identified for Class II, Division 1 (probably dust ignitionproof) or be totally enclosed pipe-ventilated; in Class II, Division 2 areas they may generally be totally enclosed nonventilated, totally enclosed pipe-ventilated, totally enclosed water-air-cooled, totally en-

Figure 5.3(a)

An explosion-proof junction box with a screw-type cover.

Source: *NEC Handbook*, 2005, Exhibit 501.5. (Courtesy of O. Z. Gedney Co.)

closed fan-cooled, or dust ignitionproof. Many of the Class II, Division 2 options also apply to Class III applications, but the surface temperatures are limited to lower values in Class III areas. Dust ignitionproof motors and motors listed for Class I, Division 1 areas are not suitable for Class III areas unless they also meet the requirements in 503.125. Although many motors suitable for Class I or Class II areas may also be suitable for Class III areas, the temperature limits for Class III may be exceeded on equipment that was designed and identified only for Class I or Class II areas.

Similar issues apply to luminaires. Many luminaires are available for both Class I and Class II areas, and some applications, such as coal handling, will involve both flammable gas and combustible dust, where both ratings are required in accordance with 500.8(A)(6). In some cases, heat-producing equipment such as a luminaire may present a greater design challenge for a Class II area than for a Class I area. A luminaire for a Class II, Division 1 area must exclude dust, maintain lower surface temperatures than Class I in many cases, and control surface temperatures even when blanketed with dust, with the dust possibly acting as a thermal insulator. That is, a luminaire for a Class II, Division 1 area must generally be dust ignition-proof. Luminaires for Class III areas must also be dust-tight, a more easily met requirement than dust ignitionproof, but the temperature restrictions are more severe. In fact, luminaires that are specifically identified for Class III areas are generally not readily available, but luminaires that are totally enclosed and meet the surface temperature requirements of 503.130 may be used. Examples of luminaires for Class I and Class II areas are shown in Figures 5.3(b) and 5.3(c). The markings that are required on the nameplates on these luminaires, including temperature markings, are covered in 500.8(B).

The identification and temperature limits discussed do not apply in every area where combustible materials are used. They only apply in those areas that are classified according to the definitions in 500.5. The materials that exist in an area are not the only consideration. Knowledge about quantities of material and their likelihood of being present in ignitible quantities is critical to classification. The likelihood of an ignitible quantity being present may

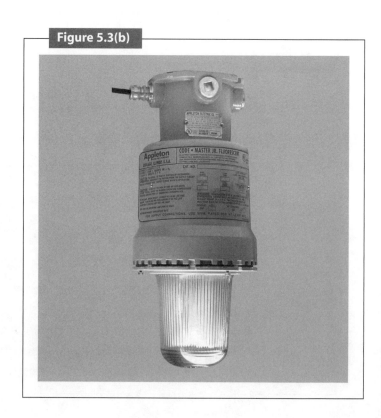

Figure 5.3(b)

A luminaire for use in Class I, Group C and D locations.

Source: *NEC Handbook*, 2005, Exhibit 501.20. (Courtesy of Appleton Electric Co.)

Figure 5.3(c) A luminaire for use in Class II, Division 1 locations.
Source: *NEC Handbook*, 2005, Exhibit 502.5. (Courtesy of Cooper Crouse-Hinds.)

be dependent on factors such as dust collection, ventilation, and housekeeping procedures. Equipment is required to be identified in accordance with 500.8 only where it is applied in an area that requires classification as defined in 500.5 for areas classified by division.

This discussion has included a brief comparison of some protection techniques used in different classified areas. A more complete listing of the protection techniques and the areas where they are applicable can be found in 500.7 for areas classified by division, and in 505.8 and 506.8 for areas classified by zone. As noted previously, ordinary equipment may be used in some cases. In other cases, ordinary equipment may be used in conjunction with other protection techniques. These other techniques that allow the use of ordinary equipment are based on two concepts: (1) reducing the available energy below the level required for ignition of a specific material, or (2) isolating the equipment in such a way that it will not be exposed to the materials in question. Intrinsically safe and nonincendive circuits are based on reduced energy levels, and purged and pressurized enclosures provide for isolation of the electrical equipment from the flammable, combustible, or ignitible materials. Intrinsically safe systems are covered by Article 504, and purging and pressurization techniques are covered by NFPA 496, *Standard for Purged and Pressurized Enclosures for Electrical Equipment*. (Nonincendive circuits are similar to intrinsically safe circuits but are based only on normal energy levels and do not include the redundant energy-limiting features of intrinsically safe circuits. Nonincendive circuits are therefore limited to Division 2 areas, where the hazardous condition is not likely to exist normally.)

At the end of Section 500.7, 500.7(L) provides for the use of "other protection techniques." This section refers to protection techniques that are a design feature of equipment that is evaluated by a third-party testing organization for use in a hazardous (classified) area. Some operating protocols that are approved by the authority having jurisdiction may be other protection techniques. For example, combustible gas detection systems were listed as a protection technique in the 2002 *NEC*, but not in previous codes, so the use of a combustible gas detection system prior to the 2002 *NEC* would have been an "other protection technique."

This brings us back to the issue of "Approval for Class and Properties," which is the title of 500.8(A). This section says the following:

> Suitability of identified equipment shall be determined by any of the following: (1) Equipment listing or labeling, (2) Evidence of equipment evaluation from a qualified testing laboratory or inspection agency concerned with product evaluation, (3) Evidence acceptable to the authority having jurisdiction such as a manufacturer's self-evaluation or an owner's engineering judgment.

In all cases, the equipment must be acceptable to the authority having jurisdiction—that is, "approved"—in accordance with 110.2. Most authorities prefer to approve equipment based on the first criterion, listing or labeling. The other criteria may involve the more common protection techniques listed in 500.7(A) through (K), or may utilize one of the other protection techniques mentioned in 500.7(L).

Another list of protection techniques and their applications can be found in 505.8 for Class I zone-classified areas. In fact, these alternate protection techniques are a primary reason for classifying an area by zone, and the language of 500.8(A) is mirrored in 505.9(A) for identifying suitable equipment. Section 505.8 does not include "other protection techniques" like 500.7 does, but at least one alternate technique is permitted in zone-classified areas. According to the exception to 505.20(A), Exception No. 1 to 505.20(B), and Exception No. 3 to 505.20(C), equipment listed for appropriate division-classified areas may also be used in zone-classified areas. This means, for example, that explosionproof (Class I, Division 1) equipment with the appropriate ratings for the same gas or vapor may also be used in Class I, Zone 1 areas even though explosionproof equipment is not explicitly mentioned as a protection technique in 505.8.

Question 5.4

Is there any length limitation on flexible conduit runs or flexible fittings in Class I areas?

Keywords

FMC
LFMC
Class I
Division 1
Division 2
Grounding
Bonding jumper

NEC References

250.102, 250.118, Article 348, Article 350, 501.10, 501.30, 501.140, 502.30

Related Questions

- Does the *NEC* limit the use of "explosionproof flex"?
- Are flexible wiring methods intended only for motors in hazardous locations?
- Is flexibility in wiring methods limited only to motors requiring flexible wiring due to movement and vibration in Class I areas?
- Are flexible wiring methods also permitted for applications such as providing wiring protection in tight quarters or replacing short runs that require complex or difficult conduit bends?
- What is the longest run allowed between an instrument panel enclosure and the instrument itself, using either Type FMC or Type LFMC in a Class I, Division 2 area?

Answer

The permissions to use a flexible wiring method in Class I locations are based on the need to provide for movement within the wiring system. Some installations require the use of a wiring method with a degree of flexibility because of excessive equipment vibration or limited movement of equipment that is permanently connected (hardwired). Section 501.10(A)(2) states, "Where necessary to employ flexible connections, as at motor terminals, flexible fittings listed for Class I, Division 1 locations or flexible cord in accordance with the provisions of 501.140 shall be permitted." Figure 5.4(a) shows an example of a flexible fitting, often called "explosionproof flex."

The language "as at motor terminals," in the wording of 501.10(A)(2), is provided as an example rather than as a limiting factor. Therefore, the permission to use wiring methods that provide a degree of flexibility is not limited to motors only, but is limited to those installations within the Class I, Division 1 location for which there is a need to provide for movement or to limit vibration at permanently installed equipment. The uses of flexible cord in 501.140 are strictly limited to portable equipment or equipment that is treated as portable in the rule, including electric submersible pumps.

Connecting stationary equipment that is not subject to operational movement or vibration is typically not where "limited flexibility" of the wiring system is needed. Flexibility for installation convenience is not the criterion. The permission to use wiring methods that allow a degree of flexibility was not intended to open the door to using flexible methods instead of rigid wiring methods just because it might be difficult to install the rigid wiring methods. The permission to use the flexible conduits and flexible cords recognized the fact that wiring systems are at times subject to movement or substantial vibration. This use is intended only where absolutely necessary. The determination of the need to provide flexibility is the responsibility of the authority enforcing the code, because this assessment can only be made through on-site case-by-case evaluation. (Not all permitted wiring methods are rigid. Type MC-HL and ITC-HL cables are relatively flexible, perhaps more flexible than the flexible fittings or explosionproof flex shown in Figure 5.4(a), but these cable types are permitted only in industrial applications, according to 501.10(A)(1)(c) and (d).)

In Division 2 areas, the wiring methods permitted for flexibility are expanded in 501.10(B)(2) to include flexible metal fittings, flexible metal conduit (FMC) with listed fittings, liquidtight flexible metal conduit (LFMC) with listed fittings, liquidtight flexible nonmetallic conduit with listed fittings, or flexible cord listed for extra-hard usage and provided with approved bushed fittings. Types FMC and LFMC are not permitted for wiring in Class I, Division 1 areas. According to Section 501.10(B)(2), FMC and LFMC are permitted "where provision must be made for limited flexibility" in Class I, Division 2 areas. There is no specific length limitation for these conduits in trade size 1/2 and larger, at least not in Article 348 or Article 350. However, the provisions of 501.10(B)(2) must be met, and as quoted earlier, these flexible conduits are only permitted where provision for limited flexibility "must be made."

The fact that motor terminals are not mentioned with regard to Class I, Division 2 locations is not meant to broaden the application of flexible connections in Division 2 locations, even though the permitted methods are broadened and, for that matter, are more flexible than

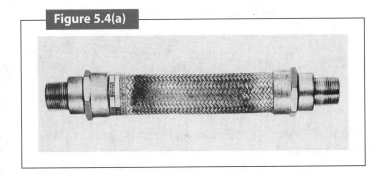

Figure 5.4(a)

An explosionproof flexible fitting.

Source: *Electrical Installations in Hazardous Locations*, 1997, Figure 4-13.

A flexible connection made with liquidtight flexible metal conduit in a Class I, Division 2 area (or a Class II area) must include an internal or external bonding jumper.

Source: Based on *NEC Handbook*, 2005, Exhibit 501.13.

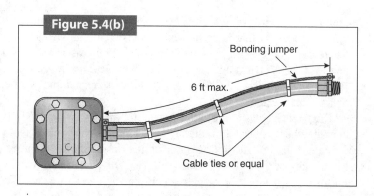

Figure 5.4(b)

the so-called explosionproof flex that is required in most Division 1 applications. In most cases, limited flexibility is not needed for more than a few inches or a few feet, so the length limitation is based on the flexibility needed, and should be justifiable.

In ordinary areas, FMC and LFMC may be used for grounding where the total ground-fault return path is not over 6 feet through the flexible conduit. This use has other restrictions, found in Section 250.118(5) and (6), and those restrictions become more severe in Class I, Division 2 areas. According to Section 501.30(B), bonding jumpers must be installed in FMC or LFMC where it is used as permitted in 501.10(B). A bonding jumper installed on the outside of the flexible conduit is limited to 6 feet according to 250.102(E), as shown in Figure 5.4(b). An exception to the requirement for a bonding jumper applies only to LFMC with listed fittings. This exception allows up to 6-foot lengths for circuits protected at not over 10 amperes where the circuit is not used to supply power utilization equipment. This rule could well be applied to an instrument circuit in a Class I, Division 2 or Class II, Division 2 location. (See 502.306(B) for Class II areas.)

In summary, the *NEC* does not provide a specific length limit for flexible wiring methods in Class I areas. The length that is allowed is the length that is needed, and it should be as short as possible. Unlimited lengths or flexible methods used primarily for ease of installation are not permitted.

Question 5.5 Is "explosionproof flex" permitted between an enclosure and a required seal?

Keywords

Explosionproof
Flex
Flexibility
Coupling
Fitting, seal

NEC References

501.10, 501.15, 501.130

Related Questions

- Is "explosionproof flex" considered to be a "coupling" as mentioned in Section 501.15(A)(1)?
- Does the classification of an item depend on what the manufacturer calls it or on how it is listed?

Answer

Section 501.15(A)(1) says, in part, "only explosionproof unions, couplings, reducers, elbows, capped elbows, and conduit bodies similar to L, T, and Cross types that are not larger than the trade size of the conduit shall be permitted between the sealing fitting and the explosionproof enclosure." Figure 5.5 shows the types of fittings that are permitted between an enclosure and a seal. The only type of fitting permitted between a seal and the enclosure that is not represented in Figure 5.5 is an elbow. The permitted distance between the seal and the enclosure is the same with or without any of these fittings.

The flexible fitting referred to in the question is described in the Underwriters Laboratories (UL) product standard (UL 886) and the guide card information as a "flexible connection fitting." According to UL, it has not separately evaluated any "explosionproof couplings." The original proposal for this requirement was referring to the couplings that are attached to and considered as an integral part of rigid metal and intermediate steel conduit. No other fittings have been evaluated as explosionproof couplings. (Explosionproof three-piece couplings are called "unions" in this section.) Based on this information, a flexible connection fitting is not one of the fitting types permitted between a seal fitting and an explosionproof enclosure per Section 501.15(A). This interpretation is based on how the product testing organization classifies and lists this fitting, and not on the terminology employed by equipment manufacturers.

The fitting shown in Figure 5.5 is often called "explosionproof flex" by installers. This fitting is called a "flexible fitting listed for Class I, Division 1 locations" in 501.10(A)(2), where the subject is "Flexible Connections" under the heading "Wiring Methods." According to the UL book *General Information for Electrical Equipment*, "Prospective users should first ascertain from authorities having jurisdiction under what conditions these flexible connection fittings will be accepted. The use of flexible fittings should be avoided wherever possible. They should be used only when conditions are such that threaded rigid conduit cannot be used." For many users, the cost of these fittings often serves as a practical deterrent to their use as well.

The term "fitting or flexible connector" is used in Section 501.130(A)(3), where flexibility for pendant luminaires is the subject, but this is a separate type of fitting, called a "flexible fixture fitting" under UL standards.

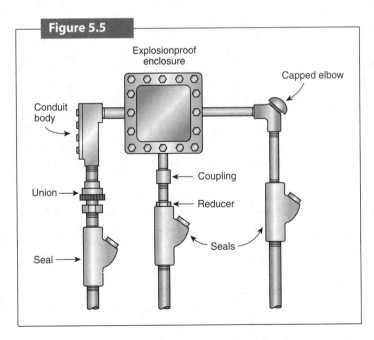

Figure 5.5 The types of fittings permitted between a seal and an enclosure that is required to be explosionproof are very limited.

Question 5.6 — What types of facilities are and are not covered by Article 511?

Keywords

Classified
Repair garage
Parking garage
Sump
Class I

NEC References

500.5, 511.1, 511.3, NFPA 88B

Related Questions

- Does Article 511 apply to areas where only lubrication and tire services are performed?
- Is the sump in a car wash area a classified location?
- Is a car wash treated differently from a repair garage or a parking garage?
- Are service areas for compressed natural gas (CNG) vehicles covered by Article 511?
- Are general-purpose receptacles permitted in battery charging areas?

Answer

Article 511 applies to specific types of facilities that often include hazardous (classified) locations. According to 511.1, Article 511 covers "locations used for service and repair operations in connection with self-propelled vehicles . . . in which volatile flammable liquids or flammable gases are used for fuel or power." This definition is based on two primary points: the type of operations and the type of vehicles. Based on the scope and further information in 511.3, areas not used for service or repair, such as parking garages, are not classified. Also, since the scope only applies to vehicles using flammable liquids and gases, service areas for vehicles using only combustible liquids for fuel, such as diesel-powered trucks, are not covered. However, unless a garage strictly limits its work to diesel vehicles only, Article 511 will usually be presumed to apply, but areas where only diesel vehicles are serviced may still be unclassified according to 511.3(A)(6). (Combustible and flammable liquids are defined in other standards, such as NFPA 497, *Recommended Practice for the Classification of Flammable Liquids, Gases, or Vapors and of Hazardous (Classified) Locations for Electrical Installations in Chemical Process Areas*.) Since the scope of Article 511 includes service areas for vehicles fueled by flammable gases, CNG (compressed natural gas) and LPG (liquified petroleum gas) vehicle service areas are covered.

Figure 5.6(a) illustrates the areas that are classified in a commercial garage where LPG, gasoline, and CNG vehicles are serviced. This diagram shows that most of the classified areas may be eliminated and the classification of pit areas may be reduced with adequate ventilation as described in 511.3. Without adequate ventilation, CNG vehicles create classified areas near the ceiling because the fuel vapors are lighter than air, but the areas classified for LPG vehicles are the same as for gasoline vehicles. Notice that both parts of Figure 5.6(a) include some ventilation because a certain amount of ventilation will be needed to keep exhaust vapors from accumulating.

Section 511.3(A)(4) addresses service areas where only tires are serviced or where only lubrication and oil changes are performed, that is, where Class I liquids are not transferred. In addition, Table 514.3(B)(1) mentions these areas as a possible part of a motor fuel dispensing facility. According to Table 514.3(B)(1), a lubrication or service room without dispensing (of flammable liquids or gases), like that shown in Figure 5.6(b), is unclassified if adequate ventilation is provided.

Where lubrication and oil changes are the only services provided, the liquids that are dispensed are typically combustible liquids rather than Class I flammable liquids. Section

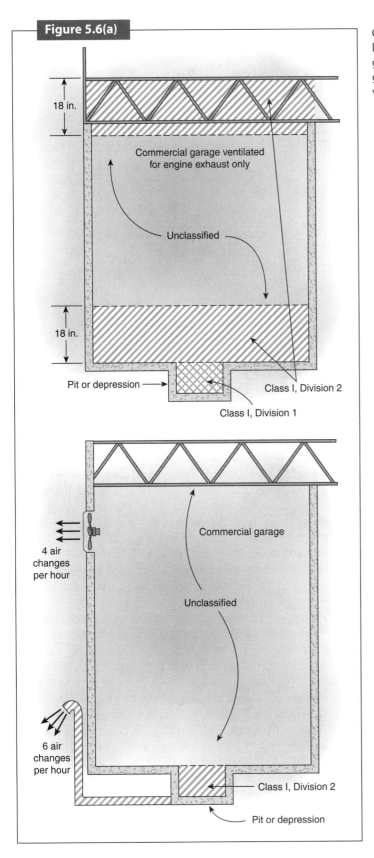

Figure 5.6(a) Classification of locations in commercial garages where LPG, gasoline, and CNG vehicles are serviced.

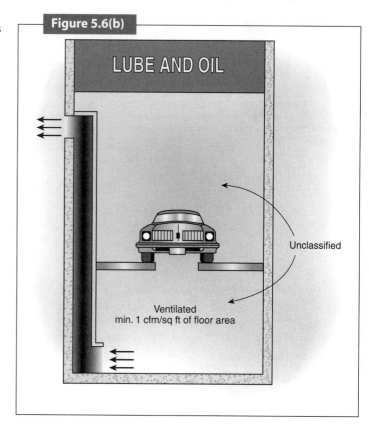

Figure 5.6(b) A "lubritorium" that offers only oil, filter, and lubrication services may be unclassified if properly ventilated.

511.3(A)(4) says the pit areas may be unclassified if they are adequately ventilated, and 511.3(A)(6) says the above-floor areas may be unclassified if Class I liquids are not transferred.

Areas where *only* tires are serviced may have no sources of flammable liquid other than the vehicle fuel tanks. Nevertheless, these areas are commercial garages as defined in Article 511 because they do involve service and repair of vehicles; thus, without proper ventilation, parts of these areas would also be classified unless a different determination were made by the authority having jurisdiction based on the specific application. Many shops that sell and repair tires do other types of service work as well.

According to 511.10, battery charging equipment may not be located in the classified locations specified in 511.3. Since the battery charging equipment may only be located in unclassified areas, general-purpose receptacles may be used to supply the battery chargers. Battery charging equipment for electric vehicles must comply with Article 625, which also requires ventilation in some cases.

If a sump were located in a commercial garage, the sump would be a Class I, Division 1 or 2 area according to 511.3(B)(3), as shown in Figure 5.6(a). However, since there is usually no vehicle repair performed in a dedicated wash area, that portion of a facility does not qualify as a commercial garage per Section 511.1 of the *NEC*. Vehicle washing is not defined as a repair operation by the *NEC* or NFPA 88B, *Standard for Repair Garages*. Based on the Section 500.5 definitions of Class I, Division 1 and 2 locations, car and vehicle washing areas are not treated as hazardous (classified) locations. Although the possibility of a leaky vehicle fuel tank exists, experience with these types of facilities has shown that area classification is unnecessary, as illustrated in Figure 5.6(c). Typically, electrical equipment in vehicle washing areas needs no special rating other than being suitable for the wet locations.

A car wash area involves large quantities of water that may mix with comparatively small quantities of fluids that leak from vehicles. Some of these fluids may be flammable, but they

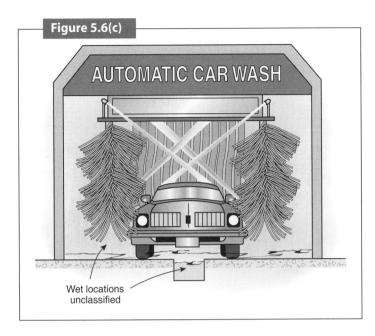

Figure 5.6(c)

Car wash areas are not considered part of a commercial garage as defined in the *NEC* and NFPA 88B.

are highly diluted and therefore highly unlikely to form flammable mixtures in air, even in the sump. On the other hand, even though 511.3 says a parking garage is not a classified area, as illustrated in Figure 5.6(d), the sumps in some vehicle parking garages are treated as classified areas, perhaps because there are many more vehicles (and therefore more leaks) and less water to dilute the leaking fluids, or perhaps because the sump is also used for sewage. Other areas below floor level in a parking garage, such as an elevator pit, are not classified areas in most cases.

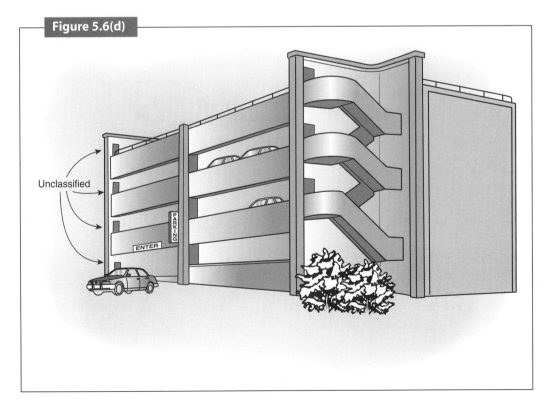

Figure 5.6(d)

Parking garages are not covered by Article 511 and typically do not include classified areas.

Question 5.7: What type of portable lighting is permitted in garages and aircraft hangars?

Keywords

Garage
Hangar
Explosionproof
Portable lighting
Hand lamp
Trouble light
Class I
Division 1
Division 2

NEC References

501.130, 511.3, 511.4, 513.3, 513.10

Related Question

- Where do I find the requirements for portable lighting in classified areas?

Answer

Where portable lighting equipment is used in areas that are designated as Class I, Division 1 or 2, they must be provided with construction features that enable them to be used safely where there is or could be an ignitible environment. The *National Electrical Code* covers portable lighting equipment used in areas where ignitible concentrations of gases or vapors exist under normal operating conditions (Division 1) or could exist under abnormal conditions (Division 2). The general requirements are found in Section 501.130(A)(1) for Division 1 and in 501.130(B)(1) and (B)(4) for Division 2. Both require identification for Class 1, Division 1 generally. A portable "hand light" or "trouble light" of this type is shown in Figure 5.7.

Figure 5.7

An explosionproof hand lamp and cord reel for use in Class I locations.

Source: *NEC Handbook*, 2005, Exhibit 501.21. (Courtesy of Appleton Electric Co.)

Before considering the specific requirements for portable lighting, we must first consider whether garages and aircraft hangars are Class I, Division 1 or 2 locations. Garages, covered in Article 511, are either major repair garages or minor repair garages as defined in 511.2. The classification of the spaces in either type of garage depends on whether the garage is used for repair of vehicles that use flammable liquids or gases (or hydrogen) as fuel. This concept was introduced to Article 511 in the 2008 *NEC*. The classification of specific spaces in the garage is covered in 511.3 and is too complex to repeat here, but, in general, spaces within 18 in. of the floor in a major repair garage used for repair of vehicles that use gasoline as fuel are Class I, Division 2 locations, unless ventilation of at least 1 cfm per square foot of floor area is provided. Spaces within 18 in. of the floor in a major repair garage used for repair of vehicles that use a lighter-than-air fuel are unclassified, but spaces within 18 in. of the ceiling are Class I, Division 2. Other spaces in the garage may be classified based on specific conditions. In minor repair garages, floor areas are usually unclassified, but pits and subfloor spaces may be classified. Aircraft hangars

More specific requirements for commercial (repair) garages and aircraft hangers are found in 511.4(B)(2) and 513.10(E)(1). Section 511.4(B)(2) describes and requires portable lighting of the sort illustrated in Figure 5.7 (suitable for Class I, Division 1) unless the equipment is installed and arranged so that the portable light cannot be used in the classified area. For example, if a cord reel similar to the one shown in Figure 5.7 is hung below the classified area near the ceiling (where there is one) and arranged so that the lampholder assembly cannot reach into the classified area near the floor, the reel and lampholder assembly are not required to be suitable for a classified area. The other aspects of the description in 511.4(B)(2) would still apply, however. Similar, but less specific, requirements apply in aircraft hangers, according to 513.10(E)(1).

Classified areas may not exist in some garages, depending on the types of vehicles serviced and the adequacy of the ventilation (see 511.3(A)). However, the classified areas in aircraft hangers are described in 513.3 without exceptions that would allow the areas to be unclassified due to ventilation.

A product-testing organization such as Underwriters Laboratories (UL) can provide information regarding specific construction and performance requirements contained in product standards for these types of equipment, both for classified areas and unclassified areas.

Question 5.8

Is there a way to use ordinary equipment in a hazardous area?

Related Questions

- May I use an ordinary luminaire for a spray booth?
- How do I determine where ordinary equipment may be used and where special equipment for hazardous locations is required?

Answer

Areas are classified or considered hazardous under the *NEC* due to the presence of flammable or explosive materials that could be ignited by electrical equipment. To prevent the ignition of gases, vapors, dust, fibers, or flyings in an area where such a hazard is likely, electrical equipment must either be removed from the area, or special equipment must be used.

In order for the function of many types of electrical equipment to be realized, the equipment often must be located in the classified area, may not be in separate enclosures, and may not operate on limited-energy sources. In these cases, special equipment is required. The type of special equipment depends on the classification of the area, and equipment must be selected that is suitable for the specific classification of the area and the flammable or combustible materials upon which that classification is based. Descriptions of classified areas are found in Section 500.5, and requirements for equipment in those areas are found in 500.8.

The *NEC* defines classified areas, but generally does not say what areas are classified or the extent of those areas. Other National Fire Protection Association (NFPA) standards, building codes, and industry standards are usually the basis for area classification, and the *NEC* is used for installations in those areas once the classification has been determined. Nevertheless, for some common types of facilities, where area classifications have been well established

Keywords

Classified area
Hazardous area
Equipment
Luminaire
Spray booth

NEC References

500.5, 500.8, 516.3

and well documented, area classification information is included in the *NEC*. Aircraft hangars, commercial garages, and spray operations are examples of common facilities for which area classification information is available in the *NEC*. For instance, Figure 5.8(a) shows area classification drawings for spray booths that are included in the *NEC*. The interiors of the booths are Class I or Class II, Division 1 areas. The dotted outlines in the drawings of Figure 516.3(B)(2) show the extent of Division 2 areas in two situations as described in 516.3(B)(2). A smaller classified area is achieved by interlocking the spray equipment with the exhaust system so that spraying may not take place unless the exhaust ventilation system is operating. Any equipment inside the classified areas must be suitable for those areas as required by Section 500.8.

An important point to remember in all classified areas is that not all the electrical equipment that serves classified areas must be located in those areas. Let's consider the example of the spray booth. Certainly some lighting is required for the spray operation, and motors will be required to operate an air compressor and an exhaust fan. Controls for the lights and motors will be needed, and perhaps interlock circuits for providing fan operation. But do any of these items have to be located within the spray booth or within the classified areas outside the spray booth?

The compressor may be remote and may operate automatically to maintain air pressure in a remote tank. Only the air piping must come to the classified area. Fans may also be de-

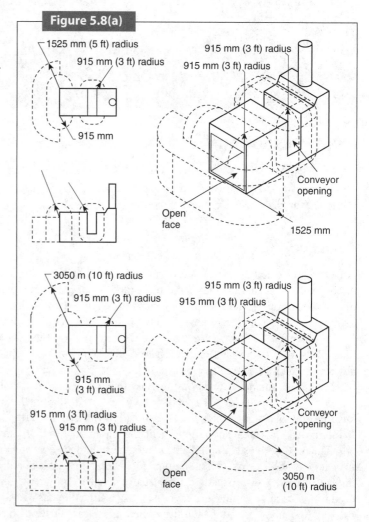

Class I or Class II, Division 2 locations adjacent to a closed top, open face, or open front spray booth or room.

Source: *NEC*, 2005, Figure 516.3(B)(2), and NFPA 33, 2003, Figures 4.3.2(a) and 4.3.2(b).

Figure 5.8(b) An enclosed spray booth illuminated through translucent panels by fixtures mounted on the outside of the booth.

Source: *Electrical Installations in Hazardous Locations*, 1997, Figure 3-12.

signed so that the motor is outside the exhaust duct and outside the spray areas. Controls may certainly be located near but not inside the classified areas, or they could be designed to use intrinsically safe or nonincendive circuits. Even the lighting may be located outside the area and directed through the open front of the booth. For many spray booths, this idea is used routinely, often by installing translucent panels in the walls or ceilings of the spray booths and locating the light sources outside the booth, as shown in Figure 5.8(b).

This spray booth example illustrates the advice found in the Fine Print Note (FPN) of Section 500.5:

> Through the exercise of ingenuity in the layout of electrical installations for hazardous (classified) locations, it is frequently possible to locate much of the equipment in a reduced level of classification or in an unclassified location and, thus, to reduce the amount of special equipment required.

Such an exercise in ingenuity not only saves money on installation and equipment but may also reduce the hazard.

Question 5.9

Are optometrists' or psychologists' offices patient care areas?

Related Questions

- How are health care facilities defined?
- What is a patient care area?
- Are medical and dental clinics and offices covered by Article 517?

Keywords

Health care
Medical
Dental
Ophthalmology

Keywords

Optometry
Psychology
Patient care area

NEC References

517.2, 517.13, NFPA 99

Answer

The definition of a health care facility in 517.2 does include medical and dental offices. NFPA 99, *Standard for Health Care Facilities*, defines a medical/dental office as follows:

A building or part thereof in which the following occur:

(a) Examinations and minor treatments/procedures are performed under the continuous supervision of a medical/dental professional.
(b) Only sedation or local anesthesia is involved and treatment or procedures do not render the patient incapable of self-preservation under emergency conditions.
(c) Overnight stays for patients or 24-hour operation are not provided.

A dictionary definition of "medical" is "of, relating to, or concerned with physicians or the practice of medicine; requiring or devoted to medical treatment." The definition of "medicine" is "the science and art dealing with the maintenance of health and the prevention, alleviation or cure of disease." The definition of "optometry" is "the art or profession of examining the eyes for defects and faults of refraction and prescribing correctional lenses or exercises, but not drugs or surgery." The difference between an optometrist and an ophthalmologist is that the ophthalmologist treats diseases of the eye, whereas optometrists generally do not. Based on these definitions, it appears that an optometrist's office is not covered by Article 517. The same line of logic can be applied to the practice of psychology: It is not a medical practice by definition; thus, it is not covered under Article 517. However, many other medical and dental facilities are covered by Article 517.

Another consideration, especially when the psychologist's or optometrist's practice is in a clinical setting, is whether the examination rooms used by these professionals are for patient care purposes. Even in hospitals, the lobbies, hallways, waiting rooms, private offices, and similar areas are not considered patient care areas. Nevertheless, many designers and installers will treat all rooms of a hospital or clinic as patient care areas simply because they may eventually be used that way. From the standpoint of the code, however, only those areas intended for use in patient care are required to meet the provisions of Section 517.13, which restricts wiring methods and imposes special requirements for grounding methods.

A patient care area is defined in Section 517.2, which includes an explanatory Fine Print Note (FPN):

Any portion of a health care facility wherein patients are intended to be examined or treated. Areas of a health care facility in which patient care is administered are classified as general care areas or critical care areas, either of which may be classified as a wet location. The governing body of the facility designates these areas in accordance with the type of patient care anticipated and with the following definitions of the area classification.

FPN: Business offices, corridors, lounges, day rooms, dining rooms, or similar areas typically are not classified as patient care areas.

Section 517.2 goes on to define general care areas, critical care areas, and wet locations as subcategories of patient care areas. A patient care area does not necessarily involve a bed, and examining rooms and clinics are specifically mentioned in the definition of a general care area. Patient contact with appliances is an issue that makes the patient care area subject to special rules.

Back to the question regarding optometrists and psychologists: The final answer to the question of whether optometrists' and psychologists' offices are medical facilities that include patient care areas should probably be based on the way that these two professions are regulated within a jurisdiction. The government agency that regulates the practice of medicine in the jurisdiction will have to determine whether optometry and psychology (or other practices) are considered to be medical practices. If they are, Article 517 applies; if not, it doesn't. This decision may also rest on the method of examination or treatment. For example, optometrists use equipment much like what a dentist uses, in the sense that electrically powered equipment is used to make contact with the patient. This is much less likely in a psychologist's office.

Question 5.10

Is special wiring or other special equipment required for patient care areas?

Related Questions

- What wiring methods are permitted in patient care areas?
- When is an insulated grounding conductor required in health care facilities?
- Is "redundant grounding" a requirement for all patient care area wiring?
- Is Type MC or Type AC cable permitted for wiring in patient care areas?
- May Type NM cable be used in examination and treatment rooms?
- Are isolated grounding receptacles required in patient care areas?
- What is a "hospital-grade" receptacle?
- Where are hospital-grade receptacles required?
- Do the rules for patient care areas modify the requirements for emergency wiring in health care facilities, or the other way around?

Keywords

Patient care area
Patient vicinity
Health care facility
Grounding
Redundant grounding
Equipment grounding conductor
Type MC
Type AC
Type NM
Metallic raceway
Receptacles
Hospital grade
Isolated grounding

NEC References

250.148, 517.2, 517.10, 517.13, 517.16, 517.18, 517.19, 517.30, 517.32, 517.33, 517.45, 517.80

Answer

Part II of Article 517 covers patient care areas in all health care facilities, according to 517.10(A). This includes patient care areas in all the health care facilities described or listed in the definition of "health care facility" in 517.2, including medical and dental offices. Section 517.13 provides specific requirements for grounding within patient care areas. Although the subject of the section is grounding, 517.13(A) also restricts the types of wiring methods that may be used in a patient care area, because it requires the wiring method to include an equipment grounding return path in the form of the raceway or cable armor itself. This means that only metallic raceways and metallic cables may be used, and where metallic cable is used, the cable armor or sheath must be suitable as an equipment grounding conductor.

In addition to the requirement of 517.13(A) for a wiring method that includes an inherent grounding path, 517.13(B) says the wiring method must also include an insulated equipment grounding conductor for grounding "the grounding terminals of all receptacles and all non-current-carrying conductive surfaces of fixed electric equipment likely to become energized that are subject to personal contact, operating at over 100 volts." Thus, *most* line-voltage electrical equipment in a patient care area must be supplied with two grounding paths: one through the raceway or cable armor or sheath, and one through an insulated equipment grounding conductor contained within the wiring method. This special grounding requirement is often called "redundant grounding."

Equipment is considered to be "likely to become energized" if the equipment is conductive, contains insulated conductors, and failure of the insulation could energize the equipment. This, of course, includes virtually all metallic electrical equipment, including most appliances; in the case of a patient care area, it may include a powered chair or table for examinations or procedures or a bed.

Taken as a whole, Section 517.13 permits the use of Type MC (metal-clad) or Type AC (armored) cable, but they must be of special types. The most common type of MC cable has interlocked armor that is not suitable as a grounding means but also includes an insulated equipment grounding conductor. Type MC and AC cables that are intended for use in patient care areas often have the additional marking "HC," for health care. The type of MC cable that may be used in patient care areas must have a continuous sheath that is suitable for use as an equipment grounding path in addition to the insulated equipment grounding conductor. This distinguishes "MC-HC" from the more common types of MC cable that have interlocked armor. The most common type of AC cable has a shorting strip inside the cable that makes the armor suitable for grounding, but it does not contain an insulated equipment grounding conductor, so such a conductor must be added to AC cable for it to be the "AC-HC" type used in patient care areas. Nonmetallic raceways, such as ENT (electrical nonmetallic tubing) or RNC (rigid nonmetallic conduit), or cables with nonmetallic sheaths, such as NM cable, may not be used to supply power or lighting circuits in patient care areas in any type of medical or dental facility.

The requirement for an insulated equipment grounding conductor has two exceptions. Section 517.13(B), Exception No. 1, permits a metal faceplate to be grounded by means of metal screws that are used to attach the faceplate to a box or wiring device such as a switch or receptacle. According to Exception No. 2, luminaires that are more than 7.5 feet above the floor in a patient care area and switches that are located outside the "patient vicinity" are not required to have insulated equipment grounding conductors. This does not affect the choice of wiring method, because the exception only applies to the insulated equipment grounding conductor and not to the requirement to use a metal raceway or cable.

Again, the concern with personal contact comes up in 517.2 in the definition of "patient vicinity" (or "patient care vicinity," as the term is used and defined in NFPA 99, *Standard for Health Care Facilities*). A patient vicinity is described with regard to a bed location as the space within a room that is within 6 feet of the patient bed and within 7.5 feet of the floor. This is the area in which a patient is likely to make contact with equipment or other persons who are in contact with equipment and where shock hazards to the patient are significantly greater than in other situations. Figure 5.10(a) is a diagram of a patient bed. This bed may or may not be an inpatient sleeping bed or a "patient bed location." The area where such a bed or an examining table or treatment chair is located is a patient care area where special wiring and grounding methods must be used.

A patient bed location is defined in 517.2 as "the location of an inpatient sleeping bed; or the procedure table used in a critical patient care area." This is a specific type of patient care area, most commonly found in hospitals, where additional requirements and restrictions apply to wiring methods.

Patient bed locations are required by 517.18 and 517.19 to be supplied by branch circuits from both the normal system and the emergency system. According to 517.30(C)(3), the wiring of the emergency system of a hospital must be mechanically protected by installation in nonflexible metal raceways or Type MI (mineral insulated) cable. This eliminates even special types of AC or MC cable for the emergency circuits in patient bed locations, but not for circuit wiring from the normal branch. (Some flexible raceways or cables are permitted for emergency wiring in headwalls in patient care areas or for other specific and limited uses according to 517.30(C)(3)(3).) Emergency circuits must still have redundant grounding if installed in patient care areas, because the requirements for patient care areas as well as the requirements for emergency systems apply.

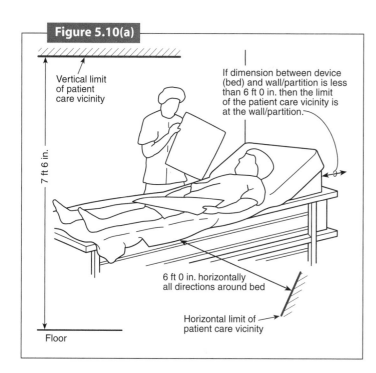

A patient bed, one of many kinds of patient care vicinities.

Source: *Health Care Facilities Handbook*, 2005, Exhibit 3.8.

Although the requirement for raceway protection for emergency circuits in Section 517.30(C)(3) refers specifically to hospitals, and only to hospitals, these same requirements also apply in other health care facilities if electrical life support equipment is used, if critical care areas are present, or if generator systems are used to supply essential electrical systems. These similar requirements can be found in 517.45(B), (C), and (D).

The requirements for raceways for emergency circuits or redundant ground paths in patient care areas are sometimes interpreted as applying also to signaling and communications systems within patient care areas. Section 517.80 is often cited to support this interpretation. However, 517.80 mentions only insulation and isolation, not wiring methods. In patient bed locations, many signaling systems, such as nurse calls, fire alarms, and some communications, are required to be on the emergency system, either on the life safety branch or the critical branch, in accordance with 517.32 and 517.33. Even in these emergency systems, raceways are not required for secondary circuits of Class 2 or Class 3 communications or signaling systems because no special requirements are established for these systems in 517.30(C)(3)(5). Chapter 7 does not require conduit for these systems unless the failure of the circuit would *create* a direct life hazard, which is not the case with nurse calls, thermostat controls, or communications systems (including television systems). Since raceways are generally not required for signaling and communications systems that are part of the emergency system, it does not make sense to say that the less vital systems, such as patient phones or TV cables, should require raceways. Also, as we saw in 517.13(B), redundant grounding does not apply to systems under 100 volts. Figure 5.10(b) summarizes the requirements for raceways and wiring methods in patient bed locations. The wall switch in Figure 5.10(b) is assumed to be within the patient care vicinity as defined, otherwise it would still be subject to the restrictions on wiring methods, but it, like the luminaires, would not require an insulated equipment grounding conductor according to 517.13(B), Exception No. 2.

Receptacles at general care or critical care patient bed locations are required to be "hospital grade" in accordance with 517.18(B) and 517.19(B)(2). Hospital-grade receptacles are characterized by having high-impact-resistant faces and bodies and higher spring tension in

212 Chapter 5 Special Occupancies

Summary of wiring method and redundant grounding requirements in patient care areas that include a patient bed location in a hospital.

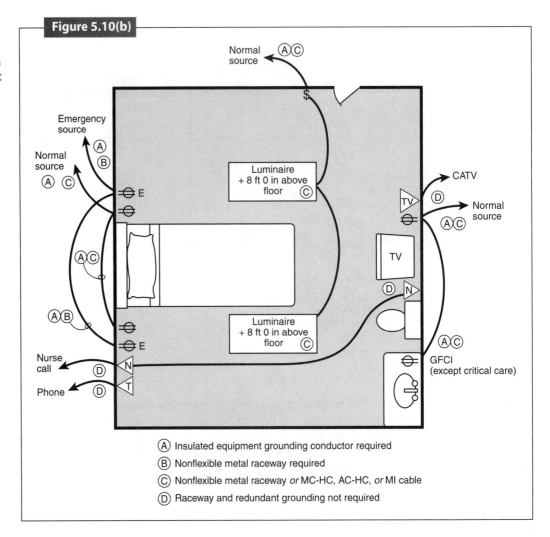

Figure 5.10(b)

Ⓐ Insulated equipment grounding conductor required
Ⓑ Nonflexible metal raceway required
Ⓒ Nonflexible metal raceway *or* MC-HC, AC-HC, *or* MI cable
Ⓓ Raceway and redundant grounding not required

the receptacle contacts, among other features, and are identified by a green dot on the receptacle face. (See Figure 2.21(b) in Chapter 2 of this book.)

If insulated equipment grounding conductors are required in patient care areas, are isolated grounding receptacles also required? The insulated equipment grounding conductors required by 517.13 are not required to be isolated, just insulated. In fact, most metallic electrical equipment is required to be connected to both the wiring method and the insulated equipment grounding conductor, and equipment grounding conductors are required to be connected to boxes in accordance with 250.148. These interconnections provide the "redundant grounding" mentioned previously.

Isolated grounding receptacles are permitted by 517.16, but they are not required, and their use should be limited to applications where isolated grounding conductors are absolutely necessary. Where an isolated grounding receptacle is used, an *insulated* equipment grounding conductor that is *isolated* from boxes and panelboards is needed. This insulated conductor is necessary to ground equipment connected to the isolated grounding receptacle as required by 250.146(D) and 406.2(D). The isolation of this equipment grounding conductor reduces some of the benefits of a redundant grounding path. The Fine Print Note to 517.16 provides important information and cautions with regard to isolated grounding in patient care areas.

Question 5.11

What loads are permitted on the equipment branch of an essential electrical system in a hospital?

Related Questions

- What is meant by "other selected equipment" in 517.34(B)(8)?
- If standby loads do not qualify for the life safety, critical, or equipment branch, may they be supplied from an essential system generator?
- Is another transfer switch required for nonessential loads on an essential power system?

Answer

Section 517.30(B)(5) prohibits loads from being placed on an essential system (including equipment system) *transfer switch* unless those loads are specifically named in Article 517. The loads that are named are found in Sections 517.32 through 517.34. The loads permitted on the equipment system are named in Section 517.34. Therefore, equipment that is not specifically mentioned in Section 517.34 is required to "be served by their own transfer switches" in a way that will assure that the generating equipment will not be overloaded. This requires monitoring of the load on the generator and automatic load shedding when the generator becomes overloaded. It does not prohibit the use of the generator (or other source) for nonessential loads.

To take a closer look, Section 517.34(B) contains a list of specific equipment permitted to be connected to the equipment system. Quoting from Section 517.34(B)(8): "Other selected equipment shall be permitted to be served by the equipment system." Clearly, the phrase "other selected equipment" refers to nonspecific equipment or nonspecific loads and is actually a generalized phrase that does not "name" anything. Therefore, because item (8) of the list is nonspecific, and 517.30(B)(5) requires equipment to be specifically named, equipment included in item (8) is permitted to be served by the equipment system, but must "be served by their own transfer switches." The text of item (8) is extracted from Section 3.4.2.2.3(e) of NFPA 99, *Standard for Health Care Facilities*. The appendix of NFPA 99 includes the following commentary:

> Consideration should be given to selected equipment in kitchens, laundries, and radiology rooms and to selected central refrigeration.
> It is desirable that, where heavy interruption currents can be anticipated, the transfer load be reduced by the use of multiple transfer devices. Elevator feeders, for instance, might be less hazardous to electrical continuity if they are fed through an individual transfer device.

The equipment that is selected should be reviewed to be certain that it qualifies as an essential system load by evaluating the effect on patient safety if power to the equipment in question were to be interrupted for an extended period.

The *NEC* language addressing this issue was changed in the 1996 *NEC* to clarify that other loads that do not qualify as essential system loads (life safety branch, critical branch, or equipment branch) may still be connected to the generating equipment provided the necessary safeguards are in place to prevent generator overloading. The revised wording allows the use of automatic load-shedding systems found in large health care facilities with multiple generators that are connected to automatic paralleling and load control systems. It also allows

Keywords

Essential electrical system
Equipment branch
Standby
Generator
Transfer switch

NEC References

517.30, 517.34, 701.6, NFPA 99

Addition of nonessential system loads and transfer means to an essential electrical system in a hospital.

Source: Adapted from *NEC*, 2005, FPN Figure 517.30, No. 1.

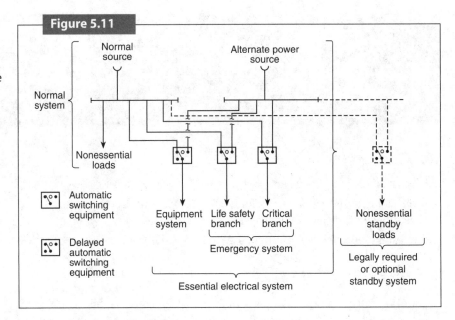

Figure 5.11

manual transfer of standby loads as long as there is an automatic system to shed these loads. As noted previously, this language is also consistent with NFPA 99.

Figure 5.11 is a modification of FPN Figure 517.30, No. 1, in the *NEC* that shows how additional transfer switches might be added to comply with the requirements of 517.30(B)(5) and 517.34(B)(8). The additional transfer means could supply both optional standby and legally required standby loads "where automatic selective load pickup and load shedding is provided as needed to ensure adequate power to the legally required standby circuits" in accordance with 701.6 since 701.7 does not restrict the use of the transfer switch like 700.6 does for emergency systems. This creates a hierarchy of needs that requires the loads to be prioritized and power to be supplied in the following order:

1. Emergency system (life safety and critical branches)
2. The equipment system
3. The legally required standby system
4. The optional standby system

Power can be supplied to all of these systems at the same time if the power source is adequate.

Question 5.12 What requirements apply to temporary wiring at trade shows and conventions?

Keywords

Temporary
Trade show

Related Questions

- What article covers conventions and trade shows?
- Is GFCI protection required for temporary wiring for trade shows and conventions?

Answer

Most trade shows and conventions are held indoors in convention centers (see Figure 5.12). Convention centers usually include a variety of conference rooms, auditoriums, and exhibition halls, all of which are covered by Article 518, "Places of Assembly." For theatrical areas, Article 520, which covers, in part, theaters, performance areas, and similar locations, applies. In those cases where all or part of a convention or trade show is held in outdoor areas, such as stadiums, fair grounds, or similar locations, Article 525, "Carnivals, Circuses, Fairs, and Similar Events," covers the portable wiring and equipment in or on structures. Section 525.3 refers to other articles for wiring for permanent structures (Articles 518 and 520), for audio wiring (Article 640), and for pools and fountains (Article 680).

A significant similarity in the rules for portable and temporary wiring under these articles is that flexible cord is permitted. A significant difference is that ground-fault circuit-interrupter (GFCI) protection is generally required for temporary and portable outdoor wiring under 525.23 if convenience-type receptacles are used, but GFCI protection is not required for temporary wiring in places of assembly, according to 518.3(B). Section 518.3(B) refers to Article 590 for temporary wiring, but specifically states that the GFCI requirements of 590.6 do not apply. Another difference is that although Article 518 refers to Article 590, "Temporary Installations," for temporary wiring, Article 525 provides its own rules for events that are within the scope of Article 525. For the most part, Article 518 covers permanent buildings for assembly of persons, and Article 525 covers portable structures, such as portable tents, in mostly outdoor installations. If a circus or fair is held partially or entirely in permanent structures such as indoor sports arenas, Articles 518 and 520 apply to the permanent buildings.

According to 90.3, Articles 518, 520, 525, and 590 have the authority to modify other articles of the *NEC*. These modifications may be somewhat less restrictive, somewhat more restrictive, or both. For example, in 518.3(B), Exception, hard usage or extra-hard-usage flexible cord is permitted to be installed in a cable tray dedicated to that purpose, a less restrictive rule because this is not a typical use of cord or cable tray. In 518.4, however, the

Convention
Places of assembly
Performance area
Circus
Fair
Flexible cord
GFCI

NEC References

210.8, 400.8, 518.3, 518.4, 525.23, 590.6

Figure 5.12

Trade shows and conventions may involve aspects of Articles 518, 520, 525, and 527 as well as other articles.

wiring methods used for fixed wiring are generally more restrictive than in other occupancies. As another example, in Sections 525.23 and 590.6, most 125-volt, 15- or 20-ampere receptacles are required to have GFCI protection even if they are not in any of the locations where GFCI protection is required by 210.8(B), but hard-usage or extra-hard-usage flexible cord is permitted for both feeders and branch circuits in these applications, and the flexible cord is permitted or required to be fastened to structures, contrary to 400.8.

Chapter 6

Special Equipment, Conditions, and Systems and Related Standards

Question 6.1

When does Article 645 apply?

Keywords

Information technology
IT equipment
Computer room
Disconnecting means
Smoke detectors
Air-handling space
Other space for environmental air
Plenum-rated
Fire-rated construction

NEC References

300.22, 400.8, 645.1, 645.2, 645.5, 645.10

Related Questions

- Must a room used for computer or IT equipment comply with Article 645?
- Is there any advantage to complying with Article 645?
- How many disconnecting means are permitted at exit doors?
- When are smoke detectors required in raised floor spaces?
- How many smoke detectors are required by Article 645 and where should they be mounted?

Answer

Although Article 645 is titled "Information Technology Equipment," not all information technology (IT) equipment or computer equipment is covered by this article. IT equipment is used widely, in almost all occupancies, but few such locations are the "information technology equipment rooms" that, according to 645.1, are the focus of Article 645. Section 645.2 says that Article 645 applies if five conditions are met. Section 645.2 does not require any of the conditions to be met, but if they are, the article applies; if any conditions are not met, the regular rules of the code apply without modification by Article 645.

The five conditions of 645.2 can be summarized as follows:

1. Special means for disconnecting power to IT and heating, ventilating, and air-conditioning (HVAC) equipment are provided in accordance with 645.10.

2. A separate HVAC system is provided, or HVAC system ductwork is separated by fire dampers from other occupancies.

3. The room is used for listed IT equipment.

4. Room occupants are restricted to those necessary for operation and maintenance of the IT equipment.

5. The room is separated from other occupancies by fire-rated construction.

A sixth condition was listed in some previous code editions, but it was deleted because it was an unnecessary statement and outside the scope of the *NEC*: "The room must comply with all building codes." This is not an insignificant requirement, however, because in some cases local building codes may be more restrictive than the *NEC*. For example, smoke detectors connected to HVAC fan shutdown systems were required in many areas before the requirement was added to the *NEC*, and some local building codes do not permit some of the wiring methods in air-handling spaces that are permitted by Article 645.

An IT room is not required to have a raised floor, and the under-floor space is not required to be used for ventilation, but both of these are common practices. Raised floors may be used in other applications and may or may not be air-handling spaces. An under-floor space may also be used for air handling or "other space for environmental air" in other applications, subject to the requirements of 300.22.

From the list of conditions, it should be evident that many areas that may be thought to be "computer rooms" cannot be made subject to the provisions of Article 645. The computer labs that are present in many schools are a good example. These rooms are used by most or

Figure 6.1(a) An example of an information technology equipment room of the type covered by Article 645.

Source: *Limited Energy Systems*, 2002, Figure 3.32.

all students and therefore cannot comply with the fourth condition, which restricts access to the room. Other conditions are difficult to meet unless the room was originally constructed that way. For example, providing the fire separation, separate HVAC, and special disconnecting means may require extensive remodeling of a space if those conditions are to be met. Figure 6.1(a) shows an example of the type of room anticipated by Article 645. Many such rooms may be much smaller.

According to 645.10, means to disconnect the power to IT equipment and HVAC equipment must be placed at principal exit doors. For a small room with only one door, the disconnecting means might consist of regular safety switches. Where there are multiple exit doors, other devices, such as push buttons, may be used to control a contactor or shunt-trip device to disconnect power. The disconnecting means may be two devices, one for IT equipment and one for HVAC, or it may be a single means that interrupts power to both. If a push button is used, pushing the button must disconnect the power. Other devices, such as break-glass stations or switches, may also be used. These disconnecting means are not required to disconnect power for room lighting.

If the floor area is used for ventilation, smoke detection systems must automatically shut down HVAC fans if smoke or fire is detected in the raised floor space. The *NEC* does not cover the spacing or mounting of these devices. In most cases, they will have to be part of a fire alarm system, wired in accordance with Article 760 and designed (including spacing) in accordance with NFPA 72, *National Fire Alarm Code*. Typically, they would be installed on the underside of the floor structure, facing down, as shown in Figure 6.1(b). Few, if any, smoke detectors are intended and listed for face-up mounting, and smoke is more likely to rise and be more readily detected on the underside of the raised floor than in other locations.

Article 645 is a "permissive article" because it says "if you meet these conditions, you may do these things." But with all the conditions placed on its use, why would anyone want to use Article 645? The article provides a number of modifications of the rules in Chapters 1 through 4 of the *NEC* that can be advantageous to the user and designer of IT equipment. The use of flexible cords is an example. Section 645.5(B)(1) permits flexible cords to be used in lengths of up to 15 feet, and this cord may pass through the floor and connect with an attachment plug to a receptacle located in the space beneath a raised floor. In general, flexible cords may not pass through floors, walls, or ceilings according to 400.8.

Two other examples of the benefits of Article 645 are found in 645.5(D) and (E). Sections 645.5(D)(2) and (5) permit nonmetallic wiring methods and non-plenum-rated cables for data wiring in air-handling spaces. This is permitted because of the requirements for separated HVAC systems with smoke detectors, fire-rated construction and isolation from other occupancies, and the restricted access to the spaces, all of which reduce the risk of exposure of people to smoke

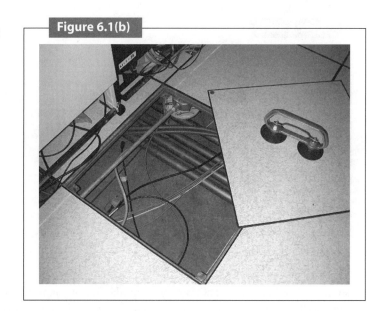

Wiring for IT equipment in a raised floor with a smoke detector.
Source: Low Voltage Seminar.

and reduce the risk of fire or smoke being spread throughout a building in case a fire were to start in the IT room. Section 645.5(E) increases the flexibility of the installation and makes changes in equipment and wiring much easier by allowing many cables and much of the wiring in the floor space to be installed without being secured in place if it is "listed as part of, or for, information technology equipment." Figure 6.1(b) shows both fixed wiring and flexible cabling in a raised floor space. The conduits shown must be fastened in place in accordance with 645.3(D)(2) because they likely supply branch circuits within the space, and they do not fall under the provisions of 645.5(E) because they are not supplied with or specifically listed for IT equipment.

Question 6.2 Why are special boxes required for connections to pool luminaires?

Keywords

Box
Junction box
Deck box
Swimming pool
Spa
Hot tub
Underwater luminaire

NEC References

680.23, 680.24, 680.42, 680.43

Related Questions

- How are junction boxes for pool lighting different from regular boxes?
- Is a special-purpose "deck box" required where there is no "deck" or where the box is not in the deck?
- Are special boxes required for connections to underwater lights in spas or hot tubs?

Answer

Section 680.24 covers the requirements for so-called deck boxes. The rules in 680.24 apply to junction boxes connected to conduits where the conduits extend directly to forming shells of wet-niche or dry-niche luminaires or to mounting brackets of no-niche luminaires. The first requirement in 680.24(A)(1) is very clear in requiring such boxes to be listed as a swimming pool junction box. Boxes that are listed for this purpose should be labeled

"Swimming Pool Junction Box" and be marked "Listed," along with the identification of the applicable listing agency.

In simple terms, the difference between a swimming pool junction box and other boxes is in the respective product standards. Swimming pool junction boxes are evaluated under the provisions of Underwriters Laboratories (UL) 1241, and the evaluation and listing of other boxes is based on UL 50. The UL *Electrical Equipment Construction* directory does not provide a lot of detail on what the product standard provides, because it primarily is about how to identify products that meet the standard. However, Section 680.24(A)(1) does describe some of the required features of listed swimming pool junction boxes. These features include provisions for connection of threaded conduit or nonmetallic conduit to the box. In addition, the box must be constructed of corrosion-resistant materials, include integral means for terminating grounding conductors, and provide means of ensuring continuity between metal conduits that attach to the box. Another provision that relates to the listing of the box is found in 680.24(D). This section says that the integral grounding provisions must include terminals for one more grounding conductor than the number of conduit entries. Although most equipment grounding conductors in pool lighting will be 12 AWG or perhaps as small as 16 AWG in cords, any box that is to be used with nonmetallic conduit and connecting to a wet-niche luminaire must have provisions for terminating the 8 AWG copper conductor required by 680.23(B)(1).

Comparable enclosures used for other purposes, such as those used to house ground-fault circuit-interrupter (GFCI) devices or transformers, must meet requirements similar to those for junction boxes, with one added requirement: They must be capable of being sealed to prevent circulation of air between the enclosure and attached conduits. (The GFCI devices mentioned in this context are not necessarily GFCI receptacles. They might be GFCI receptacles if the box is far enough away from the pool or on the other side of a permanent barrier, in accordance with both 680.22(A) and 680.24(B)(2).)

Installation requirements are also pretty much the same for junction boxes and other types of enclosures for swimming pools. The boxes must be at least 4 feet from the inside edge of the pool, and the inside bottom of the box must be at least 4 inches above the pool deck or 8 inches above the maximum water level, whichever provides a greater elevation. This requirement is illustrated by Figure 6.2(a), which shows only the 8-inch dimension because the maximum water level in this case is assumed to be the lip of the pool, which is higher than

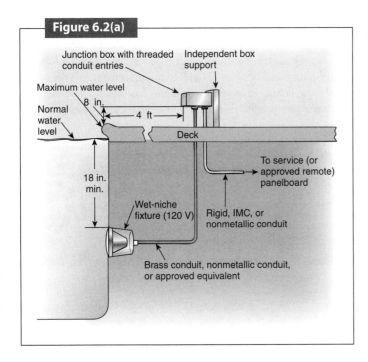

Junction box for a swimming pool shown with a wet-niche luminaire (lighting fixture).

Source: Based on *NEC Handbook*, 2005, Exhibit 680.1.

Flush junction or deck box used with a wet-niche forming shell on a lighting system of 15 volts or less.

Source: Based on *NEC Handbook*, 2005, Exhibit 680.7.

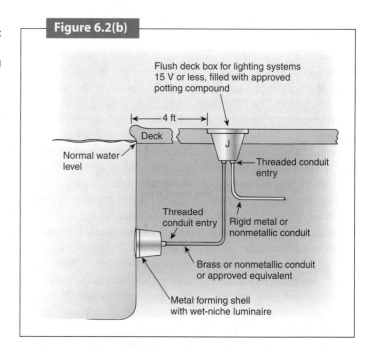

Figure 6.2(b)

the deck. In many cases, the maximum water level might be lower than this, depending on the placement and elevation of drains, and the 4-inch dimension above the pool deck might be the higher elevation. The boxes can be closer than 4 feet from the pool if separated from the pool by a permanent barrier such as a wall or fence. Permitted wiring methods as illustrated in Figure 6.2(a) are covered by 680.23(B)(2) for the connections to the forming shell, and by 680.23(F)(1) for the branch circuit supplying the deck box.

Flush deck boxes may only be used for junction boxes and only with lighting systems operating at 15 volts or less. Flush deck boxes must meet the requirements for other types of boxes, except for elevation; in addition, they must be filled with an approved potting compound to prevent the entrance of moisture. Figure 6.2(b) illustrates the use of a flush deck box. The drawing shows that the 4-foot horizontal dimension still applies to these boxes even though the 4-inch or 8-inch vertical dimensions do not.

Because Sections 680.42 and 680.43 (Part IV, "Spas and Hot Tubs") require compliance with Part II of Article 680 and the rules for boxes are not modified in Part IV, the requirements discussed earlier also apply to boxes connecting to underwater lights in spas and hot tubs.

Requirements for reliable grounding and bonding of electrical equipment in swimming pools and similar areas are critical to personnel safety in those areas. Listing and labeling provide a way of ensuring that an appropriate product is selected to meet these requirements.

Question 6.3 Where are receptacles required for pools, spas, and hot tubs?

Keywords

Receptacle

Related Questions

- What is a "general-use receptacle"?

- May a receptacle installed to meet the requirement for outdoor receptacles be used for a swimming pool, spa, or hot tub?
- What can be done if the deck area around a pool or hot tub is smaller than the distance required for a receptacle?
- Why must I have a receptacle within a certain distance of a pool, spa, or hot tub?
- Is a receptacle near a pool permitted on the other side of a fence or wall?
- Must receptacles required by Article 680 be readily accessible?

Pool
Spa
Hot tub
Hydromassage bathtub
Therapeutic tub
General use
GFCI
Accessible
Readily accessible

Answer

The requirements for receptacles at pools are found in Part II of Article 680. Section 680.22(A) covers receptacles specifically and primarily addresses how close receptacles may be installed to pools, and which receptacles require ground-fault circuit-interrupter (GFCI) protection. The most general requirement for receptacles is that they be at least 10 feet from the inside walls of pools. However, for circulating and sanitation equipment they may be as close as 5 feet if they are grounding, locking, GFCI-protected, single receptacles. The requirements for grounding and GFCI protection are largely for shock protection, and the use of single receptacles with a locking configuration will mean there are only enough receptacles for the pool equipment, and thus the receptacles will not be easily used for portable electrical equipment in the pool area.

Section 680.22(A)(2) covers "other receptacles," that is, other than those for pool circulation or sanitation equipment, and 680.22(A)(3) through (6) cover other details of receptacle placement. Any receptacle within 20 feet of a pool has to be GFCI protected, regardless of occupancy. At dwelling units, a receptacle must be installed within 20 feet and not less than 10 feet from the inside edge of a pool. Providing a convenient GFCI-protected receptacle in the 10- to 20-foot area from a pool helps to reduce the use of extension cords, helps to ensure that any portable electrical equipment will be GFCI protected, and helps keep most ordinary appliances away from the pool because normal appliance cords are 6 feet long or less. Where pools at dwelling units are less than 10 feet from lot lines or buildings so that a receptacle cannot be placed in the 10- to 20-foot area, one receptacle is still required and only one is permitted less than 10 feet and more than 5 feet from the pool. These requirements and restrictions are shown in Figure 6.3(a).

The "other receptacles" required at dwelling units are required to be supplied from general-purpose branch circuits. Such receptacles, rated 125 volts, 15 or 20 amperes, are often called general-purpose receptacles, or convenience receptacles. The point is that a general-purpose branch circuit is one that is not dedicated to identified equipment or loads, that is, a circuit that is not a small appliance branch circuit, an individual branch circuit, or a branch circuit otherwise intended for specific equipment. According to Article 100, a general-purpose branch circuit "supplies two or more receptacles or outlets for lighting and appliances," so the circuit used for the receptacle at a pool could be a circuit for general lighting and convenience receptacles. For that matter, a receptacle installed as an outdoor outlet as required by 210.52(E), or a receptacle installed for servicing heating, air-conditioning, or refrigeration equipment as required by 210.63, could also be used to meet the requirement of 680.22(A)(3) if it can be located where it will meet more than one requirement. For example, in Figure 6.3(a), the receptacles shown on the sides or back of the house are located at the rear or as close to the rear as permitted in order to comply with the requirements of 210.52(E).

As noted, the requirement to have a receptacle in a specific area near a pool applies to dwelling units, whether single-, two-, or multiple-family. There is no general requirement for any such general-use receptacles at commercial pools or pools in schools, gyms, hotels, or other public areas. However, if any 125-volt receptacles are installed within 20 feet of a pool, they must have GFCI protection. In addition, any receptacles "rated 15- or 20-amperes, 125 through

NEC References

210.8, 210.52, 210.63, 406.8, 680.22, 680.30, 680.40, 680.42, 680.43, 680.61, 680.70, 680.71, 680.72, 680.73

Acceptable receptacle locations within 20 feet of a permanently installed pool at a dwelling unit.

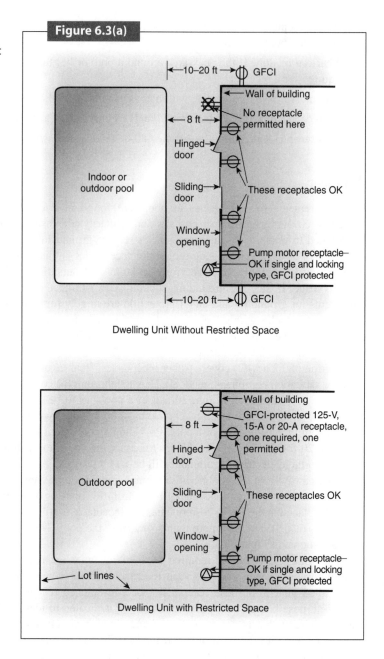

250 volts, single phase" that supply pool pump motors must be GFCI protected without regard to their proximity to a pool. These rules are found in 680.22(A)(5) and apply to all occupancies.

Section 680.22(A)(6) explains how the distance of a receptacle from a pool is measured. Permanent barriers such as walls, floors, and ceilings, with or without doors or windows, create separate spaces. A receptacle on the other side of such a barrier is usually not counted as being within 20 feet of a pool, even though the wall or barrier may be within 5 or 10 feet of the pool. The measurement is taken as the shortest path a cord would follow from a receptacle "without piercing a floor, wall, ceiling, doorway with hinged or sliding door, window opening, or other effective permanent barrier." Thus, receptacles in a structure, like those shown inside the building in Figure 6.3(a), are permitted without GFCI protection. Where cords could pass through openings without windows or doors, over short walls, or around partitions (essentially, where cords could reach a receptacle without violating the restricted uses for cords in 400.8), a receptacle on the other side of the barrier is subject to 680.22(A) and requires GFCI protection if the receptacle is within 20 feet of the inside walls of the pool as measured along the cord.

The rules discussed earlier apply primarily to swimming pools because they are in Part II of Article 680, which is titled "Permanently Installed Pools." However, these rules may also apply to other installations covered by Article 680, depending on the specific language of the other parts. Part III covers storable pools, and according to 680.30, only Parts I and III apply to storable pools. Storable pools may not be installed in precisely the same location every time they are set up, so rules about specific placement of receptacles do not apply. However, 680.32 does require GFCI protection for all electrical equipment used with storable pools, including power supply cords.

Similarly, Part IV applies to spas and hot tubs, and 680.40 says only Parts I and IV apply to these installations. However, according to 680.42, spas and hot tubs installed outdoors must also comply with Part II. Part II is modified by 680.42 with regard to wiring methods, but the provisions for receptacle locations apply to outdoor spas and hot tubs without modification. With regard to receptacles, Part II does not distinguish between indoor and outdoor pools, but Part IV does distinguish between indoor and outdoor spas and hot tubs. Specifically, the requirements for indoor spas and hot tubs found in 680.43 invoke and then modify the receptacle requirements of Part II. For indoor spas and hot tubs, a convenience receptacle must be located not less than 5 feet and not over 10 feet from the inside walls of the spa or hot tub, rather than the 10 to 20 feet required for pools and outdoor spas and hot tubs. (The special rule for limited spaces allowing a receptacle closer than 10 feet also applies to outdoor spas and hot tubs, and is reasonably also applicable to spas or hot tubs located outdoors on decks where space is limited by railings, but the literal language refers only to lot lines.) Receptacles for indoor spas and hot tubs must be GFCI protected if within 10 feet, rather than within 20 feet as required in Part II.

Part V of Article 680 applies to fountains, but the provisions of Part II only apply to fountains that have water common to a pool. GFCI protection is usually required for fountain equipment.

Part VI of Article 680 applies to therapeutic tubs and pools, and Part VII applies to hydromassage bathtubs. According to Section 680.61, Part II (and the receptacle rules discussed previously) applies only to permanently installed therapeutic tubs—that is, those that are installed in the ground or on the ground or that cannot be readily disassembled. Hydromassage bathtubs must only comply with Part VII, according to 680.70. Section 680.72 says electrical equipment, including receptacles, shall be installed according to the rules for bathroom areas. This means that receptacle outlets must be GFCI protected as required by 210.8 and that receptacles may not be installed in the tub space, per 406.8(C). Section 680.71 also requires all 125-volt receptacles rated 30 amperes or less and within 5 feet of the tub to be GFCI protected. However, no receptacles are required around hydromassage tubs unless the tub is in the same room as a bathroom basin. Receptacles may also be required for cord-and-plug-connected hydromassage bathtub equipment.

Although Section 680.22(A)(3) does not specifically say that required convenience receptacles must be readily accessible, a requirement for ready access is implied because the receptacles must not be over 6 feet above the platform serving the pool. As noted, this requirement applies to pools, spas, hot tubs, and permanently installed hydrotherapeutic tubs, but not to storable pools or hydromassage bathtubs. Receptacles at basins in bathrooms must be readily accessible, but electrical equipment for hydromassage bathtubs, including receptacles for cord-and-plug-connected hydromassage tub pumps, must only be accessible according to 680.73, not readily accessible. This rule is illustrated by Figure 6.3(b). Although the illustration shows the access in the same room as the tub, the access could be arranged to be on the other side of a wall adjoining the tub, and is sometimes on the other side of an outside wall.

Access panel for hydromassage tub electrical equipment including receptacles.

Source: Based on *NEC Handbook*, 2005, Exhibit 680.19.

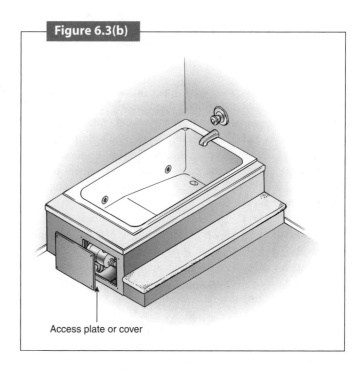

Figure 6.3(b)

Access plate or cover

Question 6.4 — How is bonding to be accomplished for hydromassage bathtubs?

Keywords

Hydromassage bathtub
Bonding
Equipotential
Pool
Spa
Hot tub
Solid conductor
Double-insulated pump

NEC References

680.21, 680.26, 680.31, 680.42, 680.43, 680.70, 680.74

Related Questions

- How do the bonding requirements for hydromassage bathtubs compare with those for pools, spas, and hot tubs?
- Are isolated metal parts such as valves in plastic piping required to be bonded?
- Is stranded wire permitted for bonding at pools, spas, hot tubs, and hydromassage bathtubs?
- Must an 8 AWG bonding conductor extend to a service or grounding electrode?
- What is to be bonded in a double-insulated pool or hydromassage pump?

Answer

The purpose of bonding metal parts of a hydromassage bathtub is the same as the purpose of bonding in the vicinity of pools, spas, and hot tubs. This purpose is stated in Section 680.26(A). Bonding is intended to eliminate voltage gradients in the pool area. This is bonding to create an equipotential plane, as opposed to the bonding required by Article 250 that is intended to provide fault current paths. Bonding for equipotential need not extend beyond the area of the equipotential plane, whereas bonding for fault current must extend back to the source that supplies the fault current.

The requirements for pools are more extensive than for other systems covered by Article 680. Section 680.26 does apply to spas and hot tubs, but Part IV of Article 680, which covers spas and hot tubs, modifies the bonding requirements somewhat. For example, at spas and hot tubs, certain items, such as the metal bands around a hot tub, are not required to be bonded.

Although the purpose of bonding at a hydromassage bathtub is the same as for pools, the rules of 680.26 do not apply to hydromassage bathtubs, according to 680.70—only the rules of Part VII of Article 680 apply. Nevertheless, the reasoning behind the similar rules in other parts can be applied to understand the rules of Part VII. The only rule about bonding that applies to hydromassage bathtubs is found in 680.74. All metal parts in pools are required to be bonded by the general rule in 680.26, but certain small isolated metal parts are exempted. For hydromassage bathtubs, only those items mentioned in 680.74 are required to be bonded.

Only metal piping systems and grounded metal parts that contact the circulating water of a hydromassage tub are required to be bonded. This will include pump motors and some piping fittings, but other metal items in the vicinity of a hydromassage bathtub are not required to be bonded to the hydromassage bathtub bonding system. For example, shower rods or handrails are not required to be bonded. Neither are metallic faucets in an adjacent basin in a bathroom. Metal fittings such as valves, faucets, and drains that are part of a hydromassage tub water piping system are only required to be bonded when they are part of a metal piping system or are otherwise required to be grounded. (Generally, only metal parts of electrical equipment are required to be grounded.) Figure 6.4 summarizes these requirements.

According to 680.74, the means of bonding metal parts of a hydromassage tub that are required to be bonded is "a copper bonding jumper, insulated, covered, or bare, not smaller than 8 AWG solid." An 8 AWG stranded conductor is not permitted as the bonding jumper. Solid conductors not smaller than 8 AWG are also required where conductors are used as the bonding means at pools, spas, and hot tubs in accordance with 680.26(C), 680.42(B), and 680.43(E). Solid conductors are specified to ensure a mechanically sound, highly reliable, corrosion-resistant bonding method. As noted previously, the required bonding jumper is intended only to eliminate voltage gradients in the tub area, not to carry fault current, so the

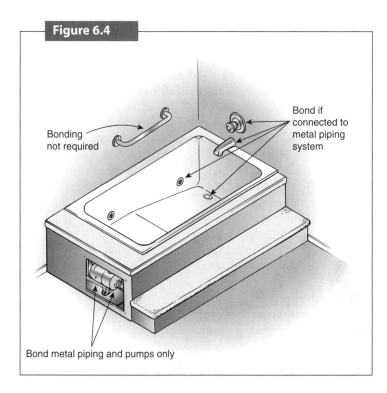

Items to be bonded at hydromassage bathtubs.

jumper is not required to connect back to a panelboard, the service, or any electrode. The bonding system may become connected to the equipment grounding system through equipment grounding conductor connections to bonded equipment, but such a connection is incidental, not required.

Note that a metallic pump for a hydromassage tub may be supplied with an external bonding means (a lug). There may be no other metallic parts that require bonding if the water piping and wiring methods are all nonmetallic. In that case, there is no requirement for a bonding jumper to bond the pump to anything else. Such a pump will have to be grounded with an equipment grounding conductor according to the ordinary rules of Article 250, meet the requirements of the installation instructions, and be supplied through a ground-fault circuit interrupter in accordance with 680.71.

Question 6.5 May the power supply to a fire pump originate in a service panel or switchboard?

Keywords

Fire pump
Service
Service disconnect

NEC References

695.3, 695.4

Related Questions

- May the power supply to a fire pump originate in the distribution section of a service switchboard or panelboard or must it originate before any main breaker?
- When is more than one disconnect permitted in the power supply to a fire pump?
- How can feeder sources be used for fire pumps if disconnects or overcurrent devices are not permitted in the fire pump power supply?

Answer

Sections 695.3 and 695.4 cover the permitted power sources and the method of connecting the fire pump to those sources. Where the power supply to the fire pump is from a utility (a service), Section 695.4(A) specifies that the source be directly connected to either a listed fire pump controller or a listed combination fire pump controller and transfer switch. In this basic arrangement there can be no disconnects between the source and the controller, as shown in Figure 6.5(a), which illustrates a single separate service as the power source for a fire pump where such a single source is approved as the required reliable power source.

As an alternative, Section 695.4(B) permits a "supervised connection" arrangement in which a single switching device is permitted between the source and the fire pump controller. However, there are a couple of other requirements that typically preclude the use of an overcurrent device within a switchboard or panelboard to supply a fire pump. First, 695.4(B)(2)(3) prohibits a disconnect in a fire pump supply from being located in equipment that supplies other loads. Second, Section 695.4(B)(2)(4) requires that the fire pump disconnecting means "be located sufficiently remote from other building disconnecting or other fire pump source disconnecting means such that inadvertent contemporaneous operation would be unlikely." These rules are meant to avoid possible disconnection of the fire pump in the case of a building emergency in which nonessential loads supplied by the same equipment are being shut down.

In general, if a service is the power source or one of the power sources for a fire pump, the service equipment and disconnecting means for the fire pump must be in a different location

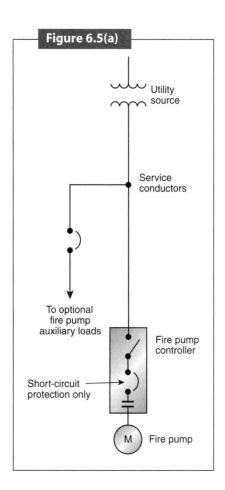

Figure 6.5(a) Typical power supply arrangement for a fire pump supplied by a single utility source.

from the regular power service. In other words, a separate service for the fire pump is usually required. However, 695.3(A)(1) says the service supply for a fire pump may be tapped from the normal service if the tap is located ahead of and not in the same "cabinet, enclosure, or vertical switchboard section as the service disconnecting means." This still requires another location for the fire pump service disconnect, but allows the fire pump to share the service conductors up to the point of the tap, as shown in Figure 6.5(b). The authority having jurisdiction must determine when a secondary source is needed to provide a reliable power supply.

Other options may be used where there is no utility (service) supply available. If there is no service supply, the power must be obtained from feeder sources. This is common in multi-building campus-style complexes where the service is perhaps a substation or otherwise remotely located from many of the buildings. In such cases, one or more feeder sources may be used, but such feeder devices must still be arranged to minimize the possibility of inadvertent disconnection of the supply. One example of this type of arrangement is shown in Figure 6.5(c), which depicts feeders derived from two separate utility sources as permitted by 695.3(B)(2). The acceptability of this arrangement depends on an evaluation of the reliability of the two sources, so in some cases the authority having jurisdiction may require that one of the two or more sources be an on-site generator in accordance with 695.3(B).

When an arrangement of two or more sources that are not utility sources is used, the connections, disconnect, and overcurrent protection in the feeder or feeders must meet the requirements for supervised connections found in Section 695.4(B). However, this rule has a significant modification where feeder sources are used in campus-style arrangements as permitted in 695.3(B). In these cases, *more than one* disconnecting means may be required in the

Fire pump service and normal service supplied from a single set of service lateral conductors with a secondary source supplied by a generator.

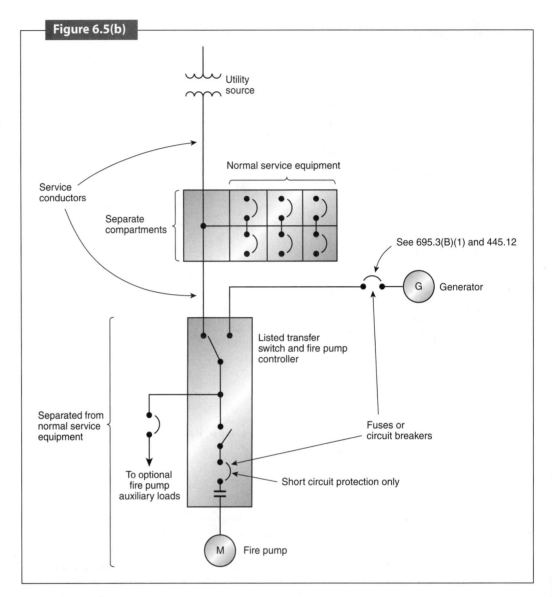

Figure 6.5(b)

fire pump supply in order to meet other code requirements. For example, additional disconnects or overcurrent devices may be needed at separate buildings or for transformers. In any case, each additional disconnect, whether there is one or more than one, must meet the requirements for supervised connections as outlined in 695.4(B)(1) through (4) and be supervised in the closed position by one of the methods in 695.4(B)(5).

To summarize the issue here, Section 695.3 is concerned with power supplies for fire pumps, and the main concern of this section is that the power source itself be highly reliable. More than one source is often needed to meet this requirement, and utility sources may not always be available. Section 695.4 is concerned with maintaining the continuity of power between the source and the fire pump, so the location and arrangement of disconnecting means and overcurrent protection is the primary concern here. These sections must be used together and coordinated to obtain maximum reliability of power at the fire pump.

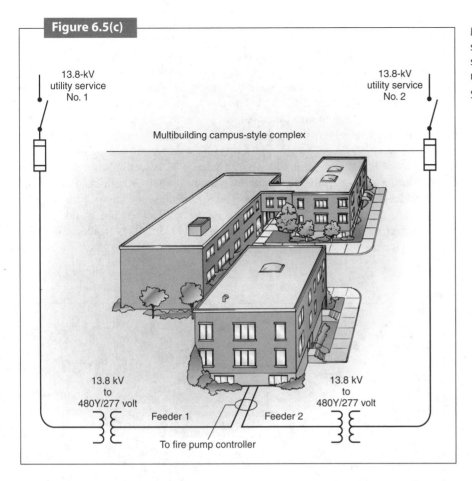

Figure 6.5(c)

Multiple feeder sources derived from separate utility sources for a campus-style application where fire pumps are required.

Source: *NEC Handbook*, Exhibit 695.3

Question 6.6

What physical separation is required for a fire pump service disconnect?

Related Questions

- Is the intent of Article 695 to provide a physically separate enclosure for the fire pump disconnect or just to provide separation from other devices via a barrier between sections?
- If a physical separation is required, by how much distance should the fire pump disconnect be separated from the disconnects for normal power?

Answer

According to Section 695.3(A)(1), a tap is permitted to be "located ahead of and not within the same cabinet, enclosure, or vertical switchboard section as the service disconnecting means," as shown in Figure 6.5(b). (See Question 6.5.) This restriction is referring to the location of the

Keywords

Fire pump
Disconnect
Service
Switchboard
Remote
Physical separation

NEC References

230.72, 695.3, 695.4, 695.12

ordinary service disconnect. This section may appear to permit the fire pump disconnect (which is also a service disconnect, by definition) to be in the same switchboard as the service disconnecting means for the normal power supply as long as it is in a separate vertical section. However, Section 695.3(A)(1) covers only the location of a tap where a separate set of service conductors is not run to the fire pump, and does not cover the location of the fire pump disconnect except by reference to Section 230.72(B). The tap connection to the switchboard is required to be arranged so as to minimize the possibility of damage from fire. The code allows judgment as to how the physical separation is to be provided for the purpose of minimizing exposure to fire.

The reference to 230.72(B) does provide some guidance. This section says: "The one or more additional service disconnecting means for fire pumps . . . shall be installed remote from the one to six service disconnecting means for normal service to minimize the possibility of simultaneous interruption of supply." This reference reinforces Sections 695.3 and 695.4 to require the disconnect for a fire pump to be located separately from the switchboard containing the normal service disconnect.

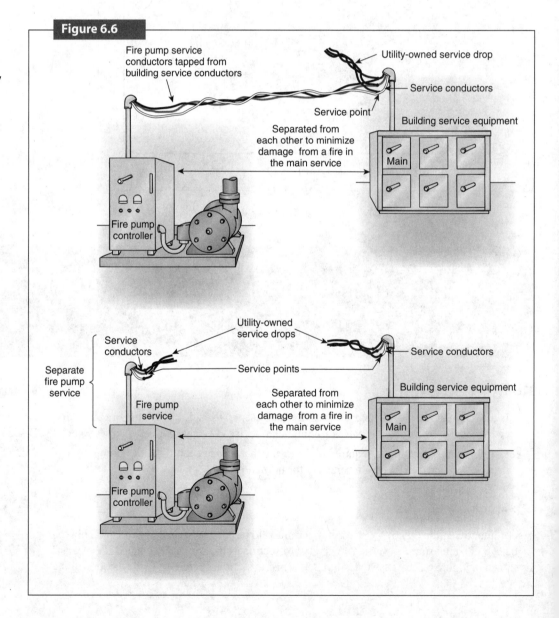

Two permitted configurations for connecting a fire pump supply to a utility service drop.

Source: Based on *NEC Handbook*, 2005, Exhibit 695.2.

Other provisions of 695.4(A) or (B) must also be met. As noted, if a direct connection is used according to (A), the service conductors for the fire pump may be supplied either from a separate service or from a tap ahead of other service disconnects, but they must connect directly "to either a listed fire pump controller or listed combination fire pump controller and power transfer switch." A fire pump controller would be used if the service alone is judged to be sufficiently reliable. A combination controller and power transfer switch would be used when another backup power supply is judged to be necessary. Either way, the fire pump disconnect must be separated from the ordinary service switchboard as shown in Figure 6.5(b) and Figure 6.6. Figure 6.6 illustrates both a separate service drop and a tap from a service drop, whereas Figure 6.5(b) shows a tap from a service lateral.

For a supervised connection complying with 695.4(B)(2), where the power source is a feeder rather than a service, a disconnect ahead of the fire pump controller must "be located sufficiently remote from other building or fire pump source disconnecting means such that inadvertent contemporaneous operation would be unlikely." In addition, 695.4(B)(2)(3) does not allow a fire pump disconnect to be within equipment that supplies other loads. Both rules require a physical separation between equipment where a tap is made and the first fire pump disconnect.

The location of the fire pump disconnect and controller is covered in another way in 695.12(A), which requires controllers and transfer switches to "be located as close as practicable to the motors that they control" and in sight from the motors. This "in sight" rule says the controller must be in the same room as and within 50 feet of the motor, but it does not say how far the controller must be from other service disconnects. Fire pumps and their associated equipment are often located in separate rooms from the main electrical service equipment, although a separate room is not necessarily a requirement of the *NEC*.

The physical spacing is a performance-based requirement that requires a judgment by the authority having jurisdiction: The fire pump controller and disconnect(s) must be far enough away so that persons disconnecting the main service would not also inadvertently disconnect the fire pump supply. The likelihood of inadvertent disconnection would be greatly increased if the fire pump disconnect were located in the same switchboard as other service disconnects.

Question 6.7

Which conductors for fire pumps require special routing and protection?

Related Questions

- What routing requirements apply to fire pump supply conductors?
- What types of protection are required for fire pump conductors?

Answer

Section 695.6 provides requirements for the power wiring to electric fire pumps. Although 695.6(A) is titled "Service Conductors," the language in this section clearly applies to supply wiring from other sources as well, because the included reference to 695.3(B)(2) only applies to feeders. The requirement that supply conductors be physically routed outside a building is a general requirement for service conductors, but Section 695.6 specifically applies this rule to most other conductors that supply fire pumps. (Service conductors are defined in Article 100

Keywords

Fire pump
Outside of building
Service conductors
Supply conductors
Fire ratings
Reliable power

NEC References

Article 100, 230.6, 230.70, 695.3, 695.4, 695.6

as the conductors up to the service disconnecting means, and the service disconnecting means is required by Section 230.70(A)(1) to be outside or inside nearest the point of entrance of the service conductors.) Like service conductors, fire pump supply conductors may be considered to be outside buildings if installed at least 2 inches under a building or encased in at least 2 inches of concrete or masonry within a building according to 695.6(B) and 230.6(1) and (2).

The intent of 695.6(A) is not to apply only to service conductors, but to require supply conductors for fire pumps to be treated like service conductors. Both the power source and the conductors supplying fire pumps must be highly reliable. The requirements for keeping conductors outside buildings in Article 230 is primarily based on the risk of having conductors without short-circuit and ground-fault protection in buildings. Although this concern also applies to service conductors that supply fire pumps, routing and protection of fire pump supply conductors are also based on reliability issues; for that reason, the rules apply to supply conductors that are not service conductors.

In many cases, perhaps most, no single source is considered to be sufficiently reliable for fire pump service as required by 695.3, so multiple sources are provided. For example, one common configuration consists of a normal utility source and a backup source supplied by an on-site generator. Section 695.3(B) permits two or more sources to be used to provide reliable power. These may be two utility sources in some cases, or other arrangements of two or more of the following: utility service connections, on-site power production facilities, on-site generators, or feeders in campus-style complexes. The final determination for what source or sources constitute a reliable source is the responsibility of the authority having jurisdiction.

When two or more sources are used, a listed combination fire pump controller and power transfer switch is needed to comply with 695.4. In such cases, the reliability of any one set of supply conductors is somewhat less critical because there is a backup set from another source. However, there is still only one set of conductors from the transfer means to the pump. Therefore, 695.6(A), Exception, says that when there is more than one source, only the conductors on the load side of the transfer means must be routed in such a way that they can be considered outside the building. (This exception modifies the requirements for feeder conductors and conductors from separate sources such as generators, but is not intended to modify the requirements for service conductors in Article 230. Service conductors are still required to be installed in such a way that they are or can be considered to be outside buildings.) The requirements of 695.6(A) and the exception are summarized in Figure 6.7(a).

In certain cases, such as where feeder sources are used, a supervised disconnecting means may be required between the source and the fire pump controller or transfer switch in order to meet other requirements of the *NEC*. Conductors on the load side of such a "final disconnecting means and overcurrent device(s)" are covered by both 695.6(A) and 695.6(B) but often must be physically routed inside a building as shown in Figure 6.7(b). Such routing is permitted only with protection that will ensure the reliability of the conductors. The options for protection are given in 695.6(B)(1) through (3) and are as follows: (1) encasement in at least 2 inches of concrete, (2) enclosure in 1-hour fire-rated construction, or (3) protection by a listed system with a 1-hour fire rating. All three of these protection methods are intended to guarantee the function of the fire pump supply for an extended time in a fire. One should understand that the last two protection methods are very different requirements from each other. A fire-rated enclosure can be designed and constructed based on the applicable building code, but a listed system must be selected, purchased, and installed according to the listing and labeling. This difference is further explained in the following quotation from the National Fire Protection Association's *NEC Handbook*:

> It is important to understand the difference between a 1-hour fire rating of an electrical circuit, such as a conduit with wires, and a 1-hour fire-resistance rating of a structural member, such as a wall. Simply stated, at the end of a 1-hour fire test on an electrical conduit with wires, the circuit must function electrically (no short circuits, grounds, or opens are permitted). The circuit and its insulation must be intact and

electrically functioning. A wall subjected to a 1-hour fire-resistance test must only prevent a fire from passing through or past the wall, without regard to damage to the wall. All fire ratings and fire-resistance ratings are based on the assumption that the structural supports for the assembly are not impaired by the effects of the fire.

Many of these requirements are not possible for the final connections between pump controllers and fire pump motors, so such local wiring is covered separately in 695.6(E), where specific wiring methods, such as liquidtight flexible metal conduit, are permitted to make final connections to fire pump motors, as illustrated in Figure 6.7(c). These special rules for the pump wiring are recognized by 695.6(B), Exception, which also makes wiring such as taps from service conductors or feeders at the location where they originate in an electrical equipment room exempt from special protection requirements.

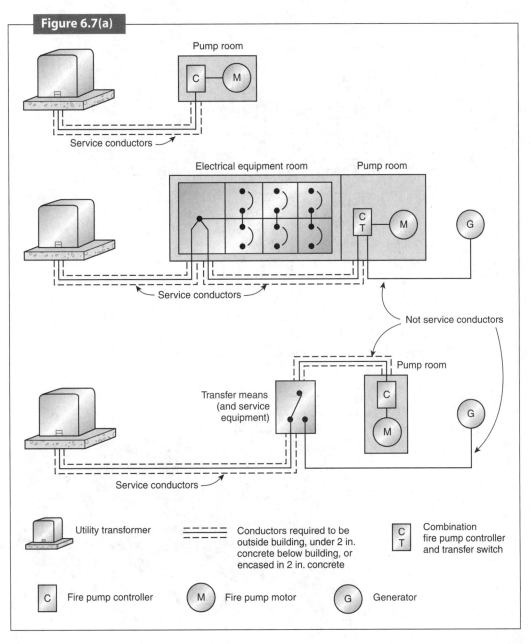

Protection of wiring methods for fire pumps supplied by service sources or combinations of services and other sources.

Protection of wiring methods for fire pumps supplied by feeder sources.

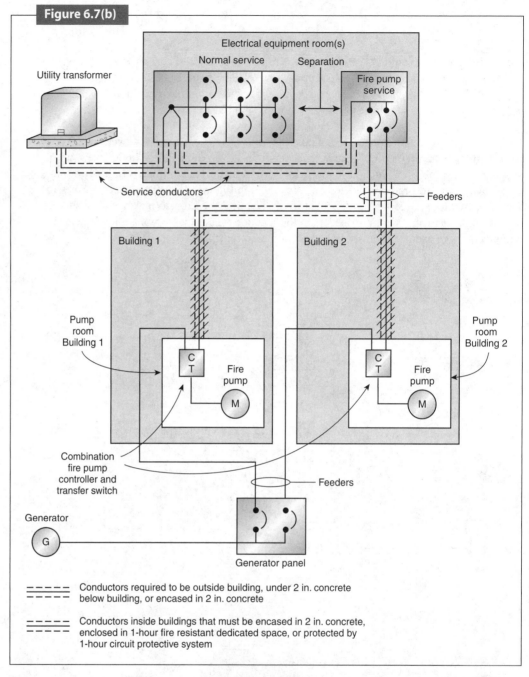

Fire pump wiring to a fire pump within the fire pump room.
Source: *NEC* Seminar.

Question 6.8

How are emergency loads identified by the *NEC*?

Related Questions

- Where does the *NEC* specify what loads are emergency loads?
- Where is emergency lighting required and how should it be located?
- May owners or designers designate loads they consider critical as emergency loads?
- May an emergency system supply any other loads?

Keywords

Emergency system
Emergency loads
Exit lighting
Egress lighting
Legally required standby
Optional standby
Separation

NEC References

700.1, 700.5, 700.6, 700.9, 700.15

Answer

Section 700.1 describes the loads that are considered to be emergency. It says, in part, "These systems are intended to automatically supply . . . power and illumination essential for safety to human life." A critical part of the description also says, "Emergency systems are those systems legally required and classed as emergency by municipal, state, federal, or other codes, or by any governmental agency having jurisdiction." In other words, the designation of what is emergency and what is not is based on standards other than the *NEC*, and only those loads essential for safety to human life are included. In many facilities, such as retail and office establishments, the only emergency loads are exit and egress lighting. In others, such as high-rise buildings, loads such as some elevators, stairwell pressurization fans, exhaust fans, and fire pumps may also be critical to life safety. Often, the emergency loads are those that are determined necessary to evacuate a building safely. In hospitals, the emergency system is subdivided into the life safety branch—the exit and egress types of loads—and the critical branch, the branch that supplies the loads necessary for preservation of life of those who cannot be evacuated or who can be evacuated only after additional preparations have been made.

In most cases, the identification of emergency loads is a function of NFPA 101, *Life Safety Code*, or the applicable building code. Other loads may be designated as either legally required standby (Article 701) or optional standby (Article 702). Loads that are not necessary for safety to human life but are necessary for fire-fighting operations, to reduce property damage, or to maintain operations during a power failure may be either legally required or optional standby. Legally required loads must be so designated by the governmental agency having jurisdiction. Optional standby loads may be whatever a user wants.

The requirements for exit lighting and egress lighting are found in NFPA 101 or the building code or both, depending on what codes are adopted by a jurisdiction. These rules are based on type of occupancy and occupant load (number of occupants in an area, usually determined by floor area). Spacing must be determined based on the applicable code and required levels of illumination, but the *NEC* does not provide this information. The function of the *NEC* is to govern the selection of power sources and the installation of the emergency wiring system once the loads have been defined by other standards. If, for example, exit and egress lighting is required, those lights will be considered to be emergency loads. If the owner of a single-family home decides to install some battery-powered exit or egress lighting, such loads are optional standby loads (unless there is some local code that requires them). Similarly, if the owner of a store decides that the frozen food cases must have a backup power supply or the owner of a casino decides the slot machines should continue to operate in a power outage, those loads are optional standby loads as well.

Chapter 6 Special Equipment, Conditions, and Systems and Related Standards

Emergency and standby loads (optional or legally required) may be supplied by a common standby source if supplied through separate transfer means.

Source: *Electrical Inspection Manual*, 2005, Figure 11.4.

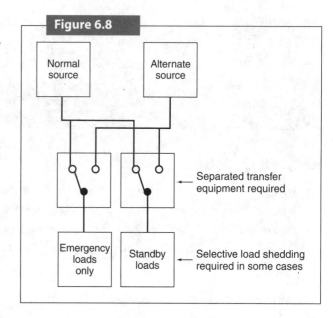

Figure 6.8

Various sections of Article 700 restrict emergency systems to supplying only emergency loads. Section 700.6 requires an automatic transfer of emergency loads from normal power to the emergency source, and the transfer means is only permitted to supply emergency loads according to 700.6(D). Section 700.9 requires the emergency wiring to be completely separated from other wiring. Section 700.15 says, "No appliances and no lamps, other than those specified as required for emergency use, shall be supplied by emergency lighting circuits." As shown in Figure 6.8, Section 700.6(B) permits the emergency power source to supply other loads or be used for other purposes, but this can be done only through separate transfer means and with entirely separate wiring systems, and the system must be designed to give the highest priority to emergency loads.

The primary purpose of all these rules and restrictions is to increase the reliability of the emergency system. Many other rules of Article 700 have the same purpose—to be sure that the emergency system will be functional when it is needed.

Question 6.9 Are transfer switch control wires considered to be emergency wiring?

Keywords

Emergency
Automatic transfer switch
Class 1
Class 2

Related Questions

- May I run the automatic transfer switch (ATS) control/start signal wires through emergency feeder pull boxes inside and outside a building?
- If the ATS control circuits are not classed as emergency wiring, does the *NEC* require the control wire to be separated from emergency power in both raceways and pull boxes?

- Does Section 725.26(B) apply to an emergency generator and its "functionally associated" automatic transfer switch?

Class 3
Functionally associated

Answer

The determination of what constitutes emergency wiring is the responsibility of the authority having jurisdiction. However, there is nothing in Article 700 or NFPA 110, *Standard for Emergency and Standby Power Systems*, that would preclude the generator start/control circuits from being classed as emergency wiring or being able to share a common raceway or enclosure with other circuits of the same emergency system. Section 700.9 requires independent raceways and enclosures for the emergency system and lists five exceptions, but it does not specifically mention ATS control wiring. The key is the classification of the control circuits, which will be emergency circuits only if they are powered by the emergency system. Typically, the ATS control circuit is a circuit powered from the generator and switched by the ATS, so in most cases, the ATS control circuit is part of the emergency system wiring.

On one hand, if the circuits are Class 1 circuits, the restriction in 725.26(B) mentioned in the question is that the Class 1 circuits may be installed with power conductors only where they are functionally associated. In the case of the feeder from the generator, the start circuit conductors are certainly functionally associated and could be installed in the same pull box or other enclosure, as shown in Figure 6.9, or even in the same raceway with the generator feeder conductors, *if* they are also classified as emergency circuits.

On the other hand, if the control circuits are Class 2 or Class 3 circuits, Section 725.55 prohibits a common raceway installation (even if they are emergency circuits), but does provide conditional permission to include these circuits in an enclosure that also contains power and lighting conductors, or in this case, emergency power and lighting conductors, where the control and power or lighting conductors connect to common equipment. This permission would apply to the transfer switch or the generator control panel, but not to a pull box. Another possibility, however, is that Class 2 or Class 3 circuits may be reclassified as Class 1 circuits if Class 1 wiring methods are used throughout the circuit. This permission is found in Section 725.52(A), Exception No. 2. If Class 2 ATS control circuits are reclassified and in-

NEC References

700.9, 725.26, 725.52, 725.55

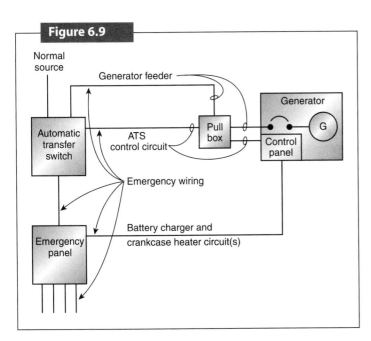

Figure 6.9

Common enclosures permitted for Class 1 and functionally associated power conductors. Both must also be classified as emergency where either is emergency.

stalled as Class 1 circuits, they would then be subject only to the requirement that they be functionally associated, and they could be installed with related power conductors in a common enclosure.

Question 6.10 Is more than one emergency system permitted in a building?

Keywords

Emergency
Source
System
Separation
Standby
Unit equipment
Transfer means
Transfer switch

NEC References

517.30, 700.5, 700.6, 700.9, 700.12, NFPA 110

Related Questions

- Is the wiring of two or more emergency sources required to be separated?
- May unit equipment be supplied from an emergency source?
- Must all emergency lighting be connected to a generator set if there is one?
- Is more than one emergency circuit permitted in the same raceway?
- May more than one type of system be used to create a single emergency power source?
- Is a single feeder from a generator permitted to supply transfer switches for emergency and other loads?

Answer

The sources of power that may be used to supply emergency loads are detailed in Section 700.12. The sources listed in 700.12 are (A) storage batteries, (B) generator sets, (C) uninterrupted power supply (UPS) systems, (D) separate services, and (E) fuel cells. Section 700.12(F) covers unit equipment such as self-contained exit lights and egress lights. Unit equipment is limited in use (to lighting) and is treated somewhat differently from the other four sources that have more general application.

Section 700.12 does not prohibit the use of multiple emergency power systems within one building, nor does it prohibit the use of individual unit equipment in a building that also has an emergency power system or systems. In fact, 700.12 clearly states that the supply system for emergency purposes "shall be *one or more* of the types of systems described in 700.12(A) through (E)" (emphasis added). It goes on to say that unit equipment may also be used as described in 700.12(F), which limits unit equipment to emergency illumination. For many buildings, emergency illumination for exit and egress is the only required emergency load, so all requirements may be met with unit equipment, and a generator could be supplied only for standby loads. Systems or sources other than unit equipment are required if there are nonlighting emergency loads.

There is no rule in Article 700 that prohibits unit equipment from being supplied by one of the sources described in Section 700.12(A) through (E). The *NEC* does not preclude (or require) multiple levels of redundancy, provided each source and the unit equipment meets its respective requirements in Section 700.12(A) through (F). Multiple levels of redundancy are often used in highly critical applications, especially where continued operation is required rather than just safe evacuation. Fine Print Note (FPN) No. 2 of 700.12 addresses the issue of "assignment of degree of reliability." A method of assessing reliability for emergency and other power systems is described in IEEE 493-1997, *IEEE Recommended Practice for the*

Design of Reliable Industrial and Commercial Power Systems ("IEEE Gold Book"). The issue of reliability is also applicable to the legally required and optional standby systems covered by Articles 701 and 702.

Section 700.9(B) permits the wiring of two or more emergency circuits from the *same source* to be installed in the same raceway or enclosure, but requires emergency wiring to be separated from the wiring of other sources. "Other sources" in this case includes other emergency sources. The rules in 700.9(B)(2) through (4) recognize, in part, that Section 700.12 permits a building to have more than one emergency power system. The purpose of Section 700.9(B) is to provide as much physical separation as possible between the normal and emergency system or between multiple emergency systems. This physical separation requirement is aimed at increasing the reliability of each of these systems so that a failure in one system is unlikely to affect another system. Section 700.9(B) includes very limited special cases

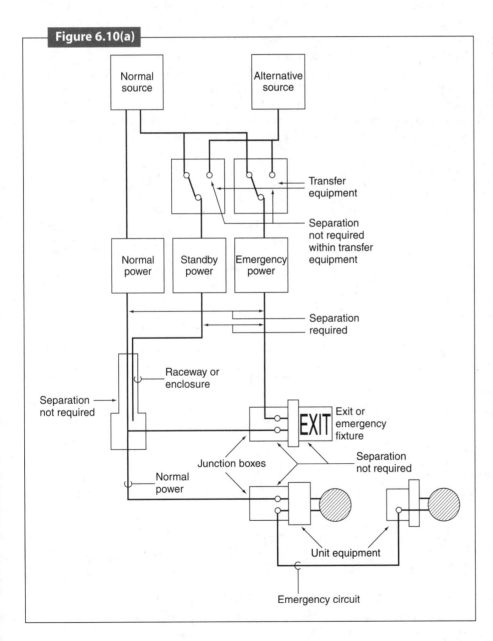

Separation of emergency system wiring from standby and normal power system wiring.

Source: *Electrical Inspection Manual*, 2005, Figure 11.5.

where emergency lighting or unit equipment is connected to multiple sources and wiring must be combined in a luminaire or junction box to supply the lighting unit.

In some cases, as in some UPS systems, more than one type of system, such as batteries and generators, may be combined to create what may be considered a single emergency source. This possibility is recognized by the language of 700.12(C), which refers to both 700.12(A) and (B), and also by the language of 700.12(B)(5), which permits an auxiliary supply to energize a system if the generator takes more than 10 seconds to develop power. This type of UPS design is used in many highly critical applications. Some of these applications may be considered to be legally required standby rather than emergency. A source used for both emergency and legally required (or optional) standby must have separated wiring and separate transfer means as required by 700.5(B) and 700.6(D). Figure 6.10(a) illustrates the requirements for wiring separations where normal, standby, and emergency systems are used.

As noted previously, the separation requirements found in 700.9(B) prohibit wiring of an emergency system from being mixed with wiring of other systems, including other emergency or standby systems. The language that prohibits this mixing says, "Wiring from an emergency source *or emergency source distribution overcurrent protection* to emergency loads shall be kept entirely independent of all other wiring" (emphasis added). Section 700.5 permits an alternate power source to supply other standby loads, but 700.6(D) permits the emergency transfer switch to supply only emergency loads. Thus, additional transfer switches are required for legally required or optional standby loads.

Nevertheless, a single feeder from a generator may supply the distribution equipment that supplies the separate transfer switches, because the separation requirement begins at the

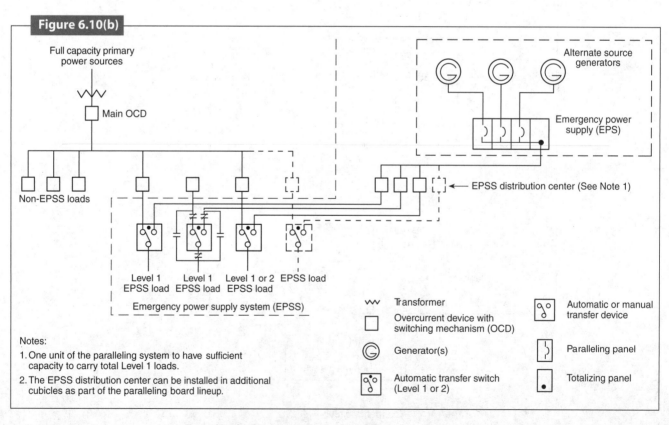

Typical multiple-unit emergency power supply system.
Source: NFPA 110, 2005, Figure B.1(b).

source *or emergency source distribution* equipment. For example, in large installations where more than one generator (or other sources) are operated in parallel to supply the emergency and standby loads or to provide the desired level of reliability, the output of the multiple generators will be combined in paralleling distribution equipment that keeps the sources in phase. This distribution equipment will also include overcurrent protection for the emergency and other feeders that supply the transfer means. In such cases, separation of emergency wiring must be maintained from the distribution point on. This scheme does not mix wiring of more than one emergency source or mix wiring of emergency systems with other systems. This type of arrangement is illustrated in Figure 6.10(b). Note that Figure 6.10(b) shows the paralleling panel separate from the emergency power supply system (EPSS) distribution center overcurrent devices, but they could be combined in one piece of equipment as mentioned in Note 2 on the drawing. (The term "emergency power supply system," or EPSS, as it is used in NFPA 110 includes standby systems other than "emergency systems" as the term is used and defined in the *NEC*. For example, a Level 2 EPSS load may be a legally required or optional standby load in the *NEC*.)

A similar arrangement with only one power source can be seen in Figure 6.10(c). Note that the alternate power source in Figure 6.10(c) supplies a bus, which typically takes the form of a distribution panel or switchboard, and all wiring from the distribution bus is separated. In a hospital application, even the critical and life safety branches of the emergency system are usually separated unless the entire load is relatively small, in which case a single transfer switch is permitted for the two branches of the emergency system and the equipment system. FPN Figure 517.30, No.2 in the *NEC* illustrates this special case that applies only to health care facilities.

The schematic arrangement of Figure 6.10(b) or 6.10(c) could also be used in other types of occupancies, but it could not be accomplished by bringing a single feeder to a junction point or wireway and then making taps to the transfer switches. For one thing, the transfer switches are required to be supplied with overcurrent protection in accordance with their listings. (This problem could be overcome by oversizing the transfer switches.) Perhaps more important and less easily resolved, however, the junction box or wireway is a wiring method in which the emergency system wiring would be separated from and then mixed with the wiring of the standby system(s) in violation of 700.9(B).

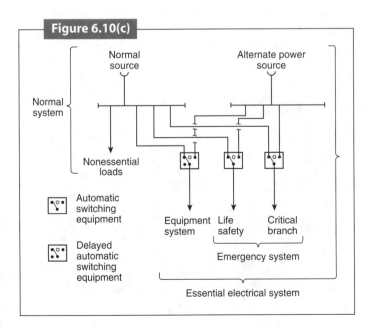

Minimum requirement for transfer switch arrangements in a hospital.

Source: *NEC*, 2005, FPN Figure 517.30, No. 1, Hospital.

Question 6.11: Is emergency power required to be supplied when a branch circuit fails?

Keywords

Emergency
Unit equipment
Generator
Emergency circuit
Normal circuit
Transfer switch
Exit
Egress

NEC References

Article 210, 700.9, 700.12

Related Questions

- Must an emergency generator start on a branch circuit failure?
- Is unit equipment required in addition to generator power?
- Are circuits supplying unit equipment considered to be emergency circuits?
- May circuits supplying unit equipment supply other loads?
- Are separate circuits permitted to supply unit equipment?

Answer

If a branch circuit provides the only illumination for an area where exit or egress lighting is required by other codes, the *NEC* requires that emergency power be available for designated lighting in that space. This does not mean that an emergency generator source must start on failure of a branch circuit. (In most cases, unless a generator is used to supply power to other than emergency loads, the generator will not start until the normal source of power, usually a feeder supplied from a utility, is lost at the transfer switch.) The emergency lighting for a specific area may be provided by sources other than generators, such as the unit equipment shown in Figure 6.11. Other methods and power sources may be used, such as using a circuit supplied by a generator along with more than one normal branch circuit for an area, to ensure that failure of a single branch circuit will not leave an area in darkness.

Emergency lighting (exit and egress lighting) is not required in all buildings or in all areas of buildings. While Section 700.12 in the *NEC* provides general requirements for sources of emergency power and some specific requirements for various types of sources, areas where emergency lighting is required, as well as types and locations of lighting units, are determined from other standards, such as NFPA 101, *Life Safety Code*, or the applicable building code. Requirements for emergency lighting may not be met by following the *NEC* alone.

Emergency circuits are those supplied by an emergency system classified as such by the authority having jurisdiction. Unit equipment may be supplied by emergency circuits,

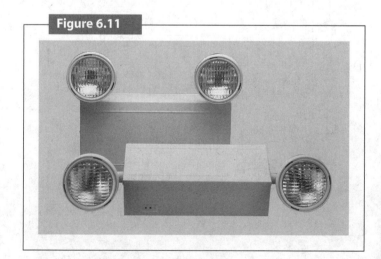

"Unit equipment" for automatic emergency lighting.
Source: *NEC Handbook*, 2005, Exhibit 700.4. (Courtesy of Dual-Lite, Inc.)

but is usually supplied by a normal lighting circuit, so that the unit equipment operates when the normal lighting source for that area is lost, and the area is not left in darkness. In fact, according to 700.12(F), "The branch circuit feeding the unit equipment shall be the same branch circuit as that serving the normal lighting in the area and connected ahead of any local switches." This section goes on to say that for any remote lighting units that are supplied by unit equipment, the wiring between the unit equipment and remote lighting units shall be treated as an emergency lighting circuit in accordance with 700.9. The circuit supplying the unit equipment is still a normal source, and unless the normal source is an emergency circuit, the supply to the unit equipment is treated as an ordinary branch circuit. By connecting the unit equipment ahead of local switches, the unit equipment will not interpret intentional and normal disconnection of the local lighting as a power failure, and the unit equipment will not be disconnected from the normal source unless there is a failure of that circuit or source.

For separate, uninterrupted areas, where at least three lighting circuits are provided for normal lighting, a separate circuit may be used for unit equipment. This rule is covered in 700.12(F), Exception. The circuit for unit equipment must be supplied from the same panelboard as the circuits for normal lighting. In effect, under the exception, three normal circuits would have to fail for the area to be left without normal lighting, and that would be most likely due to a failure of the feeder supply to the panelboard. This separate circuit for unit equipment is not an emergency circuit unless the entire panelboard supplies only emergency circuits.

Emergency circuits are restricted to supplying only emergency loads by 700.15. However, since circuits supplying unit equipment are generally required to be the same circuits that supply normal lighting, loads other than unit equipment are obviously permitted on those circuits. The load limitations on a circuit supplying unit equipment are the same as for any other branch circuit as specified in Article 210. The load of unit equipment is determined from the nameplate of that equipment. In some unit equipment, like that shown in Figure 6.11, the normal load on the branch circuit is only the current required for charging batteries. In other unit equipment, such as exit lighting, the load also includes the current required to operate the lamps that burn continuously.

According to 700.12, the "occupancy and the type of service to be rendered" by an emergency power source must be considered. In some cases, it may be determined that power for lighting must be supplied for a longer time than the 90 minutes that unit equipment is intended to operate, or that immediate restoration of lighting is needed rather than after the 10 seconds permitted for transfer to most emergency power supplies. In such cases, unit equipment may be used in conjunction with other emergency power supplies, and unit equipment may be supplied in addition to and connected to an emergency power system. However, in many cases unit equipment alone may be the only emergency power source(s) required in a given occupancy, and there will be no emergency wiring outside the unit equipment itself.

Question 6.12

Is an outside generator for an emergency or standby system a "separate structure"?

Related Questions

- Must an outside standby generator installation comply with the *National Electrical Code* Sections 225.31 and 225.32?

Keywords

Generator
Standby

Keywords

Emergency
Disconnect
Outside feeder

NEC References

225.31, 225.32, 700.6, 701.11, 702.11

- Would the generator or the structure on which the generator is placed meet the definition and be subject to the requirements for separate "structures" as referenced in Article 225?

Answer

"Structure" is defined in Article 100 as "that which is built or constructed." Based on that definition, a generator that is located outside a building is a separate structure. Since the generator is a separate structure, the answer to the questions is yes, Article 225 and Sections 225.31 and 225.32 do apply to the outside feeder conductors supplied by the generator. However, there are some special provisions for generators both in Article 225 and elsewhere in the *NEC*. Figure 6.12 illustrates the rules in 700.12(B)(6), 701.11(B)(5), and 702.11, which say that an additional disconnect is not required where the conductors supplied by the generator serve or enter a building if the generator is equipped with a readily accessible disconnect that is in sight of the building.

This "in sight" rule does not necessarily apply to other types of disconnects that supply power to but that may not be located at a building. The authority having jurisdiction must judge each case separately to determine if a disconnect in sight of a building will provide the intended level of safety or if the disconnect must be located closer to a building. However, for those generators covered by Articles 700, 701, and 702, a disconnect "in sight" is adequate, and another disconnect at the point where the wires serve or pass through a building is not required.

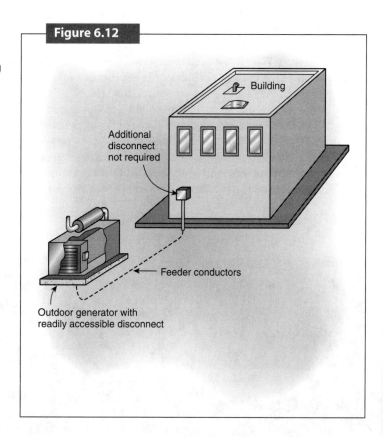

A generator disconnect may serve as a disconnect at a building when the generator is a separate structure.

Source: 2002 *NEC* Changes Seminar.

Question 6.13

How do I know if a circuit is Class 1, Class 2, or Class 3?

Related Questions

- What is the purpose of Article 725?
- What are the differences among Class 1, Class 2, and Class 3 circuits?
- How are Class 2 and Class 3 circuits defined and identified?
- What is covered by Article 725?
- When does Article 725 apply?
- Does Article 725 apply to instrumentation systems?

Keywords

Class 1
Class 2
Class 3
Remote control
Signaling
Power limited
Limited energy
Instrumentation

NEC References

90.1, 90.3, 725.1, 725.3, 725.21, 725.41

Answer

The purpose of Article 725 is the same as the purpose of the *NEC* as described in Section 90.1: "practical safeguarding of persons and property from hazards arising from the use of electricity." In the case of Article 725, the specific hazards are those associated with the use of remote-control, signaling, and limited-energy circuits. This article provides safeguards to limit the amount of energy that can be introduced into low-voltage components and circuits and also provides requirements related to the products of combustion and flame-spread ratings for signaling systems and remote-control wiring.

In the arrangement of the *NEC* as described in Section 90.3, articles in Chapter 7 may modify or amend the general wiring requirements of the articles in Chapters 1 through 4. Although the limited-energy systems covered by Article 725 typically do not require the same grade of wiring methods as systems with higher energies and higher voltages, there are still fire- and shock-protective provisions that need to be incorporated into these wiring systems in order to maintain safe operating conditions. The remote-control Class 1 circuits covered by Article 725 are not necessarily limited-energy circuits and may operate at up to 600 volts, so the ordinary wiring methods of Chapter 3 still apply to these circuits, but other rules are modified to recognize the characteristics of control circuits. Section 725.1 says:

> This article covers remote-control, signaling, and power-limited circuits that are not an integral part of a device or appliance.
>
> FPN: The circuits described herein are characterized by usage and electrical power limitations that differentiate them from electric light and power circuits; therefore, alternative requirements to those of Chapters 1 through 4 are given with regard to minimum wire sizes, derating factors, overcurrent protection, insulation requirements, and wiring methods and materials.

Circuit classification is based primarily on the circuit power supply as shown in Figure 6.13, although circuit use may also influence the circuit classification, especially with Class 1 circuits, but also in cases such as where a Class 2 or Class 3 power supply is used as a part of a listed fire alarm control panel to supply fire alarm circuits. The circuits covered by Article 725 are classified as Class 1, Class 2, or Class 3 depending on the power supply. Class 1 circuits

Classifications of remote-control and signaling circuits are determined primarily by power supplies.

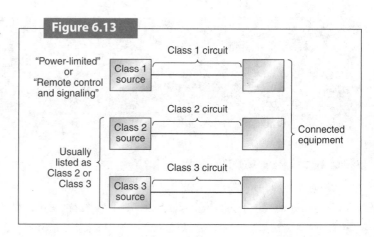

may be either Class 1 Remote-Control and Signaling, or Class 1 Power-Limited, which implies that the power-limited variety may be used for something other than remote control and signaling, but the rules for installing either type of Class 1 circuit are the same. Some Class 2 power supplies are listed as Class 2 only in dry locations; where the circuits supply equipment in wet locations, the Class 2 supply becomes a Class 3 supply because of the increased shock hazard in a wet location.

Power supply requirements are given in 725.21 for Class 1 circuits and in 725.41 for Class 2 and Class 3 circuits. Class 1 circuit power supplies are not required to be specifically listed as Class 1. Most of the power supplies for Class 2 or Class 3 must be listed or derived from other listed equipment with Class 2 or Class 3 markings. However, Class 2 or Class 3 circuits may be supplied by limited-power circuits of listed equipment such as information technology equipment or certain industrial control equipment. Thermocouples and some dry cell batteries are the only other power sources considered to be Class 2 without being specifically listed as such.

In essence, a Class 1 circuit is one that is power limited or used for remote control or signaling as described in 725.1; is not a Class 2, Class 3, or fire alarm circuit; and is not used for power or lighting applications. Class 1 circuits may operate at 30 volts or less or may be up to 600 volts. Similar circuits in industrial occupancies that are wired using Type ITC (instrumentation tray cable) are treated separately, are not considered to be Class 1 circuits, and, according to 727.5, may not be installed with Class 1 circuits.

For Class 1 circuits, usage is often the primary distinction. Most Class 1 circuits are used for remote-control applications rather than power or lighting, but some Class 1 circuits may be "low voltage" or power limited. To distinguish between Class 1 and other "low-voltage" applications, low-voltage power or lighting systems and circuits covered by other articles include the low-voltage (30 volts or less) lighting covered by Article 411 and the low-voltage (less than 50 volts) power systems covered by Article 720. Other "non-Class 1" "low-voltage" applications may be found in recreational vehicles (Article 551), pipe organs (Article 650), electroplating processes (Article 669), and solar photovoltaic systems (Article 690).

Class 2 and Class 3 circuits are defined by their power supplies. The limited energy available from these sources also tends to limit the uses of the circuits to remote-control and signaling applications, including instrumentation and data transmission. According to 725.41, Class 2 and Class 3 power sources must generally be listed as such or be supplied by other listed equipment that identifies the source as Class 2 or Class 3. Limited-power circuits supplied by listed information technology equipment or listed industrial control equipment are also considered to be Class 2 or Class 3. Class 2 power sources are typically 30 volts or less, and Class 3 sources are typically 100 volts or less, but in some cases the voltages may be higher if the current is sufficiently and reliably limited.

The power limitations of listed Class 2 and Class 3 sources are shown in Tables 11(A) and 11(B) in Chapter 9. The tables are primarily for information, however, because Class 2 or Class 3 sources cannot be constructed in the field. As noted previously, the only sources that are considered Class 2 without listing are thermocouples and some dry cell batteries.

Most industrial instrumentation circuits are covered by Article 725. Instrumentation is a form of signaling or a form of data transmission that is usually used in some control function. These circuits are covered by the requirements of Article 725. For example, a DC source may be used to create a common 4- to 20-milliampere instrumentation signal to transmit pressure information to a distributed control system or a programmable logic controller. Depending on the power supply, the circuit is usually either a Class 1 power-limited circuit or a Class 2 circuit, but could be Class 3. Determination of the type of circuit will depend on the listing of the power supply as detailed in Section 725.41. Such a circuit could be wired in the same manner as a power or lighting circuit, but under Article 725, special cable types and smaller wires can be used, and other rules may also be modified as described in the Fine Print Note to 725.1.

Question 6.14

What are the bonding and grounding requirements for control circuits?

Related Questions

- How are instrumentation and control circuits classified?
- How can I determine if equipment related to control and instrumentation circuits is required to be grounded?
- Are bonding and grounding requirements in hazardous locations different for instrumentation circuits than for power circuits?

Answer

Most instrumentation and control circuits fall under the scope of Article 725. Some may be classified as communications or may actually be fire alarm circuits, which would be covered by Chapter 8 or Article 760. In addition, Article 727 covers a specific and separate type of circuit and wiring method for instrumentation wiring in industrial occupancies. Instrumentation and control systems that use optical fiber cable are not truly electrical circuits at all, but they are covered by Article 770.

The fact that instrumentation and control circuits are covered in Chapters 7 and 8 is very significant to the choice of wiring method, installation, and grounding of these circuits, because according to Section 90.3, Chapter 7 can modify or amend Chapters 1 through 4. Also, the rules of Chapters 1 through 7 apply to Chapter 8 only when Chapter 8 says so. This means that rules that are different for controls and instrumentation are not necessarily in conflict with the more general rules, because those general rules may be changed by Chapter 7 and usually are not applicable to Chapter 8.

The wiring methods for Class 1 circuits are the ordinary methods covered by Article 300 and Chapter 3 according to 725.25, but Class 2 and Class 3 circuits may be wired using the

Keywords

Instrumentation
Control
Class 1
Class 2
Class 3
Communications
Fire alarm
Grounding
Bonding
Hazardous locations
System grounding
Equipment grounding

NEC References

90.3, 250.20, 250.86, 250.92, 250.100, 250.112, 250.162, 501.10, 501.30, 502.10, 502.30, 503.10, 503.30, 504.20, 725.1, 725.2, 725.21, 725.25, 725.41, 725.61, 800.3

alternative wiring methods that are described in Part III of Article 725, including 725.52. This does not help to answer the grounding question, even for Class 1 wiring, because very few grounding requirements are found in Chapter 3. In fact, since little is said in Article 725 about grounding, grounding of Class 1, 2, and 3 circuits and equipment must be in accordance with Article 250. Special rules do apply to grounding of Class 2, Class 3, and some Class 1 circuits, but those special rules are found in Article 250 rather than in Article 725.

To locate grounding requirements for instrumentation and control circuits, we can use the outline of Article 250. For example, if we want to know if a raceway, box, other enclosure, or equipment containing a specific circuit is required to be grounded, we can find this information in Part IV, "Enclosure, Raceway, and Service Cable Grounding." Here we see that Section 250.86 covers "Other Conductor Enclosures and Raceways." This section says, "Except as permitted by 250.112(I), metal enclosures and raceways for other than service conductors shall be grounded."

For the most part, 250.112 is a list of specific items that must be grounded, but 250.112(I) provides for certain items that are required to be grounded only under specific circumstances, that is, items that are required to be grounded only when their associated supply systems are required to be grounded. When we look at 250.112(I), we see that it specifically covers Class 1, Class 2, and Class 3 circuits along with fire alarm circuits. The equipment supplied by these circuits is required to be grounded only where the system is required to be grounded under Part II (AC systems) or Part VIII (DC systems). In other words, equipment such as raceways, boxes, and the like that contain or are supplied by a Class 1, Class 2, or Class 3 circuit must be grounded only if one of the circuit conductors of the Class 1, 2, or 3 *system* is required to be grounded. (If a Class 2 system is grounded, but not required to be grounded, the equipment is still not required to be grounded.) This series of rules is illustrated in Figure 6.14(a).

Class 1, Class 2, and Class 3 circuit equipment is required to be grounded only if the system is required to be grounded.

Source: *Limited Energy Systems*, 2002, Figure 3.6.

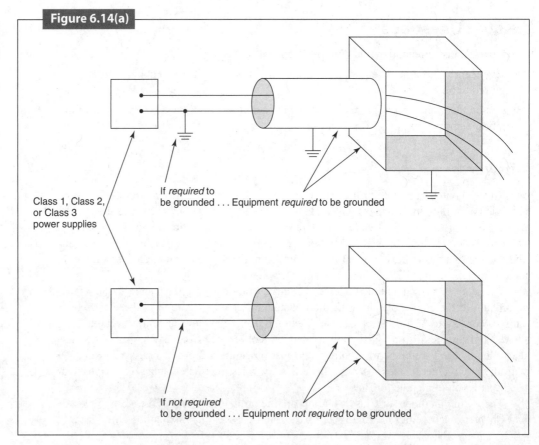

Figure 6.14(a)

The requirements for grounding DC systems are found in 250.162 and are quite easy to summarize. Although there are exceptions, generally two-wire DC systems that are 50 volts or less or over 300 volts are not required to be grounded. Most DC systems used in instrumentation are 50 volts or less, so those DC systems are not required to include a *grounded conductor*. Therefore, equipment supplied by or containing such "low-voltage" DC systems is not required to be grounded, so an *equipment grounding conductor* is not required either.

The rules for AC systems are a bit more complicated. Section 250.20(A) says AC systems of 50 volts or less (virtually all AC Class 2 systems and all AC Class 1 power-limited systems) are required to include a grounded conductor only where the circuits are installed as overhead conductors outside buildings or are supplied by transformers whose supply is over 150 volts to ground or ungrounded. (Remember that most 208-volt or 240-volt systems are 120 volts to ground.) These rules consider the possible overvoltage that could occur in the secondary circuit due to transformer failure or exposure to higher voltages such as lightning. Since the majority of Class 2 systems are derived from grounded 120-volt systems, most Class 2 systems and the equipment they supply may be ungrounded, as illustrated in Figure 6.14(b), unless they run outside, overhead. The same could be said for most Class 1 power-limited systems or other Class 1 systems that are less than 50 volts. A 24-volt system supplied by a transformer that is not listed as a Class 2 or Class 3 transformer is typically a Class 1 system, depending on the use. A Class 1 power-limited system is less than 50 volts by definition (actually 30 volts or less) and is therefore permitted to be ungrounded in many cases, depending on how it is derived.

Class 3 systems often have higher voltages, up to 100 volts, so Class 3 systems and equipment are often required to be grounded. (See 250.21(3) for special rules and exemptions for critical control systems.)

A 24-volt system could be used for low-voltage lighting or for power wiring in a recreational vehicle, neither of which is a Class 1 circuit or a circuit covered by Chapter 7. These systems have their own rules. Low-voltage lighting circuits are not permitted to be grounded, according to 411.5(A) and 250.22, but 551.10(C)(4) requires most low-voltage systems in recreational vehicles to be grounded.

Many Class 1 systems are more than 50 volts; 120-volt Class 1 control systems are very common. Such systems fall under 250.20(B)(1), which requires grounding of those systems as shown in Figure 6.14(c). Single-phase systems over 150 volts are less common, but often are not *required* to be grounded.

Again, in general, if the remote-control, signaling, or fire alarm system is *required* to include a grounded conductor, the supplied equipment and enclosures are also *required* to be

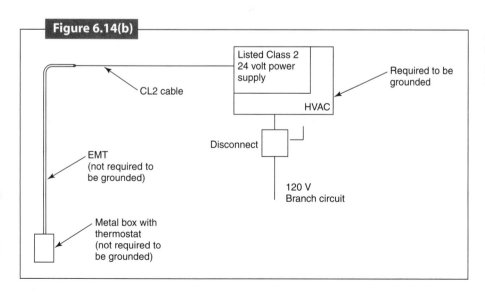

Metal raceways and enclosures that are not required to be grounded.

Source: *Limited Energy Systems*, 2002, Figure 3.31.

A Class 1 remote-control circuit where the system and equipment are required to be grounded.

Source: *Limited Energy Systems*, 2002, Figure 3.14.

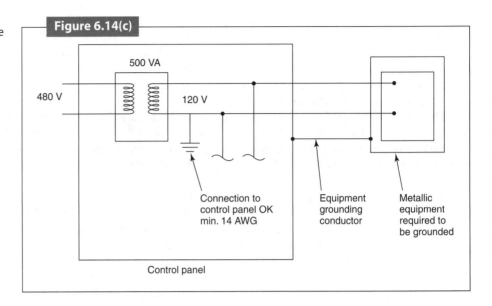

grounded. However, with regard to Class 1 circuits, this generalization of 250.112(I) only covers Class 1 circuits that are 150 volts or less and "power-limited" Class 1 circuits—those that are limited to 30 volts and 1000 VA. A Class 1 system that is over 150 volts may not be required to be grounded to comply with 250.20, but the equipment and enclosures are required to be grounded to comply with 250.110(6), which requires equipment to be grounded if it operates with any terminal over 150 volts to ground. The voltage to ground of an ungrounded system is defined in Article 100 as the maximum voltage between conductors.

As may be expected, some of these rules in Article 250 may be modified for hazardous (classified) locations. In fact, special rules for hazardous locations are found in Article 250, and are modified only for some specific situations in Chapter 5. Section 250.100 reads as follows:

> Regardless of the voltage of the electrical system, the electrical continuity of non–current-carrying metal parts of equipment, raceways, and other enclosures in any hazardous (classified) location as defined in Article 500 shall be ensured by any of the methods specified for services in 250.92(B) that are approved for the wiring method used.

The methods mentioned in 250.92(B) that are applicable are threaded couplings; threaded bosses (hubs); threadless couplings and connectors; other approved devices, such as bonding locknuts and bonding bushings; and bonding jumpers, which are often used with grounding bushings. Standard locknuts or bushings, including double locknuts, are not sufficient.

The requirements of 250.100 and 250.92 are in Part V of Article 250, titled "Bonding." Bonding is required to ensure electrical continuity and sufficient capacity to conduct fault current. In hazardous locations, the bonding path must also be connected in a manner that will prevent any arcing or sparking at connections.

As noted, many instrumentation circuits are Class 2, and thus, according to 725.2, not typically a shock or fire hazard and not capable of producing fire-starting fault currents. Therefore, in ordinary locations, bonding to provide an adequate fault path is not required. However, in many hazardous (classified) locations, a very low-energy spark could ignite the flammable gas or vapor, the combustible dust, or the ignitible fibers or flyings that define a hazardous location. Usually the shock hazard (mostly related to voltage) is not increased in a hazardous area, but because the fire hazard (mostly related to current) is much greater, a higher standard must be applied to all potential current pathways.

This special rule for hazardous locations does not change the system grounding or equipment grounding requirements discussed previously, because it does not require *systems* to be grounded to a different standard, but *equipment* that is bonded is usually also grounded. The significance is that where metal raceways and enclosures are used for control and instrumentation circuits in hazardous locations, the metal parts must be bonded to provide adequate spark-free paths for fault current, and this will usually be accomplished through a series of grounding paths that may not be required elsewhere.

This does not mean that metal raceways and enclosures are always required in hazardous areas. Permitted wiring methods are given in 501.10 for Class I areas (flammable gases and vapors), 502.10 for Class II areas (combustible dusts), and 503.10 for Class III areas (ignitible fibers and flyings). Where nonmetallic methods are permitted, the nonmetallic parts are obviously not required to bonded or grounded. Some instrumentation and control circuits may be Class 2 or Class 3 and also intrinsically safe or nonincendive. Intrinsically safe and nonincendive circuits are unable to supply enough energy in an arc to ignite the specific flammable or combustible material. (See 500.2 for further information.) Where circuits are intrinsically safe or nonincendive, any wiring method that would be suitable for a nonhazardous area is permitted, including nonmetallic methods and Class 2 or Class 3 circuit cables or their permitted substitutes. These rules can be found in 501.10(B)(3), 502.10(B)(3), 504.20, and 725.61(D). However, if, for example, a Class 2 cable is used for an intrinsically safe or nonincendive circuit and that cable is installed in metallic enclosures or raceways in a hazardous area, those metal parts must still be bonded in accordance with 250.100, because 250.100 says it applies to metal parts in classified areas "regardless of voltage."

This discussion has focused on Article 725, but the same approach (and essentially the same answers) would also apply to fire alarm circuits wired under Article 760. As noted, for communications circuits, Article 250 applies only where referenced in Chapter 8. Article 800 contains a few such references, but those references relate primarily to circuits entering buildings and to grounding of protectors. However, for a classified area, the approach for a communications circuit or any circuit reclassified as a communications circuit is a bit different, because Article 250 is referenced only indirectly. (Class 2 and Class 3 circuits are permitted, or in some cases required, to be reclassified or treated as communications circuits according to Sections 725.56(D), 725.57, and 800.133(A)(1)(b).) Section 800.3(B) says communications wiring installed in hazardous locations must comply with the "applicable requirements of Chapter 5." Sections 501.30, 502.30, and 503.30 then refer back to Article 250, including Section 250.100, which applies the special bonding requirements to communications wiring installed in hazardous locations. Communications wiring that is intrinsically safe or nonincendive is treated the same as other intrinsically safe or nonincendive circuits.

Question 6.15

How are rules for data and communications circuits applied to a Class 2 "interbuilding circuit"?

Related Questions

- Is the wiring to a camera on a pole considered to be an "interbuilding circuit"?
- Does Section 725.57 apply to outdoor CCTV security cameras when they are mounted on the same building?

Keywords

Interbuilding
Building
Structure

Keywords

Coaxial
Class 2
Communications
Protector
Electrode
Grounding conductors

NEC References

410.15, 725.41, 725.57, 800.50, 800.90, 800.93, 800.100, 800.133, 820.93, 820.100

- If a 24-volt Class 2 circuit is run as an interbuilding circuit, do the fusing or protector requirements of 800.50 and 800.90 apply?
- If a water piping system is used for grounding as permitted in 800.100(B)(1), must that connection be made within 5 feet of the building water entrance?
- If an outdoor communications cable terminates in a box that is grounded using an appropriately sized 12 AWG branch circuit grounding conductor, can that grounding conductor be used as the grounding means for the cable shields and surge protectors?

Answer

First, the *NEC* and Article 725 do not recognize "data" circuits as a separate class of circuit. Most such circuits are Class 2 according to 725.41, but they may be reclassified as communications circuits, or communications cables may be substituted for Class 2 cables. Coaxial cables listed as Type CL2 are available for use with cameras, but many such cables are listed as Type CM or Type CATV or may have multiple listings.

Section 725.57 in the *National Electrical Code* covers Class 2 or Class 3 circuit conductors extending "beyond one building and run so as to be subject to accidental contact with electric light or power conductors operating at over 300 volts to ground, or are exposed to lightning on interbuilding circuits on the same premises."

The *NEC* does not define the term "interbuilding," and most dictionaries will not define this term either. Nevertheless, "building" is defined in Article 100 as "a structure that stands alone or that is cut off from adjoining structures by fire walls with all openings therein protected by approved fire doors." Article 100 also defines "structure" as "that which is built or constructed." The prefix "inter" means "between" or "among." So "interbuilding" means "between separate structures" in this usage. Thus, the circuit for a camera on the same building is not an interbuilding circuit, but a circuit for a camera on a pole does go to a separate structure and should be treated as an interbuilding circuit, as illustrated by Figure 6.15(a).

From this discussion we can see that the requirements listed in Section 725.57 referring to Articles 800 and 820 would apply to the circuit run to a pole. Article 800 applies to communications circuits generally, but the rules from Article 820 apply specifically and only to coaxial conductors. The other important issue, however, is whether the camera circuit is "exposed" as defined in Sections 800.2 and 820.2, which refer to "a circuit that is in such a position that, in case of failure of supports and insulation, contact with another circuit may result." Most interbuilding circuits do not require primary protectors unless they are exposed to lightning or higher-powered circuits, but coaxial cables for CATV systems or the equivalent must be grounded where they enter buildings.

Section 800.90(A) requires a listed primary protector to be provided "on each circuit run partly or entirely in aerial wire or aerial cable not confined within a block." A block in this sense is "a square or portion of a city, town, or village enclosed by streets and including the alleys so enclosed, but not any street," as defined in 800.2. Section 800.90(A) goes on as follows:

> Also, a listed primary protector shall be provided on each circuit, aerial or underground, located within the block containing the building served so as to be exposed to accidental contact with electric light or power conductors operating at over 300 volts to ground. In addition, where there exists a lightning exposure, each interbuilding circuit on a premises shall be protected by a listed primary protector at each end of the interbuilding circuit.

These protectors are required to be installed according to the requirements of Article 800 as well as any listing or labeling instructions. The rules require listed protectors for most signaling, remote-control, data, or communications circuits that run between buildings, particularly if the lower-powered circuits enter enclosures such as boxes, poles, handholes, manholes, or the like that also contain conductors for power or lighting. However, such protectors would not

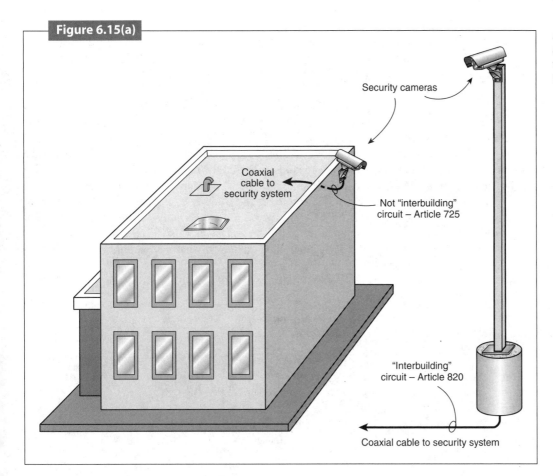

Figure 6.15(a) Application of *NEC* articles to coaxial cables that are or are not interbuilding circuits.

be required where separations are maintained and there is little likelihood of lightning. These protectors are the same type as used where communications wiring enters buildings. The protectors are required to be fused only where none of the conditions in 800.90(A)(1) can be met. Properly selected fuseless primary protectors may be used in most cases.

Section 800.100(A)(3) of the *National Electrical Code* requires a primary protector or the metallic member or members of a cable sheath that is required to be grounded by Section 800.93 to be grounded using a conductor not smaller than 14 AWG. For coaxial cables, the grounding requirements are found in Section 820.93 and 820.100, but primary protectors are not required for coaxial cables; the shields of coaxial cables are required to be grounded by a connection to an electrode.

Sections 800.100(B) and 820.100(B) list the items that may be used as electrodes for cable and primary protector grounding. In either case, the subsections refer to 250.50 and 250.52 for the electrodes to be used, and in both cases, it is the nearest of the items in the list that must be used. Where a water pipe is available, the portion that is considered for use as an electrode is just the first 5 feet of the incoming water pipe. A length restriction is needed because any portion of the interior piping between the buried pipe and the point of connection becomes an extension of the grounding electrode conductor. Also, it is possible for the metal water pipe to be interrupted by nonconductive portions or unions at meters, pressure-reducing or isolation valves, or by plastic piping. Keep in mind that the interior piping itself is not suitable as an electrode—only the directly buried portion of the metal piping qualifies as an electrode.

Considering the rules mentioned, the equipment grounding conductor for the 20-ampere branch circuit would not be permitted to be used for grounding coaxial cable shields or primary protectors. Although it may be big enough, it usually does not originate at the nearest electrode. If these grounding requirements are to be applied to an exterior pole, a driven ground rod is the most likely electrode, as required in 800.100(B)(2) or 820.100(B)(2).

From the standpoint of Article 250, a driven ground rod at a pole is a supplementary electrode. If a power circuit is also provided for a luminaire on the pole, for example, the equipment grounding conductor in the branch circuit may serve as a bonding means between the electrode at the pole and the electrode(s) in the building according to 250.54. An application like this is shown in Figure 6.15(b).

The driven ground rod in Figure 6.15(b) is not required at the pole by the *NEC*. The place where grounding is required for the coaxial cable is at the point of entry to the building. The ground rod may be provided as a design feature or for lightning protection for the pole, in which case the ground rod should be bonded to the equipment grounding conductor and coaxial shield to comply with NFPA 780, *Standard for Lightning Protection*. Equipment grounding conductors or equipment grounding terminals may also be used for secondary protectors, which are treated pretty much the same as, and may even be part of, transient voltage surge suppressors (TVSS).

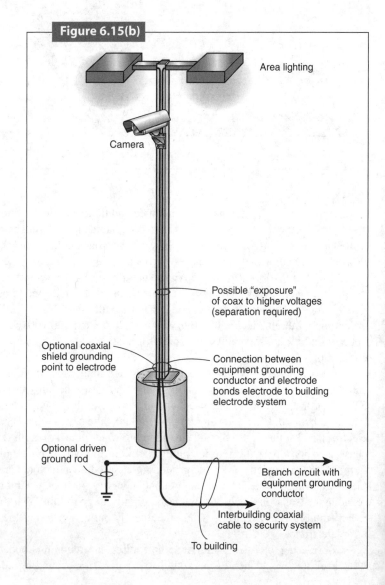

Lighting pole also used for a security camera.

One other thing: Remember that Sections 725.55, 800.133 and 820.133 require separations between the limited-energy conductors covered by these articles and any power or lighting conductors. These separations will reduce the exposure of the limited-energy circuits and may eliminate the need for protectors or other special grounding at separate structures. In Figure 6.15(b), separation could be provided in a number of ways, including installing the lighting circuit or coaxial cable in a separate raceway (such as flexible metal conduit) within the pole, using Type MC (metal-clad) or AC (armored) cable for the lighting circuit within the pole, or installing the coaxial cable in a raceway on the outside of the pole. The lighting pole is not a raceway as such, but 410.15(B) permits the pole to be used as a raceway.

Question 6.16

When or where are "low-voltage" cables required to be in conduit?

Related Questions

- What sections in the *NEC* define the type of electrical or mechanical rooms where conduit is required for low-voltage cabling?
- Is conduit required for voice and data cables that pass through a mechanical or electrical room?

Keywords

Low voltage
Cable
Conduit
Fire rating
Riser
Penetration

Answer

The *NEC* contains no specific requirement that "low-voltage" cables be installed in a raceway where they pass through an electrical or mechanical room or similar area. According to 725.8, exposed cables run on the surface of ceilings or sidewalls are required to be supported from the building structure in a manner that will not cause damage to the cables under the conditions of normal building use. Other cables are required to be supported from the structure as required by 300.11. Cables cannot be supported by being tied to electrical conduits, according to 300.11(B). They must be installed in a "neat and workmanlike manner," according to 725.8, definitely a very subjective requirement. Similar requirements are found for other low-voltage cables in Articles 760 and 800.

Generally, if low-voltage cables are properly selected for the location and type of use, they are not required to be installed in conduit. For example, a Class 2 remote-control circuit installed in a riser application (passing through more than one floor) would not have to be in conduit if it had a riser or plenum rating and marking. This and similar restrictions are covered in 725.61(B) for remote control and signaling, in 760.61(B) for fire alarm wiring, in 770.154(B) for optical fiber cable, and in 800.154(B) for communications wiring. This is one of the most significant aspects of Articles 725, 760, 770, and Chapter 8: that the rules for wiring methods are significantly altered when the energy in a circuit is sufficiently limited. The wiring methods specified for, say, Class 2 circuits in Article 725 are the methods permitted, and the other methods in Chapter 3 are not usually required for those circuits.

Some more restrictive requirements apply for certain occupancies, applications, or types of construction. These rules are most likely to apply to the application mentioned, but only in certain occupancies. For instance, 300.21 applies because it is referenced in 725.3(B).

NEC References

300.11, 300.21, 501.10, 502.10, 725.8, 725.11, 725.61, 760.61, 770.154, 800.154

According to 300.21, where the building code requires that the room or area containing electrical or mechanical equipment have a fire rating, all penetrations of the walls and ceilings must be made in a manner that will not substantially reduce the required rating of the wall or ceiling assembly or increase the spread of fire or smoke. In many cases, such penetrations are made using a conduit system or sleeve because it is easier to reseal the conduit penetration to retain the required fire rating of the wall or ceiling than to reseal a penetration with open signaling or data cables. This is only an issue of ease of installation, however, not a requirement that conduit be used in these situations, or, where it is used, that the conduit extend into areas beyond the firewall penetrations. Whether the low-voltage cables pass through the room or terminate in that room, if a fire rating is required for the room, any penetrations of the rated wall, floor, and ceiling assembly must have little or no impact on the fire rating of the assembly penetrated.

According to 725.11, if damage to low-voltage "remote-control circuits of safety control equipment" would introduce direct fire or life hazard, the circuit must "be installed in rigid metal conduit, rigid nonmetallic conduit, electrical metallic tubing, Type MI cable, Type MC cable, or be otherwise suitably protected from physical damage." This rule is based on damage or failure of safety-control circuits only, not to circuits used for such things as room temperature control or signaling. This issue (in a mechanical area) is illustrated in Figure 6.16, in which the circuits whose failure could cause a fire or life hazard are installed in the power wiring to the boiler and would have to be protected, but the Class 2 thermostat circuit could be installed using an appropriate cable type without conduit or other added protection. Some fire alarm circuits are also required to be installed in conduit for protection, but these requirements are found in NFPA 72, *National Fire Alarm Code*, not in the *NEC*.

Low-voltage cables may also be required to be installed in conduit in hazardous (classified) locations. For example, in most cases, a signal, remote-control, or communications cable will require conduit in a Class I or Class II hazardous area unless the circuit is also intrinsically safe for Division 1 areas or is nonincendive for Division 2 areas. See 501.10 and 502.10 along with the applicable article for the type of circuit for details on these requirements.

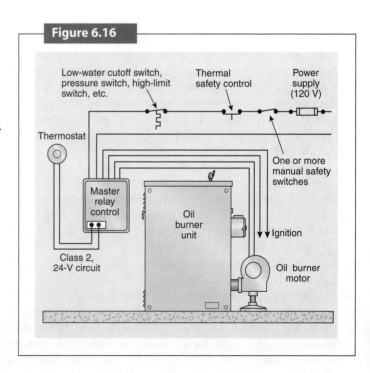

Figure 6.16

An automatic boiler control system in which the safety controls are installed in the power supply circuit, and the Class 2 circuit is not required to be reclassified.

Source: Based on *NEC Handbook*, 2005, Exhibit 725.3.

Question 6.17

Are separations required between power and instrument or control wiring?

Related Questions

- May circuits less than 600 volts from different sources be mixed in an enclosure or raceway?
- Can a 24-volt control or signal cable be run in the same raceway with a 120-volt power cable if the insulation ratings are compatible?
- What are the requirements for segregation of low-voltage cables?
- How can separations be provided between control circuits or limited-energy circuits and power or lighting circuits?

Answer

The most general rule in the *NEC* about mixing conductors of different systems is found in Section 300.3(C), which allows circuits of different systems to "occupy the same equipment wiring enclosure, cable or raceway" if the insulation on all conductors is rated for the maximum voltage applied to any of the conductors and that maximum voltage is not over 600 volts. This general rule is illustrated by Part A of Figure 6.17.

The general rule is followed by a Fine Print Note that refers to 725.55(A) for Class 2 and Class 3 circuits. Actually, the Fine Print Note may be best thought of as a reminder that Section 90.3 says that a rule in Chapter 3, such as this one, may be modified by a rule in Chapter 5, 6, or 7. In fact, the rule allowing the mixing of circuits under 600 volts is modified and becomes more restrictive for circuits covered by Articles 690, 700, 725, 727, 760, 770, and 780. A different and more restrictive rule in 230.7 also applies to service raceways and cables.

If we are considering the signaling and control circuits (including most instrumentation and data circuits) covered by Article 725, we must first determine the classification of the circuit in question. The classes of circuits—Class 1, Class 2, and Class 3—are defined in 725.2. What we see from these definitions is that the classification of a circuit under Article 725 is dependent on the power supply of the circuit. Class 1 classifications and power source requirements are given in 725.21. The requirements for Class 2 and Class 3 power sources are found in 725.41.

The rules of 725.41 make inspection and verification of Class 2 and Class 3 circuits quite straightforward: One must simply find the power supply and verify that it complies with one of the rules in 725.41.

Once the circuit has been classified, separation requirements can be found in 725.26 for Class 1 circuits or in 725.55 and 725.56 for Class 2 and Class 3 circuits, respectively. If the circuit is an industrial instrumentation circuit installed in Type ITC (instrumentation tray cable), separations are covered in 727.5.

Because Class 1 circuits are governed by different rules with regard to conductor size, conductor type, and overcurrent protection, they must generally be separated from higher-powered circuits. According to Section 725.26(A), Class 1 circuits may be freely mixed with other Class 1 circuits under the same restrictions mentioned in 300.3(C). Although these insulation requirements are repeated in 725.26(A) and not in 725.26(B), Section 300.3(C) applies to all Class 1 circuits because Article 300 is specifically referenced in 725.25. Class 1

Keywords

Separations
Circuit classification
Class 1
Class 2
Class 3
Low voltage
Instrumentation
Signaling
Control
Type ITC
Reclassification

NEC References

300.3, 725.1, 725.2, 725.21, 725.25, 725.26, 725.27, 725.41, 725.52, 725.55, 725.56, 727.5

Summary of general rule and some special rules for circuits that may be mixed in common enclosures or must be separated.

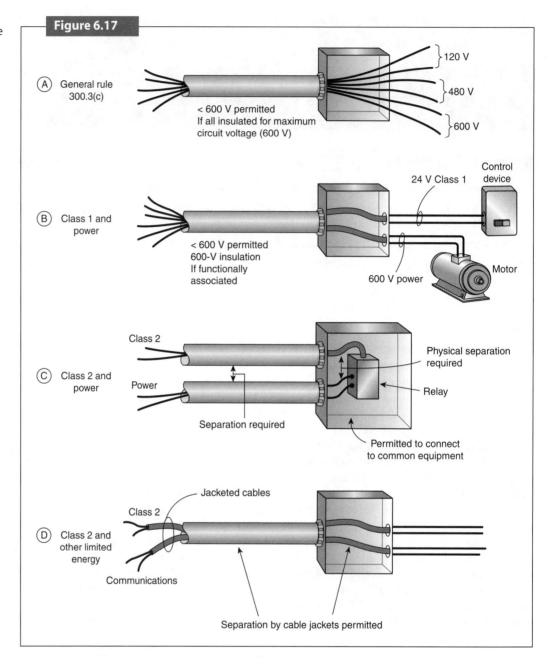

Figure 6.17

conductors are also required by 725.27(B) to be rated for 600 volts. Class 1 circuits must be separated from power and lighting circuits unless they are either functionally associated, as illustrated by Part B of Figure 6.17, or both types of circuits are in the same enclosure to connect to the same equipment in that enclosure (similar to Part C of Figure 6.17).

Connections to common equipment, such as the same relay, imply the functional association required for occupying the same raceway, but multiple power circuits may be in the same enclosure with multiple Class 1 circuits, such as in a relay or control panel, and not all of these must be functionally associated. Section 725.26(B)(2) says that conductors of power and Class 1 circuits may be installed in factory- or field-assembled control centers. In other cases, such as in manholes and cable trays, separations between Class 1 and power circuits may be provided by separate raceways, metal cable sheaths, specific separate wiring methods, or, in manholes, fixed nonconductors or secured physical separations.

Class 2 and Class 3 circuit classifications are based on the energy limitations of the power supplies. Because these power limits reduce the fire hazard and, in the case of Class 2 circuits, the shock hazard, special precautions are necessary to ensure that the power limits will not be compromised. If ordinary Class 1 wiring methods are used throughout the circuit, a Class 2 or Class 3 circuit may be reclassified as a Class 1 circuit in accordance with 725.52(A), Exception No. 2, and the separation requirements for Class 1 circuits may then be applied. Otherwise, Class 2 and Class 3 circuits, along with other power-limited circuits, must be separated from higher-powered circuits to avoid or minimize the chances of inadvertent contact with or connection to a higher-voltage system. This is especially important where Class 2 or Class 3 wiring methods have been used in any part of the circuit, because Class 2 and Class 3 cable wiring methods are more susceptible to damage and may have lower voltage ratings than Class 1 or power wiring methods.

Section 725.55 provides requirements for separations between Class 2 or 3 circuits and higher-powered circuits. The cables and circuit types that do not require separations are covered in 725.56. The general requirement of 725.55(A) reads as follows:

> Cables and conductors of Class 2 and Class 3 circuits shall not be placed in any cable, cable tray, compartment, enclosure, manhole, outlet box, device box, raceway, or similar fitting with conductors of electric light, power, Class 1, non–power-limited fire alarm circuits, and medium power network-powered broadband communications circuits unless permitted by 725.55(B) through 725.55(J).

Most of the rules in 725.55(B) through (J) cover ways of maintaining separations when the limited-energy Class 2 or Class 3 circuits have to be in the same enclosure because they connect to common equipment, as shown in Part C of Figure 6.15(a), or will be in close proximity because they must occupy a common manhole, cable tray, or hoistway. A barrier or a separate raceway is generally preferable, but physical separation is sometimes permitted; in some cases, separation may be provided by cable jackets. The possibilities are too numerous to list them all here, but a careful review of 725.55 will provide a means of separation for virtually any necessary installation.

If physical separation cannot be maintained, usually because the circuits are functionally associated, then reclassification and installation of the entire circuit as a Class 1 circuit is sometimes an option.

Sections 725.56(A) through (E) allow limited-energy circuits of the same classification to be mixed in a common cable or raceway, permit limited-energy circuits of different types to be mixed in the same cable where they are all insulated or reclassified to a higher standard, or recognize cable jackets as separations between limited-energy circuits of different classifications, as illustrated in Part D of Figure 6.17.

Question 6.18

What is the purpose of "CI" cable?

Related Questions

- Where are the requirements for fire alarm circuit integrity (CI) cable found?
- Where can I find information on survivability of fire alarm wiring?

Keywords

Circuit integrity
Fire alarm
Fire alarm cable

NEC References

725.61, 760.81, 760.82, NFPA 72

Answer

Fire alarm "circuit integrity" (CI) cable like that shown in Figure 6.18 is mentioned in Sections 760.81(F) and 760.82(G) of the *NEC*. CI Cables for related Class 2 or Class 3 circuits are mentioned in 725.61(H). The construction of the cable is such that it can sustain significant fire damage but still remain capable of continued operation without developing open or short circuits. Section 760.81 covers non-power-limited fire alarm (NPLFA) circuits and includes many notification appliance circuits, especially those that power strobes. Section 760.82 covers power-limited fire alarm (PLFA) circuits, including some notification appliance circuits (horns, speakers), most modern initiating circuits (detectors), and many signaling line circuits. However, the *NEC* only recognizes the use and listing of circuit integrity cables and provides for the CI suffix that is used for marking and identifying these cables; it does not provide any requirements for their use. The required applications of circuit integrity cable are found in NFPA 72, *National Fire Alarm Code*.

Circuit integrity cable is one method of ensuring survivability of a circuit in a fire condition. Survivability is covered by Section 6.9.4 in the 2002 *National Fire Alarm Code*. This was Section 3-8.4.1.1 in the 1999 code. The language was revised—primarily clarified—in 2002. However, the application remains essentially the same.

Requirements for survivability in a fire apply "only to systems used for partial evacuation or relocation of occupants." A high-rise is one example where partial evacuation is used, and a hospital is one example of where relocation may be needed. The requirement for survivability applies only to notification appliance circuits (horns, strobes, speakers), not to notification device circuits (detectors). Survivability may also apply to signaling line circuits or similar circuits that interconnect fire command centers with central control equipment, because these circuits may have to carry notification signals (see 6.9.4.6).

The portion of the notification appliance circuit that is required to be protected is from the control unit to the "evacuation signaling zone," which is defined as "a discrete area of a building, bounded by smoke or fire barriers, from which occupants are intended to relocate or evacuate." In any case, the methods that may be used to ensure survivability include "(1) A 2-hour rated cable or cable system, (2) a 2-hour rated enclosure, or (3) performance alternatives approved by authority having jurisdiction." Type MI (mineral-insulated) or circuit integrity cables are intended to meet the "2-hour rated cable" option. Performance alternatives may include cables in metal raceways installed in a sprinklered building (see 6.9.4.6 Exception.) Some "CI" cable must be installed in conduit as a condition of its listing and rating. Two-hour rated enclosures are also often used.

The commentary in the 1999 or 2002 *National Fire Alarm Code Handbook* and in the Annex to the *National Fire Alarm Code* is very helpful in understanding the application and

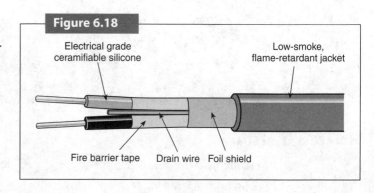

Construction of typical circuit integrity (CI) cable.

Source: *National Fire Alarm Code Handbook*, 2002, Exhibit 6.17. (Courtesy Rockbestos-Suprenant Cable Corp.)

reasons for requirements for survivability of selected fire alarm circuits. Circuit integrity cable was first recognized by the *NEC* in the 1999 edition. This cable provides one effective means of meeting survivability requirements, but it is not the only means.

Question 6.19

What wiring method should be used to interconnect residential smoke detectors?

Related Questions

- Why is Type FPL cable not suitable for interconnections for residential smoke detectors?
- Are power and signal allowed in the same cable or raceway for residential smoke detector interconnections?
- Are interconnected multiple-station smoke alarms a "fire alarm system"?

Answer

"Residential smoke detectors" as used in most dwelling units are called "multiple station smoke alarms" in NFPA 72. They are not technically "fire alarm systems" as defined in NFPA 72 and as contemplated by Article 760. Smoke alarms are treated as a separate type of alarm system in NFPA 72, which distinguishes between the "smoke detectors" typically used with a true fire alarm system and the "smoke alarms" commonly used in dwellings. According to Section 760.1, Article 760 "covers the installation of wiring and equipment of fire alarm systems including all circuits controlled and powered by the fire alarm system," so Article 760 is not really applicable to typical residential smoke alarm systems, even when they are interconnected, because they are not powered and controlled by a fire alarm panel.

More specifically, Part III of Article 760 in the *NEC* applies to the power-limited fire alarm systems with which Type FPL (fire alarm, power-limited) cable is intended to be used. Multiple-station smoke alarms are not listed as power-limited power supplies, nor are they powered by a fire alarm system; therefore, the requirements of Article 760, Part III do not apply. Since Article 760 does not apply to the interconnecting wiring, by default, Chapter 3 wiring methods must be used. Any of the Chapter 3 methods may be used, subject to local rules and any occupancy or location restrictions that apply to any particular wiring method.

Probably the most commonly used wiring method for smoke alarms in dwellings is Type NM (nonmetallic-sheathed) cable. Three-conductor (with ground) NM with 12 or 14 AWG conductors is common, where two wires are used to power the alarms, and the third wire is used to interconnect the alarms so they all sound when any one detects smoke. Most manufacturers of multiple-station smoke alarms show a three-conductor interconnection between devices. Figure 6.19 shows a smoke alarm with relay contacts and a connection for an additional wire to connect a remote notification appliance such as a bell, horn, or strobe. Where the alarms alone are used to provide the audible signal, the extra relay contact may not be used.

Keywords

Smoke detectors
Smoke alarms
Multiple-station
Fire alarm system
Interconnect
Residential

NEC References

760.1, NFPA 72

Single- or multiple-station smoke alarm with relay contacts for remote notification appliance.
Source: *National Fire Alarm Code Handbook*, 2002, Exhibit 11.4. (Courtesy Mammoth Fire Alarms, Inc.)

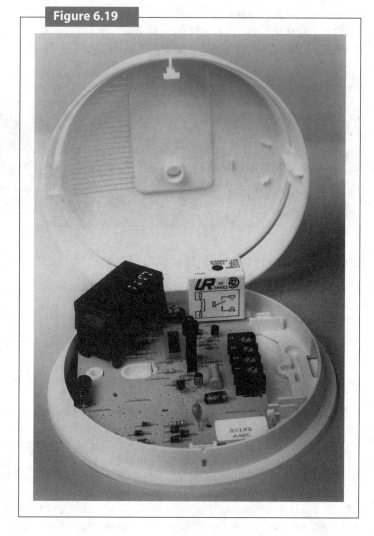

Figure 6.19

Question 6.20

Are communications cables permitted on a roof without a raceway or overhead clearance?

Keywords

Communications
Telephone
Exposed

Related Questions

- May communications cable be laid directly on a built-up gravel roof?
- Should communications distribution cabling on the roof be raised to provide the 8-foot clearance described in Section 800.44(B) in the *NEC*?

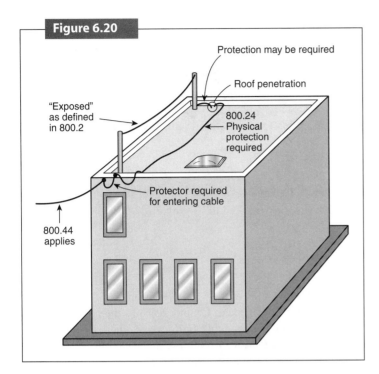

Figure 6.20 Application of rules for protection of communications conductors outside buildings.

Answer

Section 800.44 of the *NEC* pertains to overhead conductors entering buildings. It is not intended to apply to runs between the service point and end-use equipment. The clearances above roofs specified in 800.44(B) are intended to apply to the outside wiring that comes to a building, usually from a utility. However, Section 800.24 also provides requirements for installation of conductors, and provides requirements for mechanical execution of work. Conductors are not permitted to be installed where they are subject to physical damage during normal building use. Conductors lying on a roof are subject to physical damage by such things as people stepping on them, hail damage, or damage due to removing snow. Therefore, the implication is that physical protection must be provided as shown in Figure 6.20, most likely in the form of a raceway.

Another issue is that by running the conductors outside buildings, they may be exposed to higher voltages as defined in 800.2. For example, if protection is provided by elevation, the conductors become aerial conductors and may be "exposed," as illustrated in Figure 6.20. This exposure is most effectively eliminated or reduced by keeping the conductors inside the building, but the exposure may also be eliminated or reduced (at least in some cases) by installing the conductors in a raceway. A grounded raceway would probably be preferred, but 800.50(C) requires metal raceways to be grounded only where the raceways are ahead of the primary protector.

Nevertheless, where conductors are exposed to higher voltages or lightning, they are subject to the same risks as interbuilding circuits, and primary protectors may be advised where the cable reenters the building. See 800.90(A), Fine Print Notes Nos. 1 and 2.

Raceway
Aerial cable
Primary protector
Interbuilding

NEC References

800.2, 800.24. 800.44, 800.50, 800.90

Question 6.21 What raceways or equipment are required to be grounded in communications systems?

Keywords

Communications
Grounding
Electrode
Raceway
Primary protector
Cable sheath

NEC References

Article 250, 250.162, 800.2, 800.93, 800.100, 800.110

Related Questions

- Do metal raceways for communications systems require grounding?
- Are telecommunications power supplies required to be grounded?

Answer

According to Section 90.3, Chapter 8 of the *NEC*, which covers communications systems, stands apart from the rest of the *NEC*. Provisions of the first seven chapters of the *NEC* may apply to communications, but only where referenced in Chapter 8. Thus, the regular grounding requirements of Article 250 do not apply to communications except where Chapter 8 says so. For telecommunications, Article 800 is the primary reference in Chapter 8.

When we look through Article 800 for grounding requirements, we find that grounding is only required for primary protectors and some cable sheaths. This language is found in Section 800.100. Metallic sheaths of incoming cables must be grounded where they enter buildings, unless the sheath is interrupted at the point of entrance, according to 800.93. Other items that may require grounding are secondary protectors and incoming metallic conduits. However, secondary protectors themselves are not required, and metallic conduits (rigid metal and intermediate metal conduit) are required to be grounded only where they are used to extend the point of entrance into a building. (See the definition of "point of entrance" in 800.2.)

Conduits used within buildings for cable distribution, for stub-ups from outlet boxes in walls, or as sleeves or risers are not required to be grounded. Section 800.110 requires that raceways used for communications be installed according to the requirements of Chapter 3 for the specific raceway used. However, the raceway articles in Chapter 3 do not include references to Article 250 that require grounding of the raceway. Under Article 250, metal enclosures that are likely to become energized must be grounded, but these rules do not apply to communications because they are not referenced within Chapter 8 (Article 800, in this case). Furthermore, communications circuits are permitted to be wired using alternative wiring methods (not methods from Chapter 3) because they do not present a shock or fire hazard under normal conditions. Typically, even if a raceway were to become energized by a communications circuit, it would not create a shock hazard. The grounding requirements for communications equipment and raceways, as explained previously, are summarized in Figure 6.21.

The grounding requirements found in Section 800.100 are most often applied to primary protectors. The electrodes that may be used are listed in 800.100(B). These electrodes are generally part of the grounding electrode system for the power system or are other significant grounded elements of the power system. If separate electrodes are used for communications, 800.100(D) requires that the communications system electrode be bonded to the power system electrode(s). Therefore, compliance with the requirements of 800.100 results in essentially the same electrode being used for both communications and power systems.

Article 800 is based on the assumption that power is derived from a communications central station or similar systems. For the most part, these systems are rated at nominal 48 volts

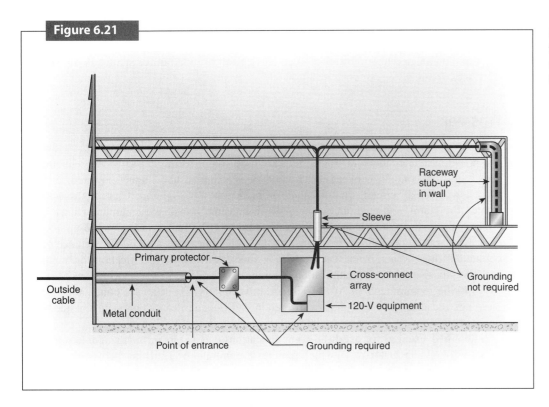

Figure 6.21 Grounding requirements for communications equipment, cables, and raceways.

DC. According to 250.162, such systems are not required to be grounded. However, where communications power supplies are supplied by a power system such as a 120-volt AC system, Article 250 does apply to the power system feeders or branch circuits that supply the communications equipment. Similarly, where switching apparatus such as a network communications rack is supplied by a building power system, the rack must be grounded in accordance with Article 250. Article 250 does not require a direct connection to an electrode, but equipment powered by 120-volt circuits or other power circuits must generally be grounded and bonded by an equipment grounding conductor as required by Part V and Part VI of Article 250, which provides an indirect connection to an electrode.

Question 6.22

Does the *NEC* specify minimum spacing between conductors for communications and power?

Related Questions

- What spacings are required between power and communications?
- Are required separations different for communications circuits inside or outside buildings?

Keywords

Communications
Separations

Keywords

Enclosures
Raceways
Framing
Type NM cable

NEC References

800.44, 800.47, 800.50, 800.133

- What is the purpose of requiring separations between communications and conductors for power or lighting?
- Are communications cables and Type NM cable permitted in the same hole in wood or metal framing?

Answer

The *NEC* does provide requirements for spacing between communications system conductors and power system conductors. These rules are based on providing for safety by reducing the possibility of contact between communications circuits and higher-powered circuits. ("Higher-powered" circuits in this context include power, lighting, Class 1, non-power-limited fire alarm, and medium-power network-powered broadband communications, all of which are circuits without inherent power limitations.) The separations required do not necessarily guarantee optimum performance of the communications system. Interference on the communications system is not seen as a fire or shock issue. Other specifications or standards, including those used by communications utilities, may address performance issues.

Spacing requirements in Article 800 are in two parts. The rules for conductors outside buildings and ahead of protectors are covered in Part II, and rules for conductors inside buildings or between protectors and equipment are covered in Part V.

Required spacings for overhead communications conductors outside and entering buildings are found in 800.44(A)(4), which requires a minimum of 12 inches between communications conductors and overhead service drops at any point in a span and at the building attachment. At least 40 inches is required between communications and power conductors at the pole. These rules are based on service conductors of not more than 750 volts where the ungrounded conductors are insulated. If communications conductors enter buildings from underground, they must be arranged to avoid accidental contact with systems over 300 volts to ground and otherwise separated by partitions or barriers in accordance with 800.47. The *NEC* does not provide a specific minimum dimension for underground conductor spacing. Communications and power conductors on buildings must be separated by having the power conductors in raceway or cable assemblies, by being firmly fixed in a nonconductor such as flexible tubing, or by at least 4 inches of spacing. These spacing requirements are illustrated in Figure 6.22. Note that the more freely the conductors may move, the greater the required spacing to avoid contact between communications and higher-powered systems.

Communications conductors between a protector and communications equipment (or communications conductors that do not require protectors) that are installed on or in buildings are covered by Section 800.133. This section does permit communications conductors in the same raceway or enclosure with certain other limited-energy systems, such as Class 2 circuits and power-limited fire alarm circuits. Generally, however, communications conductors must be separated from higher-powered circuits by installation in separate enclosures, raceways, or cable assemblies. Section 800.133(A)(2) requires at least 2 inches separation between higher-powered circuits and communications conductors, but two exceptions apply to this rule. Probably the most common application to which an exception applies is where communications cables with nonmetallic jackets or sheaths are installed in frame construction with Type NM (nonmetallic) cable. In these cases, the communications cables are not required to be installed in separate holes to maintain the 2-inch spacing according to 800.133(A)(2), Exception No. 1. The jackets or sheaths on each of the cables are considered as providing the required separation. Similarly, communications conductors in raceways or jacketed communications conductors are not required to be separated from higher-powered circuits that are in separate raceways or are in metal-sheathed or metal-clad cables.

These separation rules allow close proximity but require separations only to avoid intersystem contact, which is the primary safety issue. As noted, greater separations or shielding may be needed to improve performance and reduce electromagnetic interference in the communications system, but these performance issues are outside the scope of the *NEC*.

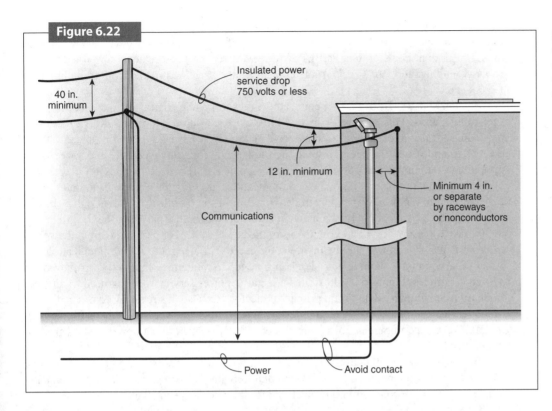

Figure 6.22 Spacing between communications conductor and power conductors outside and on buildings.

Are communications and power system grounding electrodes required to be interconnected?

Question 6.23

Related Questions

- What electrodes are permitted for grounding of communications systems?
- Is a communications system grounding electrode required where there is no electrical power system?

Answer

According to Section 90.3, "Chapter 8 covers communications systems and is not subject to the requirements of Chapters 1 through 7 except where the requirements are specifically referenced in Chapter 8." Therefore, the grounding requirements of Article 250 (Chapter 2) do not apply generally to communications. However, the articles of Chapter 8 all include rules about what may be used for grounding and what must be grounded in a communications installation. Typically, a communications installation is not required to be a grounded system; that is, it is not required to have a grounded conductor as part of the system. Protectors, arrestors, and entrance cable sheaths or shields are required to be grounded, according to Chapter 8. Some conduit is also required to be grounded if it is used to extend the point of entrance into a building, according to 800.50(C).

Keywords

Grounding
Grounding electrode
Grounding electrode system
Communications
Protector

NEC References

90.3, 250.58, 800.50, 800.100

The rules for the required grounding of all types of communications systems are similar in all four articles of Chapter 8, so for the purposes of this discussion, we will use Article 800 as an illustration. For example, Part III of Article 800 covers requirements for protectors and cable grounding, and Part IV of Article 800 covers the permitted methods of grounding protectors and cables.

Section 800.100(B) of Part IV covers the things or systems that qualify as electrodes for grounding cables. This section is divided into two lists: one for locations where there is a grounding means, and one for buildings or structures that have no grounding means. Since even ungrounded power systems are required to have a grounding electrode and a means for grounding equipment, 800.100(B)(2) typically applies only where there is no power in a building or structure; in that case, the communications grounding electrode is not required to be connected to a power system electrode. However, if the building or structure does have a power supply, as is most likely, the items in the list of things that may be used for grounding in 800.100(B)(1) are all grounded elements of the power system or part of the grounding electrode system for the power supply. In effect, the choices of electrodes for communications result in the communications electrodes being bonded (sometimes indirectly) to the power system electrodes. Figure 6.23 illustrates a metallic power service raceway used as an electrode for a communications system primary protector. Because this raceway connects directly to the service equipment and is required to be bonded using the special methods for services in 250.92(B), the communications system is effectively bonded to the power system grounding electrode.

Although Article 250 does not apply directly, the requirement of 250.58 to have only one electrode system in a building is met by compliance with 800.100. Electrode systems that are bonded together are considered one electrode for the purposes of 250.58.

One of the requirements of 800.100(A) is to keep the grounding conductor between an electrode and a primary protector as short as practicable (20 feet maximum for one- and two-family dwellings). To meet this requirement, a separate electrode, such as a driven communications ground rod, may be used, but that electrode must be bonded to the power system

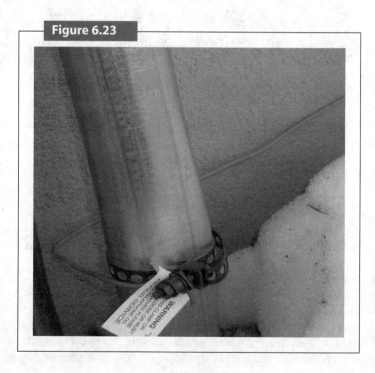

A communications system grounding electrode connection.

Source: *Limited Energy Systems*, 2002, Figure 6.7.

electrode by "a bonding jumper not smaller than 6 AWG copper or equivalent" in accordance with 800.100(D).

The reason for the requirement that a power system electrode be used for or bonded to a communication system electrode is simple. All electrodes in any given location should connect to the same "earth ground" to eliminate any difference of potential between different systems or equipment that are supposed to be grounded. One purpose of grounding is to eliminate such differences between things that could become energized, and thus to eliminate shock hazards. Using a single ground for all equipment may also help to reduce damage due to transient voltages that might not create shock hazards but are sufficient to damage electronic equipment.

Question 6.24

How is raceway fill determined for multiconductor cables?

Related Questions

- Does Table 1 in Chapter 9 apply to multiconductor instrumentation cables?
- Where can I find an example of how to calculate raceway fill for multiconductor cables?

Answer

Where multiconductor cable assemblies are installed in a raceway, each cable assembly is treated as an individual conductor based on the outside dimension of the cable assembly. This information is found in Chapter 9, Table 1, Note 9. The outside diameters that are given in a catalog or measured directly must be converted to approximate square inch areas, and then you apply the appropriate percentage of conduit fill from Chapter 9, Table 1 based on the number of cables being installed in the raceway.

The instrumentation circuits in the question are typically Class 2 or Class 3 circuits and are covered by Article 725. There is no modification in Article 725 that allows a greater fill percentage than ordinarily applies to raceways. In fact, Section 725.3(A) refers to 300.17 and applies the rules of Table 1 to all Class 1, Class 2, and Class 3 circuits. However, Section 300.17 does not have a specific rule or a reference to Table 1. Instead, the specific sections are listed in the Fine Print Note (FPN) that follows 300.17. In each of the raceway articles, there is either a rule or a reference to Table 1. Wireways and auxiliary gutters simply give a 20 percent maximum fill rule. Most other raceway articles refer to Table 1. There is no permission in the raceway articles allowing a greater fill percentage for Class 1, 2, or 3 circuits.

As an example, say we want to determine how many circular cross-section instrumentation cables can go into rigid metal conduit (RMC). The cable we want to use has an outside diameter (OD) of 0.268 inches. We are using RMC in 3/4 trade size. This example is illustrated in Figure 6.24(a).

The cross-sectional area of a cable with an outside diameter of 0.268 can be calculated using the formula for the area of a circle. This is normally given as

Keywords

Raceway fill
Instrumentation
Multiconductor cable
Class 2
Class 3

NEC References

300.17, 725.3, Table 1, Table 4, Table 5, Chapter 9, Annex C

Illustration of text example of a raceway fill calculation for multiconductor cables.

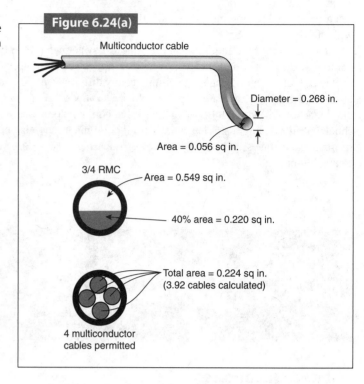

Figure 6.24(a)

$$\text{Area} = \pi r^2 \text{ or Area} = \pi d^2/4$$

where r is the radius of the circle, d is the diameter, and π is the mathematical constant pi, which is approximately equal to 3.1416. Since diameter is given, the area of one cable can be calculated as follows:

$$\text{Area} = (3.1416 \times 0.268 \text{ in.} \times 0.268 \text{ in.})/4 = 0.2256/4 = 0.056 \text{ sq. in.}$$

Now we can look for the appropriate conduit type, size, and area. In Chapter 9, Table 4, we find that the total area of a size 3/4 RMC is 0.549 square inches, but for three or more conductors, the allowable fill percentage is 40 percent. So, to eliminate one calculation, we can use the 40 percent area from Table 4 directly. From the table we see that 40 percent of the internal area of a size 3/4 RMC is 0.220 square inches. This is the area that can be occupied by cables. We get the same result if we multiply the total area by 40 percent.

Now, if we divide the area available in the raceway by the area of one cable, we get the maximum number of cables permitted. In this case, the 40 percent area is 0.220 square inches and the area of a cable is 0.056 square inches, so the number of cables permitted is

$$0.220 \text{ sq. in.}/0.056 \text{ sq. in.} = 3.92 \text{ cables}$$

According to Table 1, Note 7, we are permitted to count the 0.92 fraction as a whole number because it is greater than 0.80, so the maximum number of cables in this example is four.

Another, shorter, and somewhat easier method that works is to convert the cable size you have to one of the standard conductor types listed in Table 5. Then the tables in Annex C can be used to get a similar result without calculations. The limitation on the use of the Annex C tables in any case is that Annex C is based only on conductors that are all the same size, as stated in Table 1, Note 1.

For example, if we look at Table 5 to find a conductor type and size that is at least as big as the cable in question, we find that a size 6 AWG XHHW conductor has a diameter of 0.274

square inches, just a bit bigger (0.006 in.) than the cable we have. Now if we look at Table C8 in Annex C, we see that for RMC, the maximum number of 6 XHHW that can be installed in size 3/4 is three. (The calculated number for this particular combination is 3.73, so we cannot round up to four conductors. If we could round up, we would likely find 4 in the table.) This result is a bit more conservative than the actual calculated value, but it is a good estimation, it complies with the *NEC* if we choose a wire that is at least as big as (not smaller than) the cable we have, and it's much easier and faster than the calculation.

For another example, assume that we have a cable that is 0.156 inches in diameter. Using the alternate method for the smaller cables, we see in Table 5 that the 0.156-inch diameter is equivalent to a size 10 AWG TFE conductor, and from Table C8 in Annex C we get 11 conductors maximum in size 3/4 RMC, compared with 11.52 (11) conductors if the calculation method is used.

Where multiconductor cables with elliptical cross sections are used, Table 1, Note 9 requires that the area of a cross section of the cable be determined from the larger diameter, as if the cable were a round cable of the larger diameter. Many two-conductor and three-conductor Type TC and PLTC cables, as well as common Type NM cables, have this shape. As an example of how raceway fill would be calculated for these types of cables, consider the

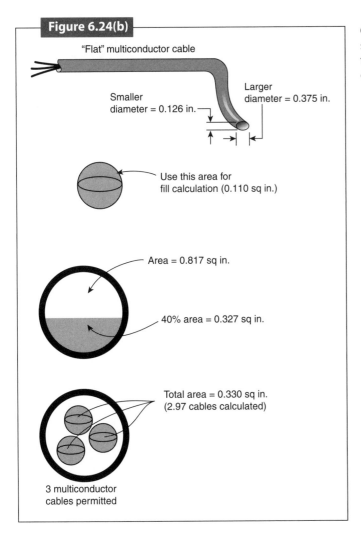

Calculation of cross-sectional area and raceway fill for a cable with elliptical cross section.

cable shown in Figure 6.24(b). This cable has a small diameter of 0.126 inches and a large diameter of 0.375 inches. If we wanted to know how many of these cables we could install in a size 1 FMC (flexible metal conduit), we would calculate the area using the larger diameter as follows:

Area of one cable = (3.1416 × 0.375 in. × 0.375 in.)/4 = 0.442/4 = 0.110 sq. in.

We will assume we can install three or more cables in this raceway. The size 1 FMC has a 40 percent area of 0.327 square inches from Table 4 for Article 348. The area available divided by the area of a single cable is 0.327/0.110 = 2.97. Therefore, we are allowed to install three cables in this raceway.

As a matter of practice, most people who have installed very much cable in raceways will choose fewer conductors or larger conduit than the *NEC* allows. This practice is essentially what the code advises in Table 1, FPN No. 1, which says: "Table 1 is based on common conditions of proper cabling and alignment of conductors where the length of the pull and the number of bends are within reasonable limits. It should be recognized that, for certain conditions, a larger size conduit or a lesser conduit fill should be considered."

Reference Index

Chapter 3, 160
Chapter 4, 160
Chapter 7, 160
Chapter 9, 271
Life Safety Code, 144
NESC, 6
NFPA 30, 190
NFPA 70E, 17
NFPA 72, 262, 263
NFPA 79, 103, 117
NFPA 88B, 200
NFPA 99, 208, 213
NFPA 110, 240
NFPA 497, 188, 190
NFPA 664, 188
Table 1, 271
Table 4, 271
90.1, 2, 97, 247
90.2, 2, 3, 5, 6
90.3, 2, 247, 249, 269
90.4, 2, 5, 89
90.5, 62
90.7, 10
Article 100, 7, 10, 28, 58, 65, 99, 123, 178, 233
Table 100, 176
110.2, 10, 89, 192
110.3, 10, 14, 151, 160, 182, 192
110.9, 12
110.10, 12
110.14, 14, 123, 142
110.16, 17
110.26, 19, 26, 28, 30, 32
110.31, 30
110.32, 19
110.33, 19
110.34, 19, 30
200.6, 36, 117, 150
200.7, 117, 150
Article 210, 244
210.1, 175
210.5, 117
210.8, 37, 46, 223
210.11, 46
210.12, 41
210.19, 54, 123, 175, 178

210.20, 54
210.21, 44, 159
210.23, 44, 46, 159
210.50, 46, 173
210.52, 37, 46, 223
210.60, 46
210.63, 37, 46, 223
210.70, 46
215.2, 54, 67, 123, 155
215.3, 54
215.12, 117
Article 220, 64
220.12, 54
220.14, 164, 175
220.43, 54
220.50, 54
220.51, 123
220.60, 54
220.61, 76
225.22, 7
225.30, 57, 58
225.31, 246
225.32, 246
225.37, 57
230.2, 57, 58
230.6, 233
230.23, 64
230.31, 64
230.40, 58
230.42, 54, 64, 123
230.53, 7
230.54, 62
230.70, 58, 233
230.71, 58, 155
230.72, 58, 231
230.79, 64
230.90, 155
230.204, 65
240.2, 67
240.4, 14, 44, 67, 121, 167
240.5, 151
240.6, 14, 121
240.20, 67
240.21, 67, 178
240.22, 67
240.24, 67

240.92, 67
Article 250, 266
250.2, 80
250.4, 80, 86, 94, 97
250.6, 76, 91, 94
250.20, 86, 249
250.21, 86
250.24, 73, 76, 91, 155
250.28, 91
250.30, 76, 80, 91, 99
250.32, 76, 80, 91
250.50, 86
250.52, 80
250.54, 86
250.58, 269
250.64, 80, 114
250.66, 80, 87, 99
250.86, 249
250.92, 73, 91, 249
250.96, 94
250.100, 249
250.102, 76, 87, 91, 196
250.104, 84, 87
250.110, 86
250.112, 86, 249
250.116, 87
250.118, 86, 89, 94, 97, 196
250.119, 117
250.122, 86, 87, 94
250.134, 94, 97
250.136, 84
250.142, 73, 76, 91
250.146, 94, 97
250.148, 209
250.162, 99, 249, 266
250.164, 99
250.166, 76, 99
300.3, 94, 259
300.5, 7
300.8, 103
300.11, 105, 106, 257
300.13, 107
300.17, 271
300.21, 111, 130, 257
300.22, 109, 111, 144, 218
310.3, 114

310.8, 119, 145
310.10, 123
310.11, 115, 119
310.12, 117
310.13, 119, 123
310.15, 14, 54, 121, 123, 129
Table 310.16, 14, 54, 123, 129
310.16, 119
Figure 310.60, 123
310.60, 129
310.62, 119
314.16, 130
314.20, 130, 133
314.21, 130
314.23, 106, 133
314.25, 130
320.15, 138
320.23, 138
320.30, 138
326.104, 114
326.116, 114
328.10, 175
330.10, 138, 145
330.23, 138
330.30, 138
334.10, 136, 144, 145
334.12, 7, 136
334.15, 138
334.23, 138
334.30, 138
336.10, 145
342.14, 142
344.10, 7
344.14, 142
344.30, 105, 141
Article 348, 196
Article 350, 196
358.12, 142
358.30, 105
368.56, 173
392.3, 144, 145
392.4, 144
392.6, 145
392.9, 145
400.2, 150
400.7, 46, 109, 150, 151, 159, 173
400.8, 46, 109, 150, 151, 159, 172, 173, 218
400.14, 173
400.22, 150
400.23, 150
406.2, 94
406.3, 154

406.4, 46
406.6, 173
406.7, 173
406.8, 46, 173, 223
408.34, 155
408.36, 155
410.4, 7, 46
410.15, 86, 254
410.16, 106
410.20, 86
410.21, 86
410.30, 159
410.101, 54
Article 411, 160
411.3, 160
Article 422, 2, 164
422.16, 162
422.31, 162
422.33, 162
422.34, 162
Article 424, 2
424.3, 123
Article 430, 2
430.6, 164, 165, 167, 175
430.7, 12, 165
430.14, 26
430.22, 54, 164, 175
430.24, 54, 164, 175
430.28, 178
430.31, 167
430.32, 167
430.51, 167
430.52, 165, 167
430.53, 167
430.62, 167
430.63, 167
430.102, 26, 162, 172
430.107, 26, 162
430.108, 173
430.109, 12, 162, 172, 173, 176
430.110, 165, 173, 176
430.221, 175
430.224, 175
430.226, 175
Table 430.247, 175
Table 430.248, 164, 175
Table 430.249, 175
Table 430.250, 175
Table 430.251(A) and (B), 165
Article 440, 2
440.3, 162, 176, 178
440.4, 178
440.12, 165

440.14, 26, 176, 178
440.41, 165
450.1, 182
450.3, 67
450.9, 182
450.11, 182
450.13, 111
450.21, 182
450.22, 182
450.27, 182
455.8, 165
480.5, 99
490.21, 65
500.1, 188
500.2, 188
500.5, 188, 190, 192, 200, 205
500.6, 188
500.7, 188, 192
500.8, 192, 205
501.10, 196, 198, 249, 257
501.11, 173
501.15, 7, 198
501.30, 196, 249
501.125, 192
501.130, 198, 204
501.140, 151, 196
502.10, 249, 257
502.30, 196, 249
502.125, 192
503.5, 192
503.10, 249
503.30, 249
503.125, 192
503.130, 192
504.20, 249
505.5, 188, 192
505.8, 192
505.9, 192
505.20, 192
506.5, 188
511.1, 200
511.3, 136, 200, 204
511.4, 204
513.3, 204
513.10, 204
516.3, 205
517.2, 7, 136, 208, 209
517.10, 209
517.13, 136, 208, 209
517.16, 94, 209
517.18, 209
517.19, 209
517.30, 209, 213, 240

517.32, 209
517.33, 209
517.34, 213
517.45, 209
517.80, 209
518.2, 136
518.4, 136
Article 590, 151
590.4, 64
640.6, 94
645.1, 218
645.2, 218
645.5, 218
645.10, 218
647.7, 94
670.1, 32
680.21, 226
680.22, 223
680.23, 114, 220
680.24, 220
680.26, 114, 226
680.30, 223
680.31, 226
680.40, 223
680.42, 220, 223, 226
680.43, 46, 220, 223, 226
680.61, 223
680.70, 223, 226
680.71, 46, 223
680.72, 223

680.73, 223
680.74, 226
695.3, 228, 231, 233
695.4, 228, 231, 233
695.6, 233
695.12, 231
700.1, 237
700.5, 237, 240
700.6, 237, 240, 246
700.9, 237, 239, 240, 244
700.12, 240, 244
700.15, 237
701.6, 213
701.11, 246
702.11, 246
Article 725, 2
725.1, 247, 249, 259
725.2, 249, 259
725.3, 111, 247, 271
725.8, 257
725.11, 257
725.21, 247, 249, 259
725.23, 247
725.25, 249, 259
725.26, 239, 259
725.27, 259
725.41, 249, 254, 259
725.52, 239, 259
725.55, 239, 259
725.56, 259

725.57, 254
725.61, 111, 144, 145, 249, 257, 262
727.5, 259
760.1, 263
760.21, 37
760.41, 37
760.61, 257
760.81, 262
760.82, 262
770.154, 257
800.2, 265, 266
800.3, 249
800.24, 265
800.44, 265, 268
800.47, 268
800.50, 254, 265, 268, 269
800.90, 254, 265
800.93, 254, 266
800.100, 254, 266, 269
800.110, 266
800.133, 145, 254, 268
800.154, 111, 257
820.93, 254
820.100, 254
830.1, 3
830.3, 3
830.179, 3
Annex C, 271
Annex E, 136
Annex G, 2

Keyword Index

125 percent factor, 54

A
AC, 129
Access, 138, 162, 223
Adjustment factor, 14, 123
Aerial cables, 265
AFCI, 41
Air conditioners, 176
Air handling, 111
Air tubing, 103
Aircraft, 5
Allowable ampacity, 129
Alternating current, 129
Aluminum, 142
Ambient temperature, 123
Ampacity, 66, 121, 123, 129
Ampacity selection, 14
Appliances, 10, 162, 163
Approved equipment, 192
Arc-fault circuit interrupter, 41
Area classifications, 188
Assembly, 136
   places of, 215
Attachment plugs, 162
Attics, 138
Automatic transfer switches, 238

B
Basements, 37, 46
Bathrooms, 46
Batteries, 98
Bedrooms, 41
Bonding
   for control circuits, 249
   for hydromassage bathtubs, 226
   for metal roofing, siding, and veneer, 87
   for service laterals, 91
   at services, 73
   water piping and, 84
Bonding jumpers, 76, 87, 196
Boxes, 130, 132, 220
Branch circuits
   AFCI protection for, 41
   for HVAC equipment, 177
   individual, 44
   motors over 600 volts and, 175
   multiwire, 107
   overcurrent protection for, 167
Broadband, 3
Buildings
   circuits between, 253, 265
   metal, 87
   separate, 57, 76
   water piping and, 84

C
Cable sheaths, 266
Cable trays, 143, 145
Cables
   aerial, 265
   ceiling grids or supports and, 106
   circuit integrity, 261
   coaxial, 254
   communications, grounding for, 266
   identification of, 115
   jacketed, 145
   listed, 115, 145, 151
   low-voltage, 257, 259
   markings on, 115, 145
   multiconductor, 145, 271
   plenum-rated, 218
CATV, 3
Ceilings, 26, 105, 106, 109
Circuit breakers, 171, 176
Circuit integrity cable, 261
Circuits. *See also* Branch circuits
   classifications for, 259
   control, 249, 259
   interbuilding, 253, 265
   normal, 244
   power-limited, 247
   short, 67, 167
   for temporary power, 63
Circuses, 215
Class 1 circuits, 238, 247, 249, 259
Class 2 circuits
   bonding and grounding requirements for, 249
   characteristics of, 247
   interbuilding, data and communications and, 254
   raceway fill and, 271
   separation requirements, 259
   transfer switch control circuits as, 238
Class 3 circuits, 239, 247, 249, 259, 271
Class A GFCI, 37
Class B GFCI, 37
Class I areas
   Article 511 on, 200
   equipment identification for, 192
   flammable gases or vapors and, 190
   flexible conduit runs or flexible fittings in, 196
   as hazardous locations, 188
   portable lighting requirements, 204
Class II areas, 188, 192
Class III areas, 188, 192
Classified areas, 188, 190, 192, 200, 205
Clearance, 26, 32, 181
Coaxial cables, 254
Code letters, 165
Color coding, 116
Combination-type AFCI, 41
Combustible properties
   boxes for noncombustible materials *vs.*, 130
   in construction, 135, 182
   in finishes, 132
   of solvents, fuel oil, or diesel, 190
   temperature ratings for luminaires and, 192
   wiring methods and, 110
Communications
   bonding and grounding requirements for, 249
   broadband, installation of, 3
   grounded elements in, 266
   grounding electrodes for, 269
   installation requirements, 2
   roof installations of, 264
   rules for Class 2 interbuilding circuits, 254
   spacing between conductors, requirements for, 267
Computer rooms, 218

Conductors. *See also* Equipment grounding conductors; Grounding electrode conductors; Service conductors
   application information for, 119
   color coding requirements for, 116
   exposed, 264
   identification of, 36
   for motors over 600 volts, 175
   parallel, 151
   secondary, 66
   single, 143
   solid, 113, 226
   sunlight resistant, 145
Conduits
   flexible metal, 89
   low-voltage cables and, 257
   rigid metal, 141
   steel, 142
Construction types, 135
Continuity, 107
Continuous load, 54
Control circuits, 2, 249, 259
Controller rating, 165
Conventions, 215
Cord and plug, 158, 162, 173
Cord set, 151. *See also* Flexible cord
Correction factor, 14, 123
Countertops, 46
Couplings, 141, 198
Covered topics, in *NEC*, 5, 6
Cranes, 32
Current ratings, 175
Current-carrying wire, 121

D

Damp locations, 7. *See also* Wet locations
Data, 2. *See also* Communications
DC currents, 129
DC systems, 98
Deadfront, 28
Deck boxes, 220
Dedicated spaces, 182
Dental offices and clinics, 207
Derating, 123
Devices, 66, 107
Diesel fuels, 188, 190
Disconnect rating, 165
Disconnecting means, 58, 162
Disconnects
   accessibility of, 162
   for air conditioners, 176
   for computer rooms, 218
   for emergency systems, 246
   for fire pump service, 231
   for HVAC equipment, 178
   for motors, 173
   for short circuits, 12
   for small motors, 171
Dissimilar metals, 142
Division 1, 188, 192, 196, 204
Division 2, 188, 192, 196, 204
Double-insulated pumps, 226
Dropped ceilings, 109. *See also* Suspended ceilings
Dry locations, 7
Dry-type transformers, 110
Drywall, 130
Ducts, 109, 110
Dust-ignitionproof properties, 192
Dwelling units, 46, 121, 138

E

Effective ground-fault circuit path, 86
Effectively grounded, 86
Egress lighting, 237, 244
Electric-discharge lighting, 158
Electrodes, 254, 266. *See also* Grounding electrode conductors
   grounding, 76, 80, 94, 269
   supplemental, 85
Emergency loads, 237
Emergency systems, 237, 240, 244, 246
Emergency wiring, 238
Enclosures, 28, 103, 142, 268
Energized work, 19, 26, 28, 30, 32
Energy-efficient motors, 165
Engineering supervision, 129
Entrances, 19
Environmental air, other space for, 109, 110, 218
Equipment, 10
   grounding, 85, 97, 249
   for hazardous locations, 192, 205
Equipment branch, 213
Equipment grounding conductors, 76, 94, 150, 209
Equipotential plane, 226
Essential electrical systems, 213
Exclusive control, 3
Exit lighting, 237, 244
Exit travel, 19
Explosionproof properties, 192, 198, 204
Explosives, 188
Exposed conductors, 264
Exposed live parts, 19, 28
Extension cord, 151
Extensions, 132

F

Fairs, 215
Fault current, 12
Fault level, 12
Feeder taps, 66
Feeder/branch-type AFCI, 41
Feeders
   for continuous loads, 57
   for HVAC equipment, 177
   for services to a building, 58
   for temporary power, 63
   wiring requirements for, 121
Finished spaces, receptacles for, 46
Fire alarm cable, 261
Fire alarms, 249, 261
Fire pumps, 228, 231, 233
Fire ratings, 233, 257
Fire resistance, 130
Fire-rated construction, 218
Fittings, 89, 142
Fittings, seal, 198
Flammable properties, 190, 192
Flash hazard, 17
Flash point, 190
Flex, 198
Flexibility, 198
Flexible cord
   for luminaire connections, 158
   plenum ceilings and, 109
   proper and improper uses of, 151
   re-identifying conductors in, 150
   for temporary wiring, 215
   use with motors, 173
Flexible metal conduits, 89
FMC, 89, 196
Framing, 268
Fuel oils, 190
Functionally-associated automatic transfer switches, 239

G

Garages, 37, 46, 204
Generators, 213, 244, 245
GFCI, 36, 154, 215, 223
Goosenecks, 62
Gray conductors, 36, 116
Green conductors, 116, 150

Grid ceilings, 26, 106
Ground faults, 67, 167
Grounded conductors, 36, 76, 91, 98, 150
Grounded outlets, 154
Grounded service conductors, 73
Grounded systems, 80, 98
Ground-fault circuit interrupters, 36. *See also* GFCI
Ground-fault current, 80
Ground-fault protection of equipment, 37. *See also* GFCI
Grounding. *See also* GFCI
  in Class I areas, 196
  for communications cables, 266
  for control circuits, 249
  electrodes, for power and communications, 269
  for flexible metal conduit, 89
  in health care facilities, 209
  for lighting poles, 85
  for metal roofing, siding, and veneer, 87
  for service laterals, 91
  sizing electrode conductors for, 80
Grounding conductors, 91, 93, 254
Grounding electrode conductors, 73, 80, 98, 114
Grounding electrode system, 269
Grounding electrodes, 76, 80, 94, 269
Grounding terminals, 94, 97
Grounding-type receptacle, 154
Grouped service or supply, 58
Guard strips, 138

H
Habitable areas, 46
HACR equipment, 46
Hallways, 46
Hand lamps, 204
Hangars, 204
Hazardous locations, 249
Hazardous locations or areas, 188, 192, 205
Headroom, 26
Health care, 207
Health care facilities, 209
Hoists, 32
Horsepower, 163
Hospital grade, 209
Hot tubs, 220, 223, 226
HVAC equipment, 2, 177
Hydromassage bathtubs, 223, 226

I
Identification
  for cables, 115
  of cables for use in cable trays, 145
  for conductors, 116
  of conductors and equipment, 10
  of equipment for hazardous locations, 192
  of grounded conductors, 150
Ignitible properties, 192
Incident energy, 17
Individual branch circuits, 44
Industrial occupancies, 143
Information technology, 218
Inrush current, 165
In-sight circuit breakers, 176
Instrumentation, 247, 249, 259, 271
Insulation class, 182
Insulation system, 182
Interbuilding circuits, 253, 265
Interconnected smoke alarms, 263
Interrupting rating, 12
Islands, kitchen, 46
Isolated equipment grounding conductor, 93
Isolated ground, 93
Isolated grounding, 97, 209
Isolated grounding receptacles, 94
Isolating switches, 65
IT equipment, 218

J
Junction boxes, 220
Jurisdiction, 5

K
Kitchens, 37, 46

L
Large equipment, 19
Legally-required standbys, 237
LFMC, 196
License, 2
Lighting, 160, 204. *See also* Luminaires
Lighting and appliance branch-circuit panelboard, 155
Limited energy, 247
Listed cords and cables, 115, 145, 151
Listed fittings, 89
Listing, for lighting systems, 160
Live parts, 28
Load calculations, 54, 164

Locations
  dry, 7
  hazardous, 188, 192, 205, 249
  for receptacles in dwelling units, 46
  for services to a building, 58
  wet or damp, 7, 136
Locked rotors, 165
Lock-out, 26
Low-voltage cables, 257, 259
Low-voltage equipment, 32, 160
Luminaires
  ceiling grids or supports and, 106
  connections for, 158
  grounding for, 86
  for hazardous locations, 192, 205
  plenum ceilings and, 109
  underwater, 220

M
Machine tools, 32
Machines, 10
Manufacturers, of cables, 115
Markings
  on cables, 115, 145
  on receptacles, 154
Mechanical code, 171
Medical offices and clinics, 207
Medium-voltage motors, 175
Metal buildings, 87
Metallic raceways, 209
Metallic wiring methods, 111
Metals, dissimilar, 142
Meter bases, 73
Meter enclosures, 73
Meter fittings, 73
Motors
  damper-type, disconnects for, 171
  disconnects for, 173
  for hazardous locations, 192
  over 600 volts, conductors for, 175
  overcurrent protection for, 167
  pool pump, 163
Multiconductor cables, 145, 271
Multiple receptacles, 44
Multiple-station smoke alarms, 263
Multiwire branch circuits, 107
Munitions, 188

N
Nameplates, 163
National Electrical Code, scope, topics covered and not covered in, 5, 6

National Electrical Safety Code (NESC), 6
Natural gray conductors, 36
Neher-McGrath method, 129
Network interface unit, 3
Neutral (conductor), 91
NIU, 3
Noncombustible construction, 135
Noncombustible finish, 132
Noncombustible properties, 130
Non-electrical equipment, 32
Nongrounding-type receptacle, 154
Nonmetallic wiring methods, 111
Normal circuits, 244
Not covered by *NEC*, 5, 6
Number of supplies, 57

O

Objectionable current, 76, 91
Objectionable currents, 94
Occupancies, 143
Ophthalmology offices and clinics, 207
Optional standbys, 237
Optometry offices and clinics, 208
Other space for environmental air, 109, 110, 218
Outdoor (outside) locations
　as damp or wet locations, 7
　for emergency generators, 246
　for feeders and branch circuits, 56
　for fire pump supply conductors, 233
　GFCI receptacles for, 37
　receptacle requirements for, 46
Outlets, 41, 45, 154. *See also* Receptacles
Overcurrent devices, 66
Overcurrent protection, 66, 155, 167, 178. *See also* AFCI; Grounding
Overhead power line, 6
Overload, 67, 167, 178

P

Panelboard, 155, 182
Panic hardware, 19
Parallel conductors, 151
Parallel paths, 76
Parking garages, 200
Patient care areas, 136, 208, 209
Patient vicinity, 209
Penetration, 257
Peninsulas, kitchen, 46

Performance areas, 215
Permanent wiring, 151
Physical protection, 114
Physical separation, 231. *See also* Separation(s)
Pin and sleeve, 173
Places of assembly, 215
Plaster ring, 130
Plenum, 109, 110
Plenum-rated cables, 218
Pneumatic tubing, 103
Point of attachment, for service heads, 62
Poles, 86
Pools, 223, 226
Portable cords and cables, 151
Portable lighting, 204
Power panelboard, 155
Power-limited circuits, 247
PPE, 17
Primary metering, 65
Primary protectors, 265, 266
Product standards, 10
Protection, 138. *See also* AFCI; GFCI
　apparel for, 17
　overcurrent, 66, 155, 167, 178
　physical, 114
Protectors, 254, 269
Psychology offices and clinics, 208

R

Raceway fill, 271
Raceways
　air tubing and, 103
　ceiling grids or supports and, 106
　for communications cables, 265, 266
　for power and communications, 268
　sleepers as support for, 104
　solid conductors in, 113
　Type NM cable and, 136
　wet locations and, 7
Raintight service heads, 62
Readily-accessible disconnects, 162
Readily-accessible receptacles, 223
Receptacles. *See also* GFCI
　AFCI protection for, 41
　disconnecting from, 162
　in health care facilities, 209
　isolated grounding, 94
　for luminaires, 158
　for motors, 173
　in plenum ceilings, 109

　for pools, spas, and hot tubs, 222
　requirements, in dwelling units, 45
　single or multiple, ratings for, 44
Reclassification, 259
Redundant grounding, 209
Re-identification, of conductors, 150
Reliable power, 233
Remote control, 247
Remote disconnects, 231
Repair garages, 200
Residential units, 138, 263
Risers, 257
RMC (rigid metal conduit), 141
Rooftops, GFCI receptacles for, 37
Running boards, 138

S

Scope, of *NEC*, 5
Seals, 198
Secondary conductors, 66
Securing exposed raceways, 104
Securing raceways to ceiling grids or supports, 105
Separate buildings, 57, 76. *See also* Buildings
Separately derived systems, 76, 97, 98
Separation(s), 231, 237, 240, 259, 267
Service
　for a building, 58
　for a fire pump, 228
　sources for a building, 56
　for temporary power, 63
　wiring requirements for, 121
Service conductors, 58, 63, 67, 177, 233
Service disconnects, 65, 73, 155, 228
Service drops, 62
Service equipment, 73
Service heads, 62
Short circuits, 67, 167
Short-circuit current rating, 12
Signaling, 247, 259
Single conductors, 143
Single receptacles, 44
Single-phase feeders, 121
Sleepers, 105
Sleeves, 114
Smoke alarms, 263
Smoke detectors, 218, 263
Snap switches, 171
Solid conductors, 113, 226
Solvents, 190

Sources, emergency, 240. *See also* Emergency systems
Spas, 220, 223, 226
Special permission, for listed fittings, 89
Spray booths, 205
Stairs, 138
Stalled rotors, 165
Standby loads, 213
Standby systems, 240, 245
Steel conduits or enclosures, 142
Steel frames, 84
Structures, 84, 253. *See also* Buildings
Subpanels, 91, 94
Substation, 6
Sufficient access, 26, 32. *See also* Access
Sumps, 200
Sunlight resistant conductors or jacketed cables, 145
Supplemental electrodes, 85. *See also* Electrodes
Supply, 58. *See also* Service
Support, 141
Support wires, 106
Supporting exposed raceways, 105
Supporting raceways to ceiling grids or supports, 105
Suspended ceilings, 105, 109
Swimming pools, 220, 223
Switchboards, 231
Switches, 12
   automatic transfer, 238
   isolating, 65
   snap, 171
   transfer, 213, 240, 244
   unit, 162
System grounding, 249

Systems, emergency, 240. *See also* Emergency systems

T

Table values, 164
Tag-out, 26
Tap conductors, 66, 177
Tap rules, 66
Taps, 177
Telephones, 264
Temperature, 192
Temporary power, 63
Temporary wiring, 214
Termination, 141
Therapeutic tubs, 223
THHN, 119
Threaded coupling, 141
Threadless coupling, 141
Three-phase feeders, 121
THW, 119
THWN, 119
Trade shows, 214
Train, 5
Transfer means, 240
Transfer switches, 213, 240, 244
Transformers, 30, 66, 110, 181
Tray cable, 145
Trouble lights, 204
Type AC (armored cable), 138, 209
Type III construction, 135
Type ITC, 259
Type IV construction, 136
Type MC (metal-clad cable), 138, 209
Type NM cable, 135, 138, 209, 268
Type PLTC (power-limited tray cable), 143, 145
Type TC (tray cable), 143, 145
Type V construction, 136

Types, of wiring and conductors, 119

U

Underground, 7
Underwater luminaires, 220
Unfinished spaces, 37, 46
Ungrounded systems, 85
Union, 141
Unit equipment, 244
Unit load, 54
Unit switches, 162

V

Ventilation, 182
Voltage ratings, 119
Voltage to ground, 28

W

Wall spaces, receptacles for, 46
Water piping, 84
Weatherheads, 62
Wet locations, 7, 136
White conductors, 36, 116, 150
Wires, types of, 119
Wiring methods, 160, 214
Woodworking, 188
Working space
   for de-energized equipment, 32
   determination of requirements for, 19
   for enclosed panels facing each other, 28
   for HVAC equipment, 26
   for transformers, 30

Z

Zones, 188, 192